AF616144

5561229
UNIVERSITY OF SURREY LIBRARY
WITHDRAWN
24 MAR 2021
UNIVERSITY OF SURREY

Studies in Surface Science and Catalysis 67

STRUCTURE–ACTIVITY AND SELECTIVITY RELATIONSHIPS IN HETEROGENEOUS CATALYSIS

Studies in Surface Science and Catalysis

Advisory Editors: B. Delmon and J.T. Yates

Vol. 68

CLASS. 541.12813
LOC. Conf -2-92
SL.
ACC. SS61229
SURREY UNIVERSITY LIBRARY

STRUCTURE–ACTIVITY AND SELECTIVITY RELATIONSHIPS IN HETEROGENEOUS CATALYSIS

Proceedings of the ACS Symposium on Structure–Activity Relationships in Heterogeneous Catalysis, Boston, MA, April 22–27, 1990

Editors

R.K. Grasselli
Mobil Central Research Laboratory, P.O. Box 1025, Princeton, NJ 08543, USA

and

A.W. Sleight
Department of Chemistry, Oregon State University, Corvallis, OR 97331, USA

ELSEVIER **Amsterdam — Oxford — New York — Tokyo 1991**

ELSEVIER SCIENCE PUBLISHERS B.V.
Sara Burgerhartstraat 25
P.O. Box 211, 1000 AE Amsterdam, The Netherlands

Distributors for the United States and Canada:

ELSEVIER SCIENCE PUBLISHING COMPANY INC.
655, Avenue of the Americas
New York, NY 10010, U.S.A.

Library of Congress Cataloging-in-Publication Data

ACS Symposium on Structure-Activity Relationships in Heterogeneous
Catalysis (1990 : Boston, Mass.)
Structure-activity and selectivity relationships in heterogeneous
catalysis : proceedings of the ACS Symposium on Structure-Activity
Relationships in Heterogeneous Catalysis, Boston, MA, April 22-27,
1990 / editors, R.K. Grasselli and A.W. Sleight.
p. cm. -- (Studies in surface science and catalysis ; vol.
67)
Includes bibliographical references and index.
ISBN 0-444-88942-6 (U.S.)
1. Catalysts--Structure-activity relationships--Congresses.
2. Heterogeneous catalysis--Congresses. I. Grasselli, Robert K.,
1930- . II. Sleight, A. W. III. Title. IV. Series: Studies in
surface science and catalysis ; 67.
QD505.A27 1990
541.3'95--dc20 91-16425
CIP

ISBN 0-444-88942-6

© Elsevier Science Publishers B.V., 1991

All rights reserved. No part of this publication may be reproduced, stored in a retrieval system or transmitted in any form or by any means, electronic, mechanical, photocopying, recording or otherwise, without the prior written permission of the publisher, Elsevier Science Publishers B.V./ Academic Publishing Division, P.O. Box 330, 1000 AH Amsterdam, The Netherlands.

Special regulations for readers in the USA – This publication has been registered with the Copyright Clearance Center Inc. (CCC), Salem, Massachusetts. Information can be obtained from the CCC about conditions under which photocopies of parts of this publication may be made in the USA. All other copyright questions, including photocopying outside of the USA, should be referred to the publisher.

No responsibility is assumed by the Publisher for any injury and/or damage to persons or property as a matter of products liability, negligence or otherwise, or from any use or operation of any methods, products, instructions or ideas contained in the material herein.

Although all advertising material is expected to conform to ethical (medical) standards, inclusion in this publication does not constitute a guarantee or endorsement of the quality or value of such product or of the claims made of it by its manufacturer.

This book is printed on acid-free paper.

Printed in The Netherlands

CONTENTS

HYDROGENATION

ZEOLITE CATALYSIS

SURFACE SCIENCE AND MODELING

PREFACE

Structure plays an important role in heterogeneous catalysis. It provides a framework for the arrangement and strategic placement of key catalytic elements, hosting them in a prescribed manner so that their respective electronic properties can exhibit their desired catalytic functions and mutual interactions. Under reaction conditions these framework structures and their key catalytic guests undergo dynamic processes becoming active participants of the overall catalytic process. They are not mere static geometric forms. The framework provides the necessary crystal structure stabilization and hence acts as a template. Non-stoichiometry and vacancy rearrangements of the solids are important factors contributing to these dynamic processes of catalytic reactions. The dynamics of catalytic structures are particularly vivid in selective oxidation catalysis where the lattice of a given catalytic solid partakes as a whole, not only its surface, in the redox processes of the reaction. The catalyst becomes actually a participating reagent. By proper choice of key catalytic elements and their host structures preferred catalytic pathways can be selected over less desired ones. However, not only in selective redox catalysis does structure play an important role, its importance is also well documented, among others, in shape selective zeolite catalysis, enantioselective hydrogenation and hydrodesulfurization.

The contributions presented in this book address the dynamic character of the solid state under catalytic reaction conditions. By relating structure to activity and selectivity in heterogeneous catalysis our understanding of such correlations has been significantly enhanced through the use of sophisticated spectroscopic means, surface science and modeling. Nonetheless, the ultimate test of the correlations remains the actual catalytic reaction.

The individual contributors who made this update of structure activity and selectivity correlations in heterogeneous catalysis possible are herewith sincerely thanked.

R. K. Grasselli
Mobil Central Research Laboratory
P. O. Box 1025
Princeton, NJ 08543

A. W. Sleight
Department of Chemistry
Oregon State University
Corvallis, OR 97331

Princeton: February 28, 1991

ACKNOWLEDGMENT

The editors gratefully acknowledge the financial support of the following corporations:

Alcoa	DuPont
Allied Signal	Exxon
Amoco	W. R. Grace
Arco	Mobil
Ashland	Monsanto
B. P. America	Petroleum Research Fund

R.K. Grasselli and A.W. Sleight (Editors), *Structure-Activity and Selectivity Relationships in Heterogeneous Catalysis*

© 1991 Elsevier Science Publishers B.V., Amsterdam

Redox Dynamics and Structure/Activity Relationships in Vanadium-Oxide on TiO_2 Catalyst

G. CENTI, M. LOPEZ GRANADOS[a], D. PINELLI and F. TRIFIRO'

Dept. of Industrial Chemistry and Materials, V.le Risorgimento 4, 40136 Bologna (Italy)

ABSTRACT

The *in-situ* evolution of crystallites of V_2O_5 on the TiO_2 surface during interaction with the o-xylene/air reagent mixture involves together with their spreading on the titania surface partial reduction and the formation of an amorphous phase characterized by a V^V:V^{IV} ratio of 2:1 and an IR band centered at 995 cm^{-1}. The phase does not form directly by reduction, but involves a preliminary reduction to a phase with a lower mean oxidation state. This reduced vanadium-oxide phase is then partially reoxidized to the active phase with a decrease in the formation of the intermediate phthalide and an increase in the selectivity to phthalic anhydride. A similar evolution in the catalytic behavior and in the mean oxidation state is observed in unsupported V_2O_5 which after about 200 hours, transforms into V_3O_7. This phase has a V^V:V^{IV} ratio similar to that present in the active phase of vanadium-oxide supported on TiO_2, but does not show its characteristic IR band centered at 995 cm^{-1}. Using a prereduced unsupported vanadium-oxide, it is possible to decrease the activation time considerably and to obtain a final catalyst, after *in-situ* treatment, whose catalytic behavior is very comparable to that of V/Ti/O, but again characterized by the presence of a phase (V_3O_7) different from that present in the V/Ti/O system. It is suggested that the catalytic behavior in o-xylene conversion to phthalic anhydride is not related to the presence of an unique special surface structure of vanadium-oxide on the TiO_2 surface, but rather to a suitable V^V:V^{IV} ratio and surface distribution. These features can also be realized in unsupported vanadium-oxide using a suitable preparation and activation procedure.

INTRODUCTION

Vanadium-oxide supported on an oxide matrix, in particular on TiO_2, is widely used for catalytic partial oxidation of hydrocarbons. Over the years a number of observations have been made suggesting the advantages to be gained by supporting the catalytically active vanadium-oxide on the surface of another oxide [1 and references therein]. However, a fundamental question arises from this concept: Are these variations to be attributed to a change in the catalytic behavior related to an increase in the available surface area, or is there a change in the local structure of the vanadium oxide species stabilized by interaction with the support ?

---

a On leave from the Institute of Catalysis y Petroleoquimica, Madrid (Spain)

It has now become clear that under suitable preparation conditions vanadium oxide may be supported on TiO_2 in a well-dispersed form with the formation of a monolayer of the active oxide on the support, and that the support appears to play a crucial role in facilitating the formation of the active structures due to geometrical or chemical factors [1-22 and references therein]. Most authors have focused their attention on this monolayer concept and on characterization of the monolayer; a variety of contradictory possible surface configurations have been suggested, such as (i) mono-oxo, hydroxyoxo, di-oxo vanadate, (ii) tetrahedrally or octahedrally coordinated vanadium species, (iii) isolated, polymeric, bidimensional or tridimensional species, and (iv) clusters or coherent lamellae, of amorphous or of paracrystalline vanadium oxide. Considerable confusion thus exists regarding the nature of the local surface configuration of vanadium oxide on TiO_2, also because sometimes the multiple molecular states that can be present simultaneously in the supported metal oxide are not sufficiently taken into account. Also problematic is the analysis of the relationship between the possible surface configurations with the activity/selectivity in o-xylene oxidation. Relatively fragmentary information exists on this fundamental aspect, and generally, very few papers analyze the evolution of the catalyst in the reaction medium, which is a fundamental aspect to assess the real nature of the active phase and the role of TiO_2 in its stabilization.

No unequivocal answer, for example, can be found in the literature on the basic problem of the difference between the catalytic behavior of supported and unsupported vanadium oxide. Does the presence of the support improve only the activity or also the selectivity? This, in turn, determines the importance of the various interpretations of the molecular structure of vanadium-oxide on TiO_2, on the catalytic behavior.

In this work the process of transformation of the catalyst in the reacting medium is studied in order to correlate the type of transformations with the catalytic behavior and the nature of the active phase. The time-on-stream evolution of structural and catalytic properties of vanadium-oxide supported or unsupported on TiO_2 were studied in a flow reactor, by characterizing the V^V:V^{IV} ratio through chemical analysis, and the surface structure by means of various physicochemical techniques.

The catalysts were prepared by solid state reaction. This method, despite its inherent simplicity, leads to a system whose catalytic performance compares well to that obtained by other methods such as wet impregnation and grafting techniques [16-19]. However, as compared to other preparation methods, preparation by solid state reaction has many advantages: (i) possible interference by other reactants is eliminated, (ii) the starting situation is clearly defined for a better correlation of the evolution of the catalytic system to its catalytic performance, and (iii) a more clear distinction of the solid state reactions occurring between the V_2O_5 and TiO_2 during calcination and during *in-situ* heating treatments in the presence of the o-xylene/air reacting mixture is possible.

EXPERIMENTAL

Catalyst Preparation The catalysts were prepared by solid state reaction of TiO_2 and V_2O_5 (Carlo Erba reagent grade). Generally, 7.7 wt.% V_2O_5 was used, an amount typical for industrial preparations. Anatase and rutile prepared by $TiCl_4$ hydrolysis were used in order to obtain highly pure TiO_2 supports and to exclude interference from doping. After mixing and gently grinding (1 min, in order to have good mixing but avoiding mechanico-chemical alterations of the samples), the powder was calcined in an oven at 500°C for 16h or longer in a static air atmosphere. The samples were then treated *in-situ* in long-run (about 500 hours) catalytic tests with a reagent mixture of 1.5% v/v o-xylene in air (reaction temperature around 320°C).

A V^{IV}-V^V mixed valence sample of unsupported vanadium-oxide was prepared by dropping a solution of V^{IV}-oxalate into an ammonia solution (pH around 9), filtration, washing and drying at 80 °C, and calcination at 280 °C.

Catalytic Tests The catalysts were tested in a conventional laboratory apparatus with a tubular fixed bed reactor working at atmospheric pressure and on-line gas chromatographic analyses of reagent and product compositions. The standard reactant composition was 1.5 % o-xylene, 20.5% O_2 and 78% N_2. The catalyst (0.52 g) was loaded as grains (0.250-0.420 mm). A thermocouple, placed in the middle of the catalyst bed, was used to verify that the axial temperature profile was within 3-5 °C.

Characterization The characterization by chemical analysis of the vanadium-oxide species present in VTiO samples and of the mean valence state of vanadium was performed as follows. The samples (about 0.5 g) were moistened at room temperature (r.t.) with 50 ml of a dilute (4 M) H_2SO_4 solution or with an ammonia solution (4 M) for fifteen minutes under stirring and then filtered. The amount of vanadium was determined separately in the filtered solution and in the residual sample dissolved in boiling concentrated H_2SO_4 (16 M). The total amounts and the valence state of vanadium were determined by a titrimetric method. In particular, a part of each fraction was titrated with 0.1 $KMnO_4$ to determine the amount of reduced vanadium species and then with Fe^{2+} to determine total amount of vanadium; another part was titrated with Fe^{2+} in order to determine the amount of V(V). From the balance it is possible to quantify the total and relative amounts of V(IV) and V(V) and of any V(III) present. The vanadium species extracted by the r.t. dilute sulphuric acid or ammonia solution will be, hereinafter, called *soluble* or *weakly-interacting* species, whereas the remaining species determined after dissolving the residue will be called *insoluble* or *strongly interacting* species. Other reactivity and spectroscopic analyses were carried out as previously reported [16-20,23,24].

RESULTS AND DISCUSSION

Formation of an Interacting V^{IV} Layer. The solid state reaction of V_2O_5 and TiO_2 in air at temperatures in the 400-500°C range, leads, in the absence of any reducing agent, to the formation of relevant amounts of V^{IV}. Chemical analysis shows, in fact, the formation of V^{IV} species that cannot be dissolved in an acid or basic aqueous medium, in contrast to other supported V^{IV}- and V^{V}-oxide species. Shown in Table I is the amount of V^{IV} formed during the calcination of V_2O_5 and TiO_2 that is approximately equal (for catalysts with a surface area of about 10 m^2/g) to the reference monolayer estimated on the basis of a geometrical coverage of the titania surface. The reference monolayer is about 0.1% w/w of V_2O_5 per m^2 of TiO_2. For the sake of comparison, the amounts of the various species of vanadium determined by chemical analysis are all expressed in Table I as % by weight of equivalent moles of $VO_{2.5}$. The formation of these V^{IV} surface sites probably occurs by a specific reaction between hydroxyl groups of TiO_2 surface and V^{V} sites and the formation of this species is the driving force for the spontaneous reduction of V^{V} in oxidizing conditions and for its surface migration. The mechanism of formation probably involves the preliminary formation of Ti^{3+} sites by dehydroxylation with consecutive electron transfer to V sites and formation of stable Ti-O-V bonds.

In contrast with what happens with rutile samples, the addition of a reducing agent during the heat treatment does not further increase the amount of insoluble V^{IV} in anatase samples (Table I). Similar results are obtained if the heat treatment is performed *in-situ* during the catalytic tests (Table I). In rutile samples, the amount of insoluble V^{IV} increases further up to a limiting value of around 3.8% w/w after long-term catalytic tests. The redox and chemical (solubility) properties of the V^{IV} sites are altered considerably by the interaction with the TiO_2 surface, in comparison with those of the V^{IV}- oxide. XRD and ESR data clearly exclude the formation of a solid solution in anatase samples, in contrast to that observed for rutile samples. In particular, ESR characterization [23] shows the presence in anatase samples of several isolated surface and unsaturated vanadyl ions in slightly different distorted octahedral environments. Some of these species may interconvert with

Table I Chemical analysis data (± 0.15%) of the distribution of vanadium oxide species in samples prepared by solid state reaction of V_2O_5 (7.7% w/w) and TiO_2.

Nature of treatment	TiO_2 Anatase (9.8 m^2/g)				TiO_2 Rutile (8.9 m^2/g)			
	% w/w of equivalent V_2O_5				% w/w of equivalent V_2O_5			
	V(IV)	V(V)	V(IV)	V(V)	V(IV)	V(V)	V(IV)	V(V)
	insoluble		*soluble*		*insoluble*		*soluble*	
a	-	-	-	7.7	-	-	-	7.7
b	0.9	-	-	6.8	0.8	-	-	6.9
c	0.9	-	6.8	-	1.8	-	5.9	-
d	0.9	-	0.3	6.5	1.8	-	0.3	5.6
e	0.9	-	3.1	3.7	1.6	-	1.2	4.9
f	0.9	-	2.0	4.8	3.7	-	0.9	3.1

(a) mixing; (b) calcination at 500°C for 24 hours; (c) calcination in the presence of a reducing agent (10% v/v of NH_3); (d) sample **c** after subsequent calcination at 400°C for 3 hours; (e) sample **a** after heating (320°C) in a flow of o-xylene/air for 24 hours or (f) for 1440 hours.

each other by the addition of suitable probe molecules, indicating the presence of Lewis acidity. In rutile samples, in contrast, no isolated surface vanadyl species could be detected, even for amounts of insoluble V^{IV} species much below those necessary for monolayer coverage. The ESR spectrum is always characterized by a broad unstructured signal with a g value of about 1.98 that is characteristic of near-lying V^{IV} paramagnetic centres. This broad signal is overlapped by another signal showing hyperfine structure which can be attributed to isolated non-vanadylic V^{4+} sites in substitutional positions in the rutile structure.

A further difference characterizes the insoluble V^{IV} sites in anatase and rutile samples (Table II). Whereas in anatase samples all the V^{IV} sites could be reduced and are accessible to gaseous reactants such as H_2, only a fraction (around 20-30%) of the insoluble V^{IV} sites in rutile samples are accessible to gaseous reactants, indicating that only a reduced fraction of these sites is localized at the surface or in subsurface layers. A further difference between insoluble V^{IV} sites in anatase and rutile samples is shown in Table II. V^{IV} sites can be more easily reduced to V^{III} in anatase samples as compared to rutile samples, but cannot be oxidized to V^{V} as occurs in rutile samples.

ESR and reactivity data thus indicate a homogeneous distribution of these V^{IV} sites on the surface of anatase, with a mean estimated distance of 4.5 Å from V centers and the presence of islands of V_2O_4 on the surface of the rutile samples.

The presence of these V^{IV} surface sites also modifies the reactivity in o-xylene oxidation of titania which is enhanced considerably as compared to pure TiO_2, even though the activation of o-xylene is relatively not selective to phthalic anhydride (around 30%). The catalytic behavior is stable in the case of V^{IV}- modified TiO_2 anatase, whereas the selectivity to phthalic anhydride improves with time on-stream in the rutile sample as a consequence of the possibility of oxidation of insoluble V^{IV} to V^{V} in these catalysts (see Table II).

Nature of the Active Layer. On the V^{IV}-modified surface of TiO_2 after calcination only V_2O_5 crystallites are present, but this phase transforms to a partially reduced amorphous phase during the consecutive *in-situ* treatment in a flow of o-xylene/air. Stable catalytic behavior may be reached in about 500 hours of time on stream. The characterization of the nature of the upper layer on V^{IV}-modified TiO_2 surfaces may be realized after its extraction with a dilute sulphuric acid solution. The analysis of this upper layer indicated (**i**) a mean valence state of vanadium of 4.71 that corresponds to a V^{V}:V^{IV} ratio of 2:1 and (**ii**) the presence of a characteristic IR band centred at 995

Table II Rates of reduction (2% H_2 in helium) and of consecutive oxidation (20% O_2 in nitrogen) at 400°C of the insoluble V^{IV} species in anatase and rutile samples after long-term catalytic tests.

sample	amount of insoluble V^{IV}	Rate of Reduction moles O removed/ mmoles V_2O_4	Fraction Reduced of insoluble V^{IV} %	Rate of Oxid. mmoles O inserted/ mmoles V_2O_4	Fraction Oxidized of insoluble V^{IV} %
anatase	0.9	10.97	100	0	0
rutile	3.7	1.13	24	0.12	23

cm^{-1} due to the symmetrical stretching mode of V^V=O. The removal of this phase decreases the activity in o-xylene oxidation but especially the selectivity, which drops from about 75% to 30%. The phase is XRD amorphous and no evidence was found of the presence of residual V_2O_5 crystallites characterized by a defined sharp band at 1020 cm^{-1} of V=O stretching mode. The shift to lower frequencies of νV=O in comparison to crystalline V_2O_5 may be attributed to the electronic effect of neighboringV^{IV} sites or to the presence of coordinatively adsorbed water that causes a weakening of the V^V=O double bond. In general, FT-IR spectra of the catalyst after different times show that a good correlation exists between the frequency of the V=O stretching band in the samples and the oxidation state of the vanadium oxide deposed on the surface of the TiO_2. The band changes position in the spectrum and decreases in frequency from 1020 cm^{-1}, corresponding to pure crystalline V_2O_5, to lower values proportional to the degree of reduction. In agreement, the consecutive oxidation of a V-TiO_2 sample after long-run catalytic tests indicates a shift to higher frequencies and the appearence of a further band at about 1010 cm^{-1}. A corresponding increase in the mean oxidation state of vanadium from 4.72 to 4.91 is observed. This suggests that the modification of initial V_2O_5 particles in the reaction medium is not completely reversible by consecutive reoxidation.

Similar results are found in rutile samples, but both the time necessary to reach a certain mean valence state of vanadium as well as the stability of the reduced catalyst against consecutive oxidation are indicative of the formation of less stable, partially reduced, vanadium-oxide species on the rutile surface in comparison to the anatase surface.

Wide line solid state ^{51}V-NMR characterization of the local coordination of V^{5+} sites in the active phase [24] indicates a significant shift of the asymmetric resonance peak due to axial shielding and a general broadening of the peak as compared to the reference signal for V_2O_5. The change is analogous to that observed in hydrated V_2O_5 and could be interpreted as a change from the nearly five-fold coordination of vanadium in the initial crystalline V_2O_5 to a nearly octahedral coordination in the active phase after long-term catalytic tests. No great differences in the ^{51}V-NMR spectra are observed in the anatase and rutile samples after long-term catalytic tests.

XPS characterization of the depth profile of vanadium in anatase samples [24] shows a considerable change in the V/Ti atomic ratio (from about 0.47 to about 0.27) after removal of the first *nm* of thickness from the catalyst using Ne^+ ions for the sputtering. The V/Ti atomic ratio then decreases at a slower rate to nearly zero with the removal of a further 12-14 *nm*. This indicates the presence of two V^V phases on the surface of TiO_2 anatase, the first corresponding to the monolayer and the second present in amorphous aggregates with estimated thicknesses of about 10-15 *nm*. In rutile samples, on the contrary, the presence of the first monolayer species is not observed, but rather only the second species.

Dynamics of in-Situ Evolution. Fig. 1 shows that the V_2O_5-TiO_2 catalyst in a pilot plant reactor undergoes a first deep reduction and then partial reoxidation with the formation of the final active

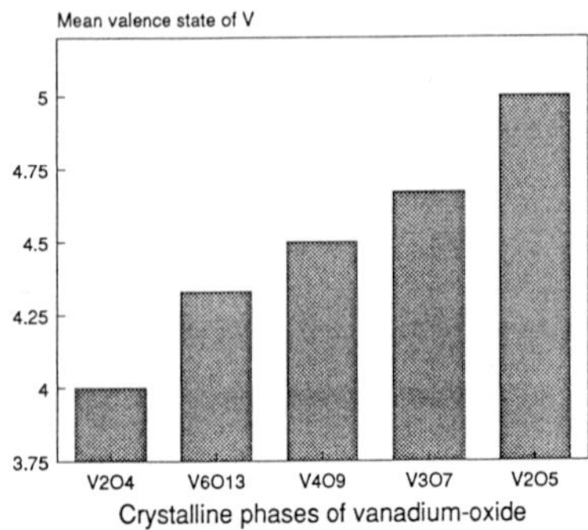

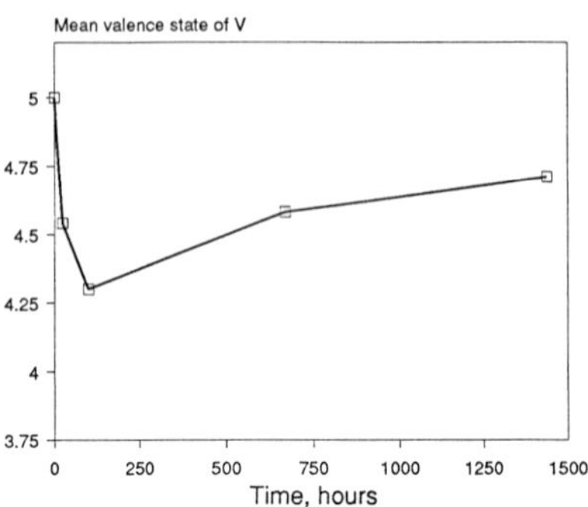

Fig. 1 Effect of time on-stream in o-xylene oxidation on the mean valence state of vanadium in the soluble part of V-oxide on TiO_2 and mean valence state of vanadium in some crystalline phases of vanadium-oxide.

catalyst whose characteristics were discussed above. The V_2O_5 is first reduced to a phase with a valence state similar to that of V_6O_{13}, however XRD analysis shows the presence of only an amorphous vanadium-oxide phase. It is thus not possible to make a definite attribution. After this stage, the catalyst starts to be progressively reoxidized and reaches a final stable mean valence state in the soluble part of vanadium similar to that present in the V_3O_7 phase. Also in this case, the vanadium-oxide phase is XRD amorphous. It should be noted that the crystalline V_3O_7 is prepared by solid state reaction of V_6O_{13} with V_2O_5 [25] and it is reasonable to hypothesize that a similar mechanism occurrs in the transformation of the V_2O_5-TiO_2 mixture to the final active catalyst. The correlation of this effect with the catalytic behavior is complex because two different catalytic effects take place at the same time: **1**) The spreading of vanadium on the TiO_2 surface, and **2**) the reduction and consecutive partial oxidation of the V-oxide upper layer and the consequent change in the nature of the supported phases.

In order to obtain a better understanding of the dynamics of these redox transformations as well as the role of titanium oxide in determining the final state of the catalyst, we carried out an analogous experiment where the evolution of an unsupported pure commercial V_2O_5 (Fig. 2) was followed. In this case, the concurrent process of spreading of vanadium-oxide on the surface of the titanium oxide is not present.

The commercial V_2O_5 also undergoes a similar reduction-reoxidation process of the supported V_2O_5 (Fig. 2), but the parallel change in the catalytic behavior may be more clearly correlated to the dynamics of phase transformation. The selectivity in phthalic anhydride is very high at the beginning, but decreases reaching a minimum value after about 40 h, and then increases by further *in- situ* treatment up to about 200 hours. A parallel evolution is observed in the activity, whereas the selectivities to CO_x and phthalide pass through a maximum. The FT-IR characterization of the catalyst after 200 hours of time on stream indicates the presence of a complex spectrum, whose characteristics are very similar to those of V_3O_7 plus some residual V_2O_5 particles (Fig. 3). This compound has a V^V:V^{IV} ratio (2:1) similar to that present in the the active phase of the V/Ti/O catalyst after 200 hours in a stream of o-xylene/air (see Fig. 2). It should be observed that the IR spectrum of V_2O_5 after *in-situ* conditioning (Fig. 3) is very different from that shown by V_2O_5

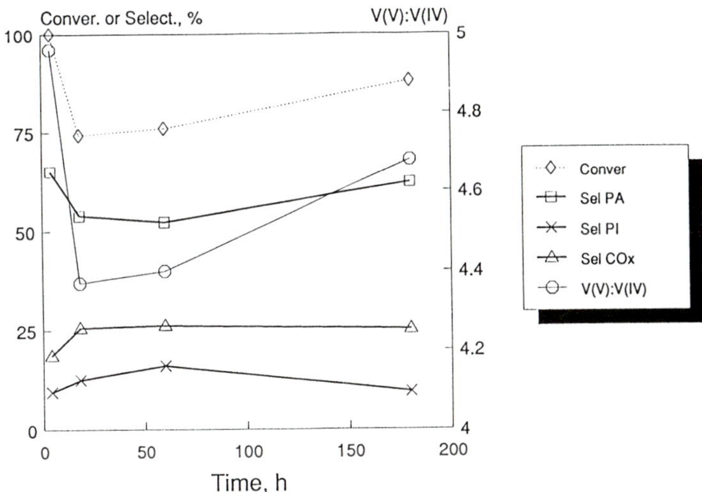

Fig. 2 Effect of time on stream on the catalytic behavior in o-xylene conversion at 327°C and on the mean valence state of vanadium of a commercial sample of V_2O_5

supported on TiO_2 after similar treatment (a broad band centred at around 995 cm^{-1}), even though in the latter case frequencies below about 900 cm^{-1} cannot be analyzed due to the stronger adsorption of TiO_2 bands. The differences in the bands in the 1100-900 cm^{-1} region may suggest that the nature of the final vanadium-oxide active phase on TiO_2 is different from that found for the unsupported V_2O_5, notwithstanding the analogous V^V:V^{IV} ratio of 2:1 and the relatively similar catalytic behavior. The selectivity to phthalic anhydride of the V-oxide supported on TiO_2, in fact, is about 75%, at 95% conversion compared to a selectivity of about 68% of unsupported V_2O_5. The lowering of the selectivity for phthalic anhydride in the latter case is due mainly to the formation of larger amount of phthalide rather than to the formation of larger amounts of CO_x. It should also be noted that the presence of coordinatively adsorbed water, whose amount is certainly proportionally higher on a vanadium-oxide phase spread on the TiO_2 surface in comparison to the same phase alone, can induce a shift of the V=O stretching frequency to lower energies and a broadening of the band in the IR spectrum. The differences in the IR spectra for supported and unsupported vanadium oxide species are thus not alone indicative of a real difference in the local structure of vanadium oxide. For example, in a hydrated V_2O_5 gel [26] a similar strong IR spectral perturbation is present in comparison to calcined V_2O_5, even though the local structure (short range order) around the vanadium centres is relatively similar.

In order to verify further the relationship between *in-situ* evolution of the catalyst and catalytic behavior, the time-on-stream evolution of a prereduced vanadium-oxide catalyst was followed. This catalyst was prepared by precipitation of V^{IV} and calcination for a short time at 280°C. The mean valence of the starting sample was 4.84, near to that found at the end of the evolution of V_2O_5-based catalysts. This sample, after an initial rapid *in-situ* modification, shows an almost constant catalytic behaviour over a long time. It exibits good activity and good selectivity in phthalic anhydride (Fig. 4) comparable to that of the V/Ti/O catalyst. The IR spectrum is very similar to that of pure V_3O_7 and the final valence state after the catalytic test is 4.62.

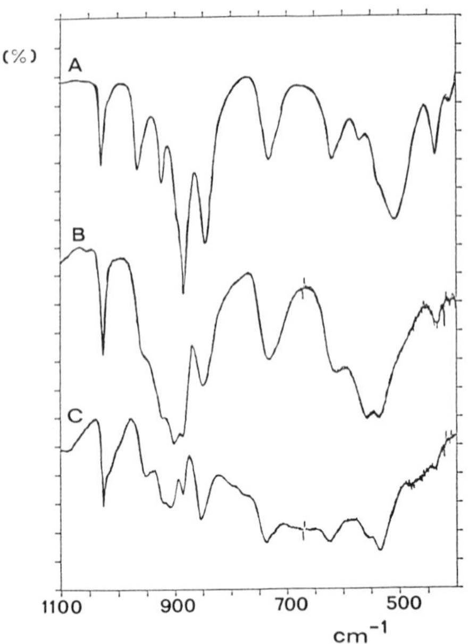

Fig. 3 FT-IR spectra of (**A**) V_3O_7, (**B**) prereduced V-oxide and (**C**) commercial V_2O_5 after 200 hours of time on-stream in o-xylene conversion.

These data further indicate that, notwithstanding the differences in the V-oxide phases present in supported or unsupported V_2O_5-based catalysts, the same selectivity and maximum yield to phthalic anhydride may be observed when a suitable starting V-oxide is used.

Relationship between surface structure on VTiO catalysts and catalytic activity. The physicochemical characterization of the catalysts prepared by solid state reaction of V_2O_5-TiO_2 and *in-situ* consecutive transformation during catalytic tests suggests the following picture of the active surface. The surface of the anatase samples is characterized by a mixed valence V^{IV}-V^{V} mono- or bi- layer in which *strongly interacting* V^{IV} species and *weakly-interacting* V^{V} species are present. A possible model of this surface, in agreement with physico-chemical characterizations [23,24], is illustrated in Fig. 5. This mono- or bi-layer covers the entire surface of the TiO_2. In addition to this phase a massive amorphous partially reduced (V^{V}:V^{IV} ratio of about 2:1) vanadium-oxide phase is also present, with an estimated thickness of about 10-15 *nm*. A slightly different picture may be proposed for the rutile samples. In this case the massive phase is still present, but the well spread mono- or bi-layer phase is absent. Chains of edge-sharing octahedra, linked to each other by shared vertices, are present in the rutile structure. In the anatase structure, the TiO_6 octaheda share four edges as compared to the two in rutile. As a consequence of such sharing, the O-O distances are short and the structure is more compact. Thus bulk diffusion of the V^{4+} sites is more difficult in the anatase structure and easier in the rutile structure, due also to the presence of cavities. This process of bulk diffusion competes with that of surface diffusion, causing the formation of islands of V(IV). These islands cover only a fraction of the titania surface. In the reaction medium, the V(IV) is partially oxidized to V(V) and thus also in this case a mixed valence vanadium-oxide is present on the surface. However, it does not cover the surface homogeneously as indicated for the anatase surface.

In conclusion, we agree with literature data [1-22] in indicating a different dispersion of the vanadium-oxide monolayer on the anatase and rutile TiO_2 surfaces, but it should be emphasized that in addition to this phase, massive (partially reduced) vanadium oxide may be present which

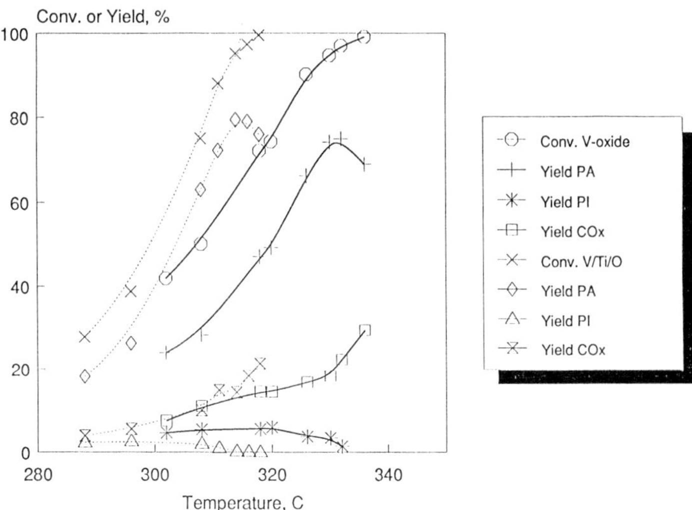

Fig. 4 Effect of reaction temperature on the catalytic behavior of VTiO samples (dashed line) and of prereduced unsupported V-oxide (solid line).

also determines the catalytic behavior. In anatase samples this massive reduced vanadium-oxide phase is more stable in reforming V_2O_5 by oxidation than the corresponding phase on rutile samples. This may be interpreted as a different stabilization due to better interaction with the V^{IV}-V^{V}-modified TiO_2 anatase surface in comparison to the direct interaction with the titania surface such as occurs in rutile samples.

The comparison of the catalytic behavior of the different samples suggests that the spreading of vanadium-oxide on TiO_2 causes an enhancement of the specific activity per gram of vanadium oxide. However, the possibility of obtaining comparable selectivities and yields of phthalic anhydride also when a suitable unsupported vanadium-oxide is used indicates that a unique special surface structure of vanadium-oxide leading to considerably superior catalytic performances is not stabilized on the TiO_2 surface. The slightly lower mean valence state in unsupported vanadium

Table III Comparison of the catalytic behavior in o-xylene oxidation of catalysts prepared by solid state reaction of V_2O_5 and TiO_2 (anatase or rutile) and of a monolayer catalyst prepared by wet impregnation with V^{IV}-oxalate of TiO_2 anatase.

Type of preparation	TiO_2	Surface area,m^2/g	% w/w of V_2O_5	T ,°C (a)	Y-max PA (b)	S-PA (c)	S-PI (c)
solid state	anatase	9.8	5.0	290	64	75	7.7
solid state	rutile	8.9	5.0	326	65	57	15.4
wet impreg.	anatase	9.8	1.3	312	63	77	5.2

(a) temperature of 50% o-xylene conversion; (b) maximum yield found of phthalic anhydride; (c) selectivity at 50% o-xylene conversion of phthalic anhydride (PA) and of phthalide (PI). Reaction conditions as reported in the experimental part.

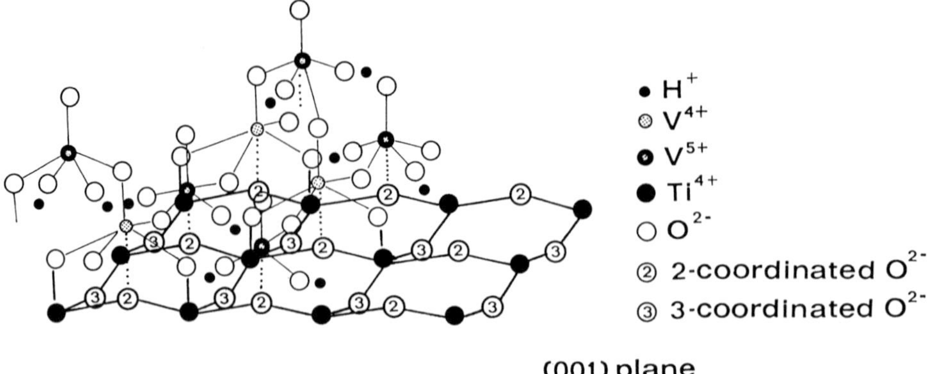

Fig. 5 Model of the surface structure of th V^{IV}-V^{V} first layer on the surface of TiO_2 anatase during o-xylene oxidation.

oxide (around 4.61) compared to supported samples (4.71) after similar times in contact with the o-xylene/air mixture (around 200 h at 330°C) may explain the slight improvment (Fig. 4) in the selectivity for phthalic anhydride in the latter sample. The role of TiO_2 thus may be related to the stabilization in the reaction conditions of a surface configuration of vanadium-oxide with a slightly higher V^{V}:V^{IV} ratio and consequently a decrease in phthalide formation and a parallel increase in the selectivity to phthalic anhydride. The comparison between the catalytic results (after long term experiments) of anatase samples prepared by solid state reaction or by impregnation, with that obtained on rutile samples (Table III) again illustrates that the presence or absence of a vanadium-oxide mono-layer on the anatase TiO_2 surface is not the determining factor to obtain selective catalysts for o-xylene oxidation. The differences in the selectivity and activity are very small and generally related to an increase in the formation of the intermediate phthalide. The differences are thus related to small differences in the number of sites (activity) more than in the stabilization of a special unique structure on the surface of TiO_2, even though second order effects of stabilization on the various vanadium-oxide phases due to interaction with the TiO_2 surface are present.

It also should be pointed out that the present data clearly show the evolution of the V- oxide species in contact with the o-xylene/air stream and that any possible correlation between structure and activity of the catalyst must take into account the nature of the *in-situ* dynamics of the active species.

REFERENCES

[1] Bond, G.C., Flamerz, S., Shukri, R., *Faraday Discuss. Chem. Soc.*, **87**, 65 (1989).
[2] Wachs, I.E., Saleh, R.Y., Chan, S.S., Chersich, C.C., *Appl. Catal.*, **15**, 339 (1985).
[3] Eckert, H., Wachs, I.E., *J. Phys. Chem.*, **93**, 6796 (1989).
[4] Bond, G.C., Brückman, K., *Faraday Discuss. Chem. Soc.*, **72**, 235 (1982).
[5] Vejux, A., Courtine, P., *J. Solid State Chem.*, **23**, 93 (1978).
[6] Gasior, M., Machej, T., *J. Catal.*, **83**, 472 (1983).
[7] Kang, Z.C., Bao, Q.X., *Appl. Catal.*, **26**, 251 (1986).
[8] Nakagawa, Y., Ono, T., Miyata, H., Kubokawa, Y., *J. Chem. Soc. Faraday Trans. I*, **79**, 2929 (1983).
[9] Inomata, M., Mori, K., Miyamoto, A., Ui, T., Murakami, Y., *J. Phys. Chem.*, **87**, 754 (1983).

[10] Miyamoto, A., Mori, K., Inomata, M., Murakami, Y. In *Proceedings, 8th Int. Congress on Catalysis*, Berlin 1984, Dechema Pub: Frankfurth AM 1984, Vol. IV, p. 285.
[11] Baiker, A., Dollenmeier, P., Glinski, M., Reller, A., *Appl. Catal.*, **35**, 351 (1987).
[12] Bond, G.C., Zurita, J.P., Flamerz, S., *Appl. Catal.*, **27**, 353 (1986).
[13] Hausinger, H., Schmelz, H., Knözinger, H., *Appl. Catal.*, **39**, 267 (1988).
[14] Kozlowski, R., Pettifer, R.F., Thomas, J.M., *J. Phys. Chem.*, **87**, 5176 (1983).
[15] Haber, J., Kozlowska, A., Kozlowski, R., *J. Catal.*, **102**, 52 (1986).
[16] Cavani, F., Centi, G., Parrinello, F., Trifiro', F. In *Preparation of Catalysts IV*, Delmon, B., Grange, P., Jacobs, P.A., Poncelet, G. Eds., Elsevier Science Pub.: Amsterdam 1987; p. 227.
[17] Busca, G., Marchetti, L., Centi, G., Trifiro', F., *J. Chem. Soc. Faraday Trans. I*, **81**, 1003 (1985).
[18] Busca, G., Marchetti, L., Centi, G., Trifiro', F., *Langmuir*, **2**, 568 (1986).
[19] Centi, G., Pinelli, D., Trifiro', F., *J. Molec. Catal.*, **59**, 221 (1990).
[20] Cavani, F., Centi, G., Foresti, E., Trifiro', F., Busca, G., *J. Chem. Soc. Faraday Trans. I*, **84**, 237 (1988).
[21] Fierro, J.L.G., Arrua, L.A., Lopez Nieto, J.M., Kremenic, G., *Appl. Catal.*, **37**, 323 (1988).
[22] Risiecka, M., Grzybowska, B., Gasior, M., *Appl. Catal., **10**, 101 (1984).*
[23] Centi, G., Giamello, E., Pinelli, D., Trifiro', F., *J. Catal.*, submitted.
[24] Centi, G., Guelton, M., Payen, E., Pinelli, D., Trifiro', F., *J. Catal.*, submitted.
[25] Waltersson, K., Forslund, B., Wilhelmi, K.-A., Andersson, S., Galy, J., *Acta Cryst.*, **30**, 2644 (1974).
[26] Repelin, Y., Husson, E., Abello, L., Lucazeau, G., *Spectrochimica Acta*, **41**, 993 (1985).

R.K. Grasselli and A.W. Sleight (Editors), *Structure-Activity and Selectivity Relationships in Heterogeneous Catalysis*

© 1991 Elsevier Science Publishers B.V., Amsterdam

MOLECULAR STRUCTURE-REACTIVITY RELATIONSHIPS OF SUPPORTED VANADIUM OXIDE CATALYSTS

G. Deo and I. E. Wachs
Zettlyemoyer Center for Surface Studies,
Department of Chemical Engineering,
Lehigh University, Bethlehem, PA 18015

ABSTRACT

The molecular structure of the surface vanadium oxide species present on different oxide supports (TiO_2, γ-Al_2O_3, and SiO_2) were determined by laser Raman spectroscopy and ^{51}V solid state NMR under hydrated and dehydrated conditions. The structure of the vanadium oxide species changes with dehydration and a four coordinated vanadium oxide species with a short terminal bond was present on all oxide supports considered. The reactivity of the supported vanadium oxide catalysts was determined via the methanol oxidation reaction. Correlation of the structure and reactivity data indicate the strength of the bridging, vanadium-oxygen-support, bond to be controlling the activity of these supported vanadium oxide catalysts. The effect of promoters/impurities on 1% V_2O_5/TiO_2 catalyst depends on their acid/base nature. Basic promoters titrate the vanadium oxide site and destroy the vanadium-oxygen-support bond of the parent 1% V_2O_5/TiO_2. Acidic promoters/impurities coordinate to the support and do not show any appreciable change to the structure, reactivity, and the vanadium-oxygen-support bond of the parent 1% V_2O_5/TiO_2.

INTRODUCTION

Supported vanadium oxide catalysts constitute an important class of oxidation catalysts which find a variety of uses in the petrochemical industry. Various studies have shown that supported vanadium oxide forms a two-dimensional metal oxide overlayer on the oxide supports which is structurally and catalytically different from bulk V_2O_5 [1-4]. The supported vanadium oxide phase is usually more active than bulk V_2O_5 during most oxidation reactions. The differences in catalytic activity of the various supported vanadium oxide catalysts have been attributed to the different structural modifications of the two-dimensional vanadium oxide overlayer. The structural modifications occur due to the vanadium oxide interaction with the surface of the oxide support. Furthermore, some researchers have attempted to correlate the activity of supported vanadium oxide catalysts, for some catalytic reactions, with the number of terminal V=O bond [5].

The present paper addresses this notion of structural differences for the supported vanadium oxide phase on different oxide supports (Al_2O_3, TiO_2 and SiO_2). The structural modifications of the supported vanadium oxide phase are studied using laser Raman spectroscopy and solid state ^{51}V NMR. The structure sensitive methanol oxidation reaction is used to probe the catalytic properties of the supported vanadium oxide phase. From these results it is possible to arrive at a conclusion regarding the structure-reactivity relationship of supported vanadium oxide catalysts.

EXPERIMENTAL

Support materials: The supports used in this study were TiO_2 (55 m^2/g) obtained from Degussa (P-25), γ-Al_2O_3 (180 m^2/g) obtained from Harshaw, and SiO_2 (300 m^2/g) Cab-o-sil.

Sample preparation: Vanadium tri-isopropoxide oxide (Alfa) was used as the precursor. The samples were made by incipient wetness impregnation of the precursor using methanol as the solvent. The impregnation was performed under a nitrogen atmosphere. The samples were then heated in nitrogen at 110 °C and finally calcined in oxygen at 500 °C. For the V_2O_5/TiO_2 samples the final calcination was done at 450 °C for 2 hrs. Details of the preparation technique have been outlined elsewhere [6]. To study the effect of promoters/impurities WO_3, Nb_2O_5, and K_2O were added, via incipient wetness impregnation of their respective precursors, to previously prepared 1% V_2O_5/TiO_2.

Laser Raman: The Raman spectra for the catalysts under ambient conditions were collected using low laser power, usually less than 20 mW. Laser induced dehydration studies were also carried out for some catalysts using higher laser powers, usually greater than 100 mW. Details of the Raman equipment have been described elsewhere [7].

NMR: Solid state ^{51}V NMR data were collected at room temperature using a General Electric Model GN-300 spectrometer. Details of the setup have been given elsewhere [4]. Dehydration experiments were performed at temperatures between 150 and 400 °C for 0.5-1 hour at 10^{-3} Torr in flame sealed containers. Within these limits the solid state ^{51}V NMR spectra showed little dependence on the dehydration conditions.

Catalytic studies: Methanol oxidation reaction was carried out in an isothermal fixed-bed differential reactor which was operated at atmospheric pressure and temperature of 230 °C. The mixture of methanol, oxygen, and helium were in the ratio 6/11/83 (molar %) and total flowrates of 25-100 sccm were employed in order to maintain $< 5\%$ conversion. The reactor was vertical and made of 6mm O.D. Pyrex glass. The catalyst was held at the middle of the tube between two layers of quartz wool. The gas flow was from the top to the bottom. Analysis of the product stream was performed using an on line gas chromatograph equipped with an FID and two TCD's. Due to the high activity of V_2O_5/TiO_2, γ-Al_2O_3, and V_2O_5/γ-Al_2O_3 these catalysts were diluted with SiO_2 to maintain conversions < 5%. For the catalytic runs, the activities and selectivities were reported as initial values. The activities for the different catalysts were converted to turnover numbers (t.o.n) which is defined as the moles of methanol converted per mole of surface vanadium atom per sec. For bulk V_2O_5 the area for a mole of $VO_{2.5}$ is known [1] and from the knowledge of the surface area of the bulk V_2O_5 material (~4 m^2/g) the t.o.n. can be determined.

RESULTS

1-20% V_2O_5/γ-Al_2O_3: The Laser Raman and ^{51}V solid state NMR spectra of these samples have been reported before. Under ambient conditions these catalysts are known to possess primarily four coordinated (metavanadate) structures at low loadings and six coordinated (decavanadate) structures at high loadings [4,6]. Dehydration of low vanadium oxide coverage catalysts (4% V_2O_5/γ-Al_2O_3) show vanadium oxide to be only four coordinated. Laser induced dehydration studies on these catalysts performed with the Raman spectrometer show the presence of a Raman band arising at ~1022 cm^{-1} which is assigned to a short terminal V=O bond. Previous *in situ* Raman studies show this terminal band to be at 1034 cm^{-1} [8].

1-8% V_2O_5/TiO_2 (Degussa): Laser Raman and ^{51}V solid state NMR spectra show that under ambient conditions four coordinated vanadium oxide structures form at low coverages which become primarily six coordinated at higher coverages similar to V_2O_5/γ-Al_2O_3 [4,9]. However, differing from vanadium oxide supported on γ-Al_2O_3 there is a preference for six coordinated vanadium oxide species on TiO_2. Dehydration of low vanadium oxide coverage V_2O_5/TiO_2 catalysts changes the vanadium oxide coordination from six to four, and the four coordinated species has a similar solid state ^{51}V NMR spectra as the dehydrated 4% V_2O_5/γ-Al_2O_3 catalyst [4]. Raman spectra of laser induced dehydrated V_2O_5/TiO_2 samples show the presence of a short V=O bond (Raman band at 1035 cm^{-1} [10]).

0.5-1.5% V_2O_5/SiO_2: From the Raman spectra of these catalysts no features of crystalline V_2O_5 were observed. Raman features of laser induced dehydrated samples show the presence of a terminal V=O band at 1033 cm^{-1}. Oyama et al. report the band position to be at 1042 cm^{-1} from *in situ* studies at 373 K after calcination of the V_2O_5/SiO_2 catalyst at 750 K [11]. Lischke et al. using uv-vis spectroscopy proposed the formation of tetrahedrally coordinated V^{5+} species upon dehydration [12].

Promoters/Impurities on 1% V_2O_5/TiO_2: The Raman spectra of

K_2O/1% V_2O_5/TiO_2, under ambient conditions, exhibit a shift of the terminal V=O band to lower wavenumbers compared to 1% V_2O_5/TiO_2. This indicates the abundance of tetrahedrally coordinated vanadium oxide species. The Raman spectra of dehydrated K_2O/1% V_2O_5/TiO_2 is stricking due to the absence of the ~1030 cm^{-1} band [13]. The Raman spectra of WO_3/1% V_2O_5/TiO_2 and Nb_2O_5/1% V_2O_5/TiO_2, under ambient conditions, exhibit a shift of the terminal V=O band to higher wavenumbers compared to 1% V_2O_5/TiO_2. This indicates a predominance of octahedrally coordinated vanadium oxide species. The Raman spectra of dehydrated WO_3/1% V_2O_5/TiO_2 and Nb_2O_5/1% V_2O_5/TiO_2 show the presence of a terminal V=O band at ~1030 cm^{-1} in addition to Raman bands of molecularly dispersed WO_3 and Nb_2O_5 [13,14].

Catalytic Studies: The main products of the methanol oxidation reaction are dimethyl ether, formaldehyde, methyl formate, methylal, CO and CO_2. Blank experiments without catalyst were performed to check the reactivity of the Pyrex tube and quartz wool. These experiments did not lead to any conversion taking place at the conditions considered. The t.o.n. gives a measure of the efficiency of each surface vanadium atom and the Raman/NMR characterization studies demonstrate that only atomically dispersed vanadium oxide was present in these samples.

Table 1 shows the activity and selectivity of the different support materials. It can be seen from this table that the activity of both TiO_2 and SiO_2 are low at these reaction conditions. The activity of γ-Al_2O_3, on the otherhand, is high and a 100% selectivity towards dimethyl ether is observed.

Table 1. Activity of Support Material for Methanol Oxidation Reaction (230 °C)

Support Material	Activity (mmole CH_3OH/g.cat./hr)	Selectivity (%)		
		HCHO	CH_3OCH_3	CO/CO_2
SiO_2	1.0	---	---	100
Al_2O_3	100.0	---	100	---
TiO_2	2.3	9.5	90.5	---

Figure 1 shows the t.o.n. (multiplied by 1E+4) of 1% V_2O_5 dispersed on different oxide supports. The 1% V_2O_5/TiO_2 catalyst is the most active and the t.o.n. is 2-3 orders of magnitude greater than bulk V_2O_5, 1% V_2O_5/γ-Al_2O_3 and 1% V_2O_5/SiO_2. The activity of the 1% V_2O_5/γ-Al_2O_3 catalyst is mainly due to γ-Al_2O_3 which greatly overshadows the effect of supported vanadium oxide and only a trace amount of methylal is observed. The amount of methylal formed during the methanol oxidation of 1% V_2O_5/γ-Al_2O_3 was used to calculate the t.o.n. for this sample.

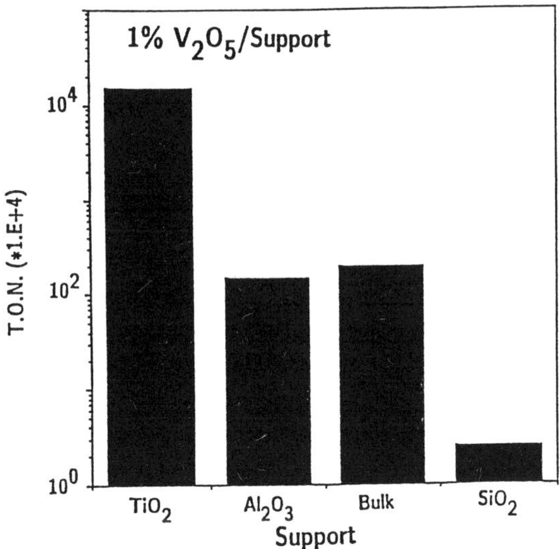

Fig. 1. Turnover Number for Methanol Oxidation Over Supported Vanadium Oxide Catalysts.

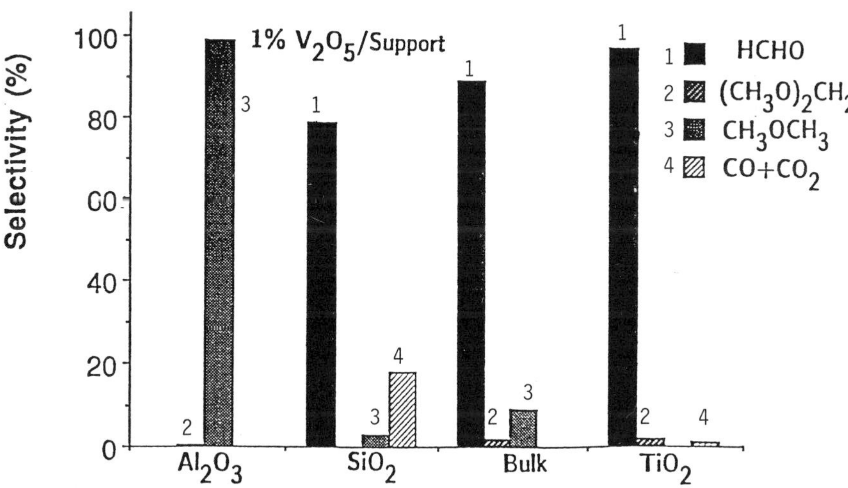

Fig. 2. Reaction Selectivities for Methanol Oxidation Over Supported Vanadium Oxide Catalysts.

The reactivity (t.o.n.) of the supported vanadium oxide catalysts exhibit the following trend:

$$1\%\ V_2O_5/TiO_2 > V_2O_5 \sim 1\%\ V_2O_5/\gamma\text{-}Al_2O_3 > 1\%\ V_2O_5/SiO_2.$$

Figure 2 shows the selectivity of the different catalysts. 1% V_2O_5/TiO_2 shows an 98% selectivity towards HCHO. On the otherhand the selectivities towards HCHO was 89% for bulk V_2O_5 and 79% for 1% V_2O_5/SiO_2. For the 1% V_2O_5/Al_2O_3 a high selectivity towards CH_3OCH_3 was observed which is typical of the support (γ-Al_2O_3). Other oxidation products were produced in minor amounts.

The reactivity 1% V_2O_5/TiO_2 during the methanol oxidation reaction is dramatically reduced with the addition of K_2O. Compared to 1% V_2O_5/TiO_2, the activity of 4% K_2O/1% V_2O_5/TiO_2 decreases by orders of magnitude. The reactivity of WO_3/1% V_2O_5/TiO_2 and Nb_2O_5/1% V_2O_5/TiO_2, on the other hand, are similar to 1% V_2O_5/TiO_2 catalyst.

DISCUSSION AND CONCLUSION

The Raman spectroscopy and solid state ^{51}V NMR studies reveal that at low vanadium oxide loadings the supported vanadium oxide phase is present as a two-dimensional metal oxide overlayer on the surface of the oxide support (100% dispersion). Under *in situ* conditions, where the oxide surfaces are dehydrated, the surface vanadium oxide phases possess tetrahedral coordination on all the oxide supports (Al_2O_3, TiO_2, and SiO_2). The ^{51}V NMR line shapes indicates the presence of a highly symmetric species [4] and the Raman signal at ~1030 cm^{-1} indicates an extremely short V=O bond. These results suggest the formation of the following surface vanadium oxide species on the different oxide supports:

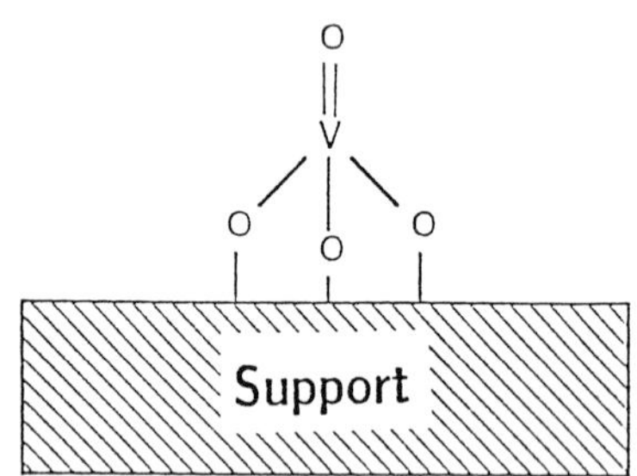

Thus, the molecular structure of the supported vanadium oxide phase is independent of the specific oxide support at low vanadium oxide loadings and the Raman band at ~1030 cm^{-1} is very indicative of this structure.

The reactivity of this highly distorted, tetrahedral

surface vanadium oxide species, however, is markedly dependent on the specific oxide support to which it is anchored. The combined structural characterization and catalytic studies suggest that the bridging oxygens, V-O-S (S=support), are responsible for the vast differences in catalytic activities since the terminal V=O bonds are not significantly influenced by the nature of the oxide support and possess Raman bands at ~ 1030 cm^{-1}. Under the chosen reaction conditions of an excess oxygen partial pressure the rate determining step is the extraction of oxygen from the surface vanadium oxide species. The influence of the oxide support on the rate of oxygen removal from the V-O-S bond would be expected to exert its greatest influence on the overall reaction rate. Indeed, this is exactly what is observed and the overall reaction rate correlates with the ease of oxygen removal from the different oxide supports since titania is significantly easier to reduce than alumina and silica [15].

The addition of promoters/impurities to 1% V_2O_5/TiO_2 has a pronounced effect on its structure and reactivity. Basic promoters/impurities titrate the surface vanadium oxide species and coordinate with the surface vanadium oxide species in 1% V_2O_5/TiO_2. This is evident from the absence of the ~1030 cm^{-1} band in the dehydrated Raman spectra of K_2O/1% V_2O_5/TiO_2. As a result, the vanadium-oxygen-support bond is destroyed and the activity of K_2O/1% V_2O_5/TiO_2 is reduced dramatically. Acidic promoters/impurities on 1% V_2O_5/TiO_2 coordinate directly to the TiO_2 support without drastically influencing the vanadium oxide four coordinated species. The direct coordination to the support of acidic promoters on 1% V_2O_5/TiO_2 catalysts is evident as the ~1030 cm^{-1} Raman band remains unaltered for the WO_3/1% V_2O_5/TiO_2 and Nb_2O_5/1% V_2O_5/TiO_2 catalysts. Hence, no appreciable change occurs to the vanadium-oxygen-support bond of these acidic promoted catalysts and the acitivity remains similar to 1% V_2O_5/TiO_2.

The reactivity of the distorted, tetrahedral surface vanadium oxide species appears to strongly depend on the strength of the vanadium-oxygen-support bond and correlates with the ease of oxygen removal from the oxide support. This conclusion is contrary to previous investigations on supported vanadium oxide catalysts which speculate that the vast differences in catalytic activities were due to the different vanadium oxide structures present in such catalysts and also to the strength and abundance of the terminal V=O bond.

ACKNOWLEDGMENT

We would like to thank Dr. H. Eckert for the helpful discussions. This study has been supported by the National Science Foundation grant # CBT-8810714

REFERENCES

1. (a) F. Roozeboom, T. Fransen, P. Mars, and P. J. Gellings,

Z. anorg. allg. Chem., 449 (1979) 25.
(b) F. Roozeboom, M. C. Mittelmeijer-Hazeleger, J. A. Moulijn, J. Medema, V. H. J. de Beer, and P. J. Gellings, J. Phys. Chem. 84 (1980) 2783.
(c) G. C. Bond, J. Sarkany, and G. D. Parfitt, J. Catal., 57 (1979) 476.
(d) G. C. Bond and K. Brukman, Faraday Disc., 72 (1981) 235.

2. (a) R. Y. Saleh, I. E. Wachs, S. S. Chan, and C. C. Chersich, J. Catal., 98 (1986) 102.
(b) I. E. Wachs, R. Y. Saleh, S. S. Chan, and C. C. Chersich, Appl. Catal., 15 (1985) 339.
3. (a) J. Haber, A. Kozlowska, and R. Kozlowski, J. Catal., 102 (1986) 52.
(b) H. Miyata, K. Fujii, T. Ono, Y. Kubokawa, J. Chem. Soc. Faraday Trans., 1, 83 (1987) 675.
(c) G. Bergeret, P. Gallezot, K. V. R. Chary, B. Rama Rao, and V. S. Subrahmanyam, Appl. Catal., 40 (1988) 191.
(d) J. Haber, A. Kozlowska, and R. Kozlowski, Proc. 9th Intl. Congr. Catal., (1988) 1481.
4. (a) H. Eckert, and I. E. Wachs, Mat. Res. Soc. Symp. Proc., 111 (1988) 459.
(b) H. Eckert, and I. E. Wachs, J. Phys. Chem., 93 (1989) 6796.
5. A. Miyamoto, Y. Yamazaki, M. Inomata, Y. Murakami, J. Phys. Chem., 85 (1981) 2366.
6. G. Deo, F. D. Hardcastle, M. Richards, and I. E. Wachs, Preprints Petrol. Chem. Div., ACS 34(3) (1989) 529.
7. I. E. Wachs, F. D. Hardcastle, and S. S. Chan, Mat. Res. Soc. Symp. Proc., 111 (1988) 353.
8. S. S. Chan, I. E. Wachs, L. L. Murrell, L. Wang, and W. K. Hall, J. Phys. Chem., 88 (1984) 5831.
9. J. M. Jehng, F. D. Hardcastle, and I. E. Wachs, Solid State Ionics, 32/33 (1989) 904.
10. C. Cristiani, P. Forzatti, and G. Busca, J. Catal., 116 (1989) 586.
11. S. T. Oyama, G. T. Went, K. B. Lewis, A. T. Bell, and G. A. Somarjai, J. Phys. Chem., 93 (1989) 6786.
12. G. Lischke, W. Hanke, H. -G. Jerschkewitz, and G. Ohlmann, J. Catal., 91 (1985) 54.
13. G. Deo and I. E. Wachs, unpublished results.
14. M. A. Vuurman, A. M. Hirt, and I. E. Wachs, to be submitted to J. Phys. Chem.
15. Y. Moro-oka, Y. Morikawa, and A. Ozaki, J. Catal., 7 (1967) 23.

R.K. Grasselli and A.W. Sleight (Editors), *Structure-Activity and Selectivity Relationships in Heterogeneous Catalysis*

© 1991 Elsevier Science Publishers B.V., Amsterdam

RELATIVE INFLUENCE OF STRUCTURE AND REACTIVITY OF V- AND Mo-CONTAINING CATALYSTS IN MILD OXIDATION OF HYDROCARBONS

E. BORDES

Département de Génie Chimique, Université de Technologie de Compiègne, B.P. 649, 60206 Compiègne Cédex, France

ABSTRACT

The structure and bulk reactivity of phases belonging to V-O and Mo-O systems are examined from the standpoint of their surface "crystal field" needed in selective mild oxidation of hydrocarbons (C_2-C_8). The reactivity (bulk and catalytic surface) of these M-O systems is modified by addition of another cation M' in variable amounts. It is shown that improved performance results when M', as a promoter or inside a support , contributes to stabilize a particular selective phase MO, or when a new compound MM'O with its own structure, reactivity and catalytic properties is obtained.

INTRODUCTION

In catalysis of mild oxidation of hydrocarbons, selectivity criteria already known concern directly or indirectly the structure and the reactivity of catalysts. The structural 3-D arrangement of metallic and oxygen atoms results from a set of interdependent properties (refs. 1, 2). Among them, acidity-basicity is involved since hard acid cations (ref. 3) ensure the activation of the organic molecule. Lattice oxygen ions behave as more or less hard base, acting as dehydrogenating sites leading to formation of water and/or oxygenating sites when O is inserted into the molecule. The small ionic radius and the high charge of the metallic ion M are responsible for the characteristic short covalent M=O and long ionic M...O bonds found when V or Mo are 6-coordinated, and layered structure often results. At the surface, short covalent M=O bonds act as sites and/or electron reservoir, while weak M...O bonds are either cut off, giving surface anionic vacancies (Φ), or restored in the presence of oxygen. The concept of catalytic anisotropy, which consists in differences of reactivity of different crystalline faces (ref. 4), was established from catalytic studies on such compounds, the planes of which offer a variety of strengths and energies of metal-oxygen bonds. Trying to determine the role of the structure leads therefore to describe the catalytic surface as needed by the transformation of a particular substrate into a particular product, that is the surface "crystal field".

Generally speaking, selective oxidations of hydrocarbons obey the kinetic scheme proposed by Mars and Van Krevelen (ref. 5) which takes into account the participation of lattice oxygen in the reaction through a redox mechanism. The own reactivity of the solid facing reactants determines the phases which are actually present at the steady state according to the rates of reduction and reoxidation. No new phase is detected in used molybdate catalysts although the participation of lattice oxygen is ascertained (ref. 6), while reduced phases are found in V-containing compounds. In the last case, the question is asked whether both oxidized and reduced phases are actually necessary in order for the catalyst to be selective, or if one of them could eventually play a negative role.

As shown in Table 1, the number of basic formulae known to be catalytically efficient in the main oxidation reactions is very small. Obviously, the more numerous the geometric and electronic constraints of the reaction, e.g. the 14-electron butane-maleic anhydride (MA) reaction, the less numerous are efficient catalysts. On the contrary, several compounds are active and selective in oxidative dehydrogenation provided that nucleophilic oxygens exist which take hydrogens off the

TABLE 1

Main oxidation reactions and corresponding selective catalysts.

REACTION		CATALYST		Additives	Ref.
Reactant	Product	KO*	K*		
Ethylene	Acetaldehyde	V_2O_5	V_4O_9	1 wt.% Pd	11
Propene	Acrolein	$Bi_2(MoO_4)_3$			12, 13
Propene	Acrylonitrile	USb_3O_{10}			14
Propene	Acrylonitrile	α-Te_2MoO_7			15
Butane	Maleic anhyd.		$(VO)_2P_2O_7$		16-21
Butene	Maleic anhyd.	$VOPO_4$	$(VO)_2P_2O_7$		2, 21-23
Butene	Butadiene	Bi_2MoO_6			24
Benzene	Maleic anhyd.	$(V_{0.66}Mo_{0.33})_2O_5$	$(V_{0.66}Mo_{0.33})_6O_{13}$		25, 26
o-Xylene	Phthalic anhyd.	V_2O_5	V_6O_{13}	TiO_2 anatase	27

* KO and K are respectively the oxidized and reduced forms of catalysts when both appear.

molecule. The influence of the surface crystal field is seen also in the fact that V-containing catalysts are able to oxygenate C_{2n} hydrocarbons while Mo-containing are better in oxidation of propene and branched isomers (isobutene,...). Recent developments on structure-sensitive reactions (refs. 1, 7-10) include a dynamic model of the selective oxidation of n-butane and butene in MA on (100) $(VO)_2P_2O_7$ (refs. 29-31), which shows that the transformation of one given molecule in a given product is related to a "cluster" of specific V-O and P-O sites. The properties of such clusters can be modified by addition of a second cation M'. According to its amount inside the original M-O system, M' will act as a promoter or as a support M'-O, or will be able to form a new compound MM'O with its own structure, reactivity and catalytic properties.

From these considerations emerges the fact that the reactivity of catalytic solids (bulk reactivity) is related with their surface reactivity during catalysis. The relations between structure and modes of reduction (and oxidation) will be first examined in order to account for the specific behavior of V-O and Mo-O systems in various catalytic reactions. In the second part the influence of a second cation in V-M'-O and Mo-M'-O systems will be studied in the same way.

1- STRUCTURE-REACTIVITY OF V-O AND Mo-O SYSTEMS

1-1. Structure and mechanism of reduction.

The crystal structures of the series $V_{2n}O_{5n-2}$ (V^{5+} to V^{4+}) and Mo_nO_{3n-1} (Mo^{6+} to Mo^{5+}) are made-up from distorted octahedra linked by corners and edges. Short vanadyl V=O and molybdenyl Mo=O bonds are present. In the case of MoO_3, the distortion is so large that its structure can be described with tetrahedra (ref. 33). The structure of suboxides originates from that of V_2O_5 and MoO_3 by Crystallographic Shear Plane (CSP). During reduction, the created anion vacancies (Φ) aggregate into a disc across which the crystal then collapses and shears, so that the vacant sites are eliminated and the previous cation coordination is restored. In oxidation a new anion plane is nucleated by aggregation of interstitial oxygen anions. For example, structures of V_6O_{13} and V_2O_5 are directly related by insertion of shear planes along [130]V_2O_5, so that the arrangement of the cleavage plane (010) V_2O_5 is retained in (001) V_6O_{13} (Fig. 1) (ref. 34). One important fact is that such

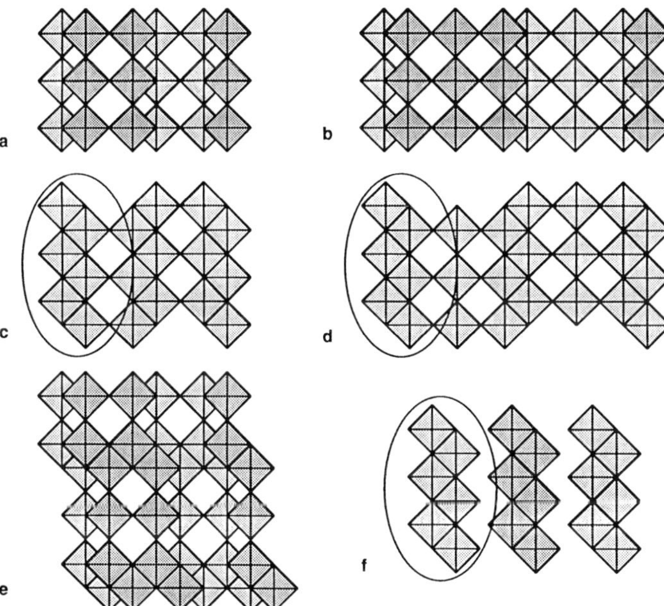

Fig. 1. Arrangement of octahedra in V-O, Mo-O and V-Mo-O related structures (idealized drawings). a) (001) plane of V_2O_5 and solid solution $(V_{1-x}Mo_x)_2O_5$; b) (010) V_2MoO_8 ; c) cleavage plane common to V_2O_5 (010) and V_6O_{13} (001) ; d) (001) V_2MoO_8 ; e) (010) V_6O_{13} ; f) (100) MoO_3 (perpendicular to (010) cleavage plane). Encircled area : framework common to these compounds.

lattices can also accommodate vacancies without reordering (point or extended defects), such as those occurring on the surface during catalysis or during partial reduction. Vacancies spread over the surface and/or diffuse into the bulk with rates depending on the solid and its morphology. The reduction of MoO_3 takes place at the surface of (010) layers but the migration of Φ is easier along the layers than perpendicularly to them (ref. 33). Therefore the extent of any reduction (or oxidation) is related to the area of the crystalline faces exposed, that is *in fine* to the morphology of the solid.

1-2. Reactivity of solid and catalytic reactivity.

When the solid faces reactants, modifications (reduction and oxidation) of surface and bulk occur according to the reducing power of the gaseous mixture and the rate of diffusion of Φ (or O). The above mechanisms apply, as Gai et al. showed by *in situ* electron microscopy experiments performed on vanadium oxides and MoO_3. For instance misfit screw dislocations, domains (leading to Mo_9O_{26}) and CSPs (leading to Mo_4O_{11}), were observed during reduction of MoO_3 by C_3H_6 up to 673 K (ref. 35). The easy release of one oxygen atom on the transformation from corner-sharing to edge-sharing octahedra has been supposed as one of the features responsible for easy insertion of oxygen into the molecule (ref. 36). For instance, the amount of energy required to remove one oxygen atom from MoO_3 is 25% smaller when accompanied by the structural rearrangement resulting in crystallographic shear (ref. 37). However, although some Mo^{5+} are detected in MoO_3 by ESR and UV-visible spectroscopy (ref. 38), the crystallization of Mo-O suboxides is generally hindered by the low rate of diffusion of vacancies through the layers, the unstability of $(MoO)^{3+}$ species and the immediate replenisment of Φ by oxygen of air. On the contrary, the same surface arrangement allows the topotactic growth of V_4O_9 and V_6O_{13} on the surface of V_2O_5, and bulk diffusion of Φ along tunnels [001] and [010]V_2O_5 is easy until the V_6O_{13} stoichiometry is reached (ref. 35). This accounts for the fact that pure V_6O_{13} and V_4O_9 are as active as prereduced V_2O_5 in oxidation of propene or of benzene (ref. 26), contrary to stoichiometric V_2O_5.

Once activated on a surface the hydrocarbon molecule has to find (at least) two kinds of lattice oxygen, one kind to be dehydrogenated and another to be oxygenated. The desired product will desorb only if it does not find surface sites to be overoxidized or decomposed. Selectivity is therefore related to a special distribution of polyhedra and adequation of valence, strength and energy of sites, in the "oxidized" and in the "reduced" states of the catalyst, which behave differently. Consequently we can assume that on V-O catalysts any reactant will find a pool of sites among which some can be active and selective for the chosen reaction. In turn, it is not easy to have the only one kind of sites which would be selective, and this is the reason why generally selectivity is best achieved when a second cation M' is present.

Obviously, activity and/or selectivity will not be observed when the natural morphology of the solid is not favorable, that is when the cleavage planes, which usually develop the largest area facing gaseous reactants, do not display the right sites. The case of MoO_3 has been largely debated, and the poor performance it exhibits in propene oxidation was attributed to the low area of the selective planes as compared to that of (010) cleavage plane (refs. 4, 8-10, 39). Special methods of preparation must be used in order to obtain samples developing the selective (101) planes yielding more acrolein from propene (refs. 10, 40). On the contrary, the cleavage (010) plane of V_2O_5 catalysts provides sites to (amm)oxidize more or less selectively various hydrocarbons which are olefins C_2-C_4 and aromatics C_6-C_8, and even 3-picoline or methanol (refs. 41, 42).

2- EFFECT OF A SECOND CATION M'

The influence of M' is different according to whether V- or Mo-based compounds are concerned. V-O structures are able to accommodate another transition metal while vanadium keeps its pseudo-octahedral coordination, even in brannerite-type MnV_2O_6 (ref. 43). Molybdenum can be

6-coordinated in mixed oxides (Bi_2MoO_6, V_2MoO_8...) or when supported in large amounts on TiO_2 anatase. In these cases layered structures made up with distorted polyhedra prevail, with various V-O (or Mo-O) bond strengths and energies, extended defects and mechanism of reduction by CSPs (refs. 1, 2, 19). Mo is 4-coordinated in almost molybdate salts and the structures and reactivity of these solids are different from the preceding ones.

2-1. Formation of definite compounds.

2-1.1. VPO phases : During the reduction of α- or β-$VOPO_4$ into $(VO)_2P_2O_7$ the single VO_6 octahedra (equatorially linked to phosphate) are paired by means of CSPs (along [110] and [211] in α and β respectively) as confirmed by electron microscopy (refs. 2, 19). The reduction of the δ and γ forms of $VOPO_4$, which appear in catalysts prepared from the precursor $VOHPO_4.0.5\ H_2O$, is performed only by means of gliding planes because pairs of edge-sharing octahedra already exist in the structures. The reduced phase found by XRD in selective catalysts after use is always $(VO)_2P_2O_7$ because possible rearrangements of the solid during reduction of $VOPO_4$ forms are limited by the presence of PO_4 tetrahedra. Phosphorus inside the V-O system modifies therefore its reactivity and particularly the thermodynamic potential of the redox couple V^{5+}/V^{4+}.

The mean maximum values of MA selectivity observed at high conversion (C = 95 mol.%) of butene and n-butane are 50 and 72 mol.% respectively. Apart from CO, CO_2, H_2O, eleven by-products, against two or three, are also respectively obtained (refs. 2, 16-23, 44). These differences cannot be solely due to the greater reactivity of butene as compared to n-butane, the more so because they are activated in the same temperature range (350-420°C) and C_4/O_2 ratio (1-1.5 % C_4/air). The active phase $VOPO_4$ contains V^{5+} (3 d^0) sites with anionic vacancies on which the π-allylic intermediate from butene is adsorbed. Since, owing to the bulk reactivity of $VOPO_4$, $(VO)_2P_2O_7$ is also found at the steady state (mean stoichiometry $VPO_{4.7}$) (refs. 2, 19), the hydrocarbons and intermediates can therefore react with several potentially active sites. On the contrary almost pure $(VO)_2P_2O_7$ is found in the best selective catalysts of oxidation of n-butane. Several workers have already suggested that MA selectivity from n-butane is related to the occurrence of (100) faces of $(VO)_2P_2O_7$ (refs. 18, 19, 45, 46). The area of (100) faces depends in turn on the crystallite size and morphology of the (100) layered precursor $VOHPO_4.0.5\ H_2O$ because the dehydration is topotactic (refs. 47, 48). Recent calculations were performed by Ziółkowski et al. using the Crystallochemical Model of Active Sites (CMAS), who presented a model of adsorption and transformation of butane and butene on $(VO)_2P_2O_7$ (refs. 30, 31). This model showed that an active "cluster", constituted by a set of sites on which adsorption, desorption, insertion of O, H, H_2O,..., species are possible, is displayed on (100) face and is necessary to oxidize selectively n-butane in MA (Fig. 2, molecule I). Partial extension of this model showed also the specific behavior of (100) as compared with other faces. In the case of butene on (100), the adsorption of C=C was found to proceed on unsaturated oxygens over vanadium. In this case the number of available oxygens to be inserted is limited, which accounts for the formation of butadiene (Fig. 2, molecule III), crotonaldehyde, dihydrofuran, acetaldehyde, etc,...and also CO, CO_2. The model therefore accounts for the formation of several by-products due mainly to the unwanted presence of crystalline $(VO)_2P_2O_7$ aside the more selective $VOPO_4$ phase.

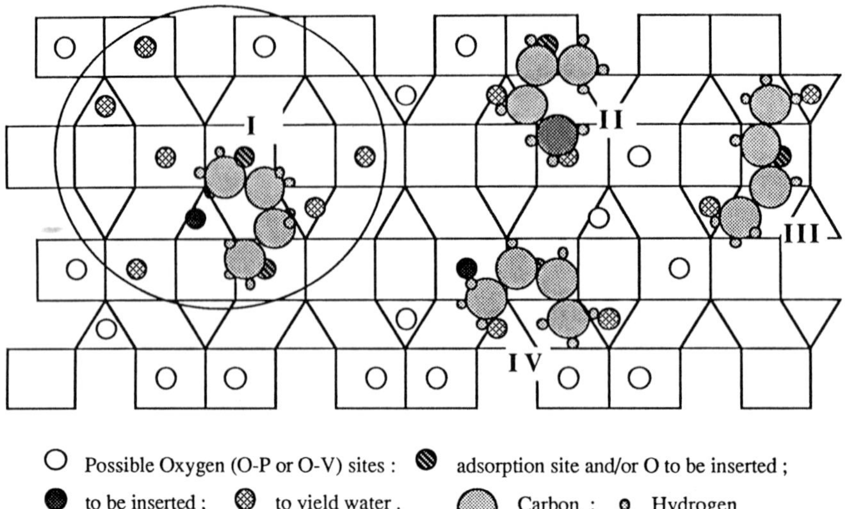

Fig. 2. Model of adsorption of C_4 on (100) $(VO)_2P_2O_7$. Molecule I = butane ; encircled area : cluster of sites involved in MA and H_2O formation. Molecules II-IV = butene ; various configurations of adsorption leading to different products by reaction with oxygen (for more details see refs. 30, 31).

2-1.2. VMoO phases : Two main phases are catalytically active, a solid solution of Mo in V_2O_5 $(V_{1-x}Mo_x)_2O_5$ (x = 0-0.33) and V_2MoO_8 (very close from $V_9Mo_6O_{40}$). The structures of these phases differ by the relative position of metallic cations in adjacent octahedra and the thickness of three octahedra instead of two for the latter, while the interconnexion of the slabs remains the same (Fig. 1). Catalytic properties are modified because the presence of Mo induces electronic changes related with a distortion of bonds intermediate in magnitude between that in V_2O_5 and that in MoO_3. Moreover VMoO catalysts are known to change markedly their chemical and phase composition during interactions with gaseous reactants because both solubility of MoO_3 in vanadium oxide and valence of both metals are modified. Germain et al. showed that the activity in benzene oxidation is maximum for the solid solution and decreases for V_2MoO_8 alone, while selectivity in MA remains constant (ref. 50). The reversible transformation of $(V_{0.7}Mo_{0.3})_2O_5$ in a mixture of V_2MoO_8 and V_2O_5, said to occur easily (ref. 49), would account for this behavior.

2-1.3. BiMoO phases : Different features are observed in Mo-M'-O systems owing to the structure and reactivity of $M'_x(MoO_4)_y$ molybdates. The diffusion of vacancies in the bulk is rate-limiting and attempts to find Mo^{5+}-O oxides in selective catalysts failed. The kind of structure makes difficult to consider any reduction by which have furthermore not been observed by electron microscopy nor in bismuth molydates or in α-Te_2MoO_7 (refs. 35, 51). Another explanation must therefore be given in order to account for the observed high mobility of lattice oxygen.

Bismuth molybdates differ, apart by the structure, not by the valence of cations as in VPO (V^{5+}, V^{4+}) or TeMoO (Te^{6+}, Te^{4+}, Mo^{6+}, Mo^{5+}) but rather by the Bi and Mo coordinations (table 2). Molybdenum accepts variable coordinations from 4 to 6 in such an extent that the structure of

TABLE 2
Structure, coordinations and polyhedra found in bismuth molybdates

Bismuth molybdate	$Bi_2Mo_3O_{12}$	$Bi_2Mo_2O_9$	α-Bi_2MoO_6
Structure	distorted scheelite	scheelite-like	koechlinite
Coord. Bi Coord. Mo	8 4	8 4	5-(6) 4-(6)
Polyhedra	isolated and distorted MoO_4	Bi_3O_2 chains unshared O-Bi irregular BiO_8 regular MoO_4	Bi_2O_2 layers unshared O-Bi MoO_4 layers (= elongated MoO_6)
Cleavage	(100)		(010)
References	51	52	51

MoO_3 itself can be described by combination of tetrahedra or of distorted octahedra. Molybdates are therefore able to restructure by modification of Mo coordination in order to accommodate oxygen vacancies created during (amm)oxidation of propene. Differences of opinions still exist as to the nature and the role of active sites (refs. 54-56), but it seems likely that oxygens linked to Bi are responsible for dehydrogenation and oxygens of molybdate groups for oxygenation, both O-Bi and O-Mo being necessarily present on the surface to obtain selectivity. During reduction and reoxidation, bismuth molybdates behave differently. Grasselli et al. showed by pulse reduction experiments under C_3H_6+ NH_3 (ref. 54) that the initial selectivity in acrylonitrile follows the order $Bi_2Mo_2O_9 > Bi_2Mo_3O_{12} > Bi_2MoO_6$, whereas for reoxidation $Bi_2MoO_6 > Bi_2Mo_2O_9 > Bi_2Mo_3O_{12}$. The layered structure of Bi_2MoO_6 and the presence of lone pairs of electrons on Bi are responsible for its quick reoxidation as compared with $Bi_2Mo_3O_{12}$ or $Bi_2Mo_2O_9$ which have a more closed-packed structure. $Bi_2Mo_2O_9$ would have a favorable balance of chemisorption sites Mo (similar to those of $Bi_2(MoO_4)_3$) and hydrogen abstraction sites O-Bi (more diversified than in Bi_2MoO_6). Moreover, chains of Bi and O atoms isolate Mo sites so as to avoid multiple oxygen insertion into the adsorbate leading to unselective products (ref. 57). However, while deeper reduction is limited to the surface for $Bi_2Mo_3O_{12}$ and α-Bi_2MoO_6, $Bi_2Mo_2O_9$ can be completely O-depleted. Reoxidation restores the initial catalytic properties except in the case of $Bi_2Mo_2O_9$ which has a lower activity than initially. In fact pure $Bi_2Mo_2O_9$ is unstable during catalytic redox cycles and can disproportionate into a mixture of $Bi_2Mo_3O_{12}$ and α-Bi_2MoO_6 (ref. 54). During reduction of (001)Bi_2MoO_6 and (010)$Bi_2(MoO_4)_3$ examined by *in situ* electron microscopy, Gai observed frequently an ordered superlattice closely related to (101)$Bi_2Mo_2O_9$ near 400°C. Microanalysis indicated the presence of $Bi_2Mo_2O_9$ in the reacted materials (ref. 51), which suggests that the disproportionation is reversible. This accounts also for the observation that the best activity and selectivity of (amm)oxidation of propene on bismuth molybdates occur for a mixture of $Bi_2Mo_3O_{12}$ and Bi_2MoO_6 phases rather than for a single phase (ref. 58).

When bismuth molybdates (Bi_2MoO_6 or $Bi_2Mo_2O_9$) are used in the oxidation of o-xylene, only o-tolualdehyde and CO_2 are obtained instead of the expected phthalic anhydride.which is selectively formed only with V_2O_5/TiO_2 catalyst *(vide infra)*. The same features are observed with MoO_3 (ref. 59). O-tolualdehyde is the first intermediate found in the rake mechanism of o-xylene-phthalic anhydride, which needs the exchange of 12 electrons, insertion of three O and formation of 3 H_2O. Like propene-acrolein, o-xylene-tolualdehyde is less demanding (4-e reaction, insertion of 1 O and formation of 1 H_2O). Therefore we can assume that bismuth molybdate lattices are not able to provide more than two lattice oxygens per reactant molecule without collapse of the molybdate structure. Since the distribution and the energy of sites is suitable for these 4-e reactions, it can be inferred that the aromatic ring is not involved during oxidation of o-xylene to o-tolualdehyde, which justifies the nucleophilic character necessary for lattice oxygens (ref. 59).

2-2. Action of promoter or support.

We have shown above (cf. §. 1) that the same structural unit found in vanadium oxides is responsible for their similar activity, but also for their lack of selectivity. These properties can be modified by means of promoters such as Pd or of a support such as TiO_2-anatase. Montarnal et al. observed that addition of Pd to V_2O_5 brings about a parallel enhancement of the catalytic conversion of C_2H_4 into acetic acid and of the reduction rate of V_2O_5 by C_2H_4 in a gas-solid reaction. While the slow step of catalysis is the reduction in the case of pure V_2O_5, it becomes the oxidation for Pd-V_2O_5 (ref. 11). The role of Pd^{2+} is not to act as an active site, as thought by authors noting the resemblance with the Wacker process, but rather to accelerate the changes (redox) in the solid state. More active sites result and, since V_4O_9 is the only reduced phase existing at the steady state besides the remaining V_2O_5, the redox couple V^{5+}/V^{4+} is well-defined and selectivity is enhanced.

Many recent studies have shown that V_2O_5 supported on TiO_2-anatase is a superior catalyst than unsupported V_2O_5 for the selective oxidation of several hydrocarbons (refs. 27, 28, 60-63). Various surface analytical methods were used to determine the actual role of TiO_2-anatase according to the amount of active V_2O_5, particularly when a monolayer of vanadia is supported. In the case of 15-20 mol.% V_2O_5/TiO_2, it is incontrovertible that above 560°C in nitrogen the reactivity of both solids is modified, since V_2O_5 is reduced in V_6O_{13} and anatase is transformed into rutile, in conditions where these transformations could otherwise not occur. Véjux et al. have proposed that this interfacial synergetic effect is due to a remarkable crystallographic fit between surface planes such as (010) V_2O_5 and (001) or (010) TiO_2 (ref. 64). The facts that, (i), the reactivity of V_2O_5 is also modified when other "supports", structurally related to anatase (e.g. $AlNbO_4$ or even rutile) are used, and, (ii), the reactivity of other oxides, structurally related to V_2O_5 (e.g., MoO_3) is modified when supported by TiO_2 anatase, substantiate this assumption (refs. 65-67). There is no need to consider that, at the atomic scale, the frameworks retain their own dimensions exactly as in the bulk. On the contrary, the low misfits (few percent) between the planes are certainly accommodated like in the case of dislocations, thereby producing an interfacial (mono)layer. Special properties arising from direct interactions V-O-Ti could account for peculiarities observed for monolayer vanadia catalysts (refs. 60-62). Cullis et al. have shown that oxygen atoms necessary to oxidize the intermediate species are supplied only by the migration of surface O in the case of monolayer while they come from V_2O_5 layers when they are thick enough (ref. 62). In the latter case the formation of V_6O_{13} is often reported (refs. 61, 63). Better performance is assumed to be due to the presence of anatase which helps to retain this particular reduced phase (same cleavage plane as V_2O_5, *vide supra*) and prevents further reduction to V^{3+}.

CONCLUSION

One active MO phase, characterized by its surface crystal field depending on its structure and its reactivity, is able to catalyze selectively one reaction, itself characterized by thermodynamics and by the molecular structure of reactant and products. In the V-O system, Pd, Mo, P, Ti (in TiO_2) modify the properties of V_2O_5, by formation of new phase(s) (P, Mo), or by modification of surface and bulk reactivity (Pd, Mo, Ti). In the last case the nucleation of one lower oxide corresponds to one well-defined V^{5+}/V^{4+} redox couple. The same ideas prevail for Mo-O system, where changes in the metal coordination modify the oxygen mobility and correspond to changes in the redox couple. When two phases are in contact, synergetic effects will be observed in catalysis only when the frameworks are structurally compatible, e.g. in the cases of VO oxides in presence of anatase or of bismuth molybdates.

REFERENCES

1 P. Courtine, ACS Symp. Series, 279 (1985) 37.
2 E. Bordes and P. Courtine, J. Catal., 57 (1979) 236 ; E. Bordes, Thèse, Compiègne (1979).
3 R.G. Pearson, "Hard and Soft Acids and Bases", Dowden, Hutchinson and Ross Inc., Stroudsburg, Penn., USA (1973).
4 J. Ziolkowski, J. Catal., 80 (1983) 263 ; ibid, 84 (1983) 317.
5 P. Mars and D.W. Van Krevelen, Chem. Eng. Sci., Suppl., 3 (1951) 41.
6 G.W. Keulks, J. Catal., 19 (1970) 232 ; ibid., 61 (1980) 316.
7 J.M. Tatibouet, J.E. Germain and J.C. Volta, J. Catal., 82 (1983) 240.
8 J.C. Volta and J.L. Portefaix , Appl. Catal., 18 (1985) 1-32.
9 J. Ziolkowski, J. Catal., 80 (1983) 263.
10 M. Abon, B. Mingot, J. Massardier and J.C. Volta, "New Developments in Selective Oxidation", G. Centi and F. Trifiro Eds, Stud. Sci. Surf. Catal., 55 (1990) 747.
11 J.L. Seoane, P. Boutry and R. Montarnal, J. Catal., 63 (1980) 182.
12 R.K. Grasselli, J.D. Burrington and J.F. Brazdil, J. Chem. Soc. Faraday Disc., 72 (1982) 203
13 A.W. Sleight and W.J. Linn, Ann. New York Acad. Sci., 272 (1976) 22.
14 R.K. Grasselli and D.D Suresh, J. Catal., 25 (1972) 273.
15 J.C.J. Bart and N. Giordano, Gazz. Chim. Ital., 109 (1979) 73.
16 F. Centi and F. Trifiro, Chim. Indust., 68 (1986) 74.
17 Papers in "Selective Catalytic Oxidation of C-4 Hydrocarbons to Maleic Anhydride", Catal. Today, 1 (1987).
18 R.M. Contractor, H.E. Bergna, H.S. Horowitz, C M. Blackstone, U. Chowdhry and A.W. Sleight, Stud. Surf. Sci. Catal., 38 (1988) 645-654.
19 E. Bordes , Catal. Today, 1 (1987) 499 ; ibid, 3 (1988) 163.
20 M.A. Pepera, J.L. Callahan, M.J. Desmond, E.C. Milberger, P.R. Blum and N.J. Bremer, J. Am. Chem. Soc., 107 (1985) 4883.
21 B.K. Hodnett, Catal. Rev.-Sci. Eng., 27 (1985) 373.
22 G. Centi, I. Manenti, A. Riva and F. Trifiro, Appl. Catal., 9 (1984) 177.
23 T.P. Moser and G.L. Schrader, J. Catal., 104 (1987) 99.
24 P.A. Batist, H.J. Prettre and G.C.A. Schuit, J. Catal., 15 (1969) 267.
25 D.J. Cole, C.F. Cullis and D.J. Hucknall, J. Chem. Soc., 1976, 2185.
26 A. Bielanski, J. Piwowarczyk and J. Pozniczek, J. Catal., 113 (1988) 334.
27 M.S. Wainwright and N.R. Forster, Catal. Rev. Sci. Eng., 19 (1979) 211.

28 M. Gasior, I. Gasior and B. Grzybowska, Appl. Catal., 10 (1984) 87.
29 J. Ziółkowski, J. Catal., 100 (1986) 45.
30 J. Ziółkowski, E. Bordes and P. Courtine, J. Catal., 122 (1990) 126.
31 J. Ziółkowski, E. Bordes and P. Courtine, "New Developments in Selective Oxidation", G. Centi and F. Trifiro Eds., Stud. Sci. Surf. Catal., 55 (1990) 747.
32 W. Thöni and P.B. Hirsch, Philos. Mag. 33 (1976) 639.
33 D.L. Kepert, "The Early Transition Metals", Academic Press, London 1972.
34 L. Fiermans, P. Clauws, W. Lambrecht, L. Vandenbroucke and J. Vennik, Phys. Stat. Sol. a, 59 (1980) 485.
35 P.L Gai, ED Boyes and J.C.J. Bart, Philos. Mag. A, 45 (1982) 531.
36 F.S. Stone, J. Sol. State Chem., 12 (1975) 271.
37 E. Broclawik and J. Haber, J. Catal., 72 (1981) 379.
38 M. Che, F. Figueras, M. Forissier, J. McAteer, M. Perrin, J.L. Portefaix and H. Praliaud, Proc. 6th Int. Cong. Catalysis (London 1976), The Chem. Soc., London, 1 (1976) 261.
39 J.C. Volta, J.M. Tatibouet, C. Pitchitkul and J.E. Germain, Proc. 8th Int. Cong. Catalysis (Berlin 1984), Dechema, Frankfurt, IV (1984) 451.
40 J.C. Volta, W. Desquesnes, B. Moraweck and G. Coudurier, React. Kinet. Catal. Lett., 12 (1979) 241.
41 A. Andersson and S.T. Lundin, J. Catal., 58 (1979) 383.
42 K. Klissurski and Y. Pescheva, React. Kin. Cat. Lett., 32 (1986) 77.
43 J. Ziolkowski and J. Janas, J. Catal., 81 (1983) 298.
44 E.Bordes and P. Courtine, Bull. Soc. Chim. Fr., 1989, 283.
45 G. Centi, F. Trifiro, G. Busca, J. Ebner and J. Gleaves, Faraday Discuss. Chem. Soc., 87 (1989) 214.
46 G. Bergeret, M. David, J.P. Broyer, J.C. Volta and G. Hecquet, Catal. Today, 1 (1987) 37.
47 E. Bordes, J.W. Johnson and P. Courtine, J. Sol. State Chem., 55 (1984) 270.
48 J.W. Johnson, D.C. Johnston, A.J. Jacobson and J.F. Brody, J. Am. Chem. Soc., 106 (1984) 8123.
49 Z.C. Kang, Q.X. Bao and C. Boulesteix, J. Sol. State Chem., 83 (1989) 255.
50 J.E. Germain and J.C. Peuch, Bull. Soc. Chim. Fr., 1969, 1844.
51 P.L. Gai, J. Sol. State Chem., 49 (1983) 25.
52 A.F. Van den Elzen, G.D. Rieck, Acta Cryst., B 29 (1973) 2433 ; ibid, 2436.
53 H.Y. Chen and A.W. Sleight, J. Sol. State Chem., 63 (1986) 70.
54 J.F. Brazdil, D.D. Suresh and R.K. Grasselli, J. Catal., 66 (1980) 347.
55 K. Brückman, J. Haber and T. Wiltowski, J. Catal., 106 (1987) 188.
56 A. Sleight, in "Advanced Materials in Catalysis", J.J. Burton and R.L. Garten, Eds, p. 181, Academic Press, New York, 1977.
57 J.L. Callahan and R.K. Grasselli, AIChE. J., 9 (1963) 755.
58 D. Carson, G. Coudurier, M. Forissier, J.C. Védrine, J. Chem. Soc., Faraday Trans. I, 79 (1983) 1921.
59 M. Gasior and B. Grzybowska, J. Catal., 52 (1978) 534.
60 G.C. Bond and K. Brückman, J. Chem. Soc., Faraday Disc., 72 (1981) 235.
61 I.E. Wachs, R.Y. Saleh, S.S. Chan and C.C. Chersich, Appl. Catal., 15 (1985) 339.
62 C.F. Cullis and D.J.Hucknall, Catal., 5 (1982) 273.
63 T. K. Mori, A. Miyamoto, Y. Murakami, J. Catal., 95 (1982) 482.
64 A. Véjux and P. Courtine, J. Sol. State Chem., 23 (1978) 93 ; ibid., 63 (1986) 179.
65 E. Bordes, J.G. Eon, A. Véjux and P. Courtine, IXth Int. Symp. Reactivity of Solids, Cracow (Poland), V-5-425 (1980).
66 J.G. Eon and P. Courtine, J. Sol. State Chem., 32 (1980) 67.
67 J. Papachryssanthou, E. Bordes, P. Courtine, R. Marchand and M.Tournoux, Catal. Today, 1 (1987) 219.

R.K. Grasselli and A.W. Sleight (Editors), *Structure-Activity and Selectivity Relationships in Heterogeneous Catalysis*
1991 Elsevier Science Publishers B.V., Amsterdam

KEY STRUCTURE-ACTIVITY RELATIONSHIPS IN THE VANADIUM PHOSPHORUS OXIDE CATALYST SYSTEM

Jerry R. Ebner, Monsanto Company, 800 North Lindbergh Avenue,
St. Louis, Missouri, 63167

Michael R. Thompson, The Pacific Northwest Laboratory[1],
Battelle Blvd., Richland, Washington, 99352

Abstract

The crystal structure of vanadyl pyrophosphate has been redetermined using single crystals obtained from a near solidified melt of a microcrystalline catalyst sample. Crystals that index as vanadyl pyrophosphate obtained from this melt are variable in color. Crystallographic refinement of the single crystal X-ray diffraction data indicates that structural differences among these materials can be described in terms of crystal defects associated with linear disorder of the vanadium atoms. The importance of the disorder is outlined in the context of its effect on the proposed surface topology parallel to (1,0,0). Models of the surface topology simply and intuitively account for the non-stoichometric surface atomic P/V ratio exhibited by selective catalysts of this phase. These models also point to the possible role of the excess phosphorus in providing site isolation of reactive centers at the surface.

Introduction

The conversion of butane to maleic anhydride on vanadium phosphorus oxide catalysts represents the only commercial process for selective functionalization of an alkane. The catalytic performance of vanadium phosphorus oxide catalysts for this reaction is unequaled by any other metal or metal-oxide system. A large number of published reports attribute catalytic activity/selectivity to the bulk crystalline phase vanadyl pyrophosphate, $(VO)_2P_2O_7$ (2-7). The catalytic performance of vanadyl pyrophosphate is strongly related to the method of preparation employed. For example, vanadyl pyrophosphate catalysts synthesized in aqueous versus organic media have significant catalytic performance differences (6,8-13). In this paper the structural complexities of vanadyl pyrophosphate are explored through single crystal X-ray structural analysis. The single crystal results seem to provide an additional perspective on the differences evidenced in the properties derived from different solvent systems. Further, the results of these crystallographic studies have led us to assemble zeroth-order models of the surface topology parallel to (1,0,0) and to gain some insight into the potential role of the non-stoichometric surface phosphorus in determining selectivity.

Results and Discussion

Structure-Activity Relationships in VPO Catalysts

The catalysts used in this study were prepared with approximately 10% excess phosphorus in aqueous and organic media according to well-established literature procedures (6). The $[VOHPO_4]_2 \cdot H_2O$ precursors from the aqueous and organic preparations were dehydrated and partially oxidized by air calcination at 400°C for 1-2 hours. The microcrystalline catalysts described here were formed and characterized after running the butane oxidation reaction at 1.4 - 2.0% butane and 1000 GHSV for approximately 750 continuous hours. Catalyst performance of the organic derived catalyst is 7 absolute yield points superior to the aqueous counterpart over the range of space velocities (Fig. 1). Many similarities exist between the two catalyst systems. Both microcrystalline catalysts have a vanadium oxidation state of 4.01 ± 0.01, a bulk phosphorus to vanadium ratio of 1.00 ± 0.025,

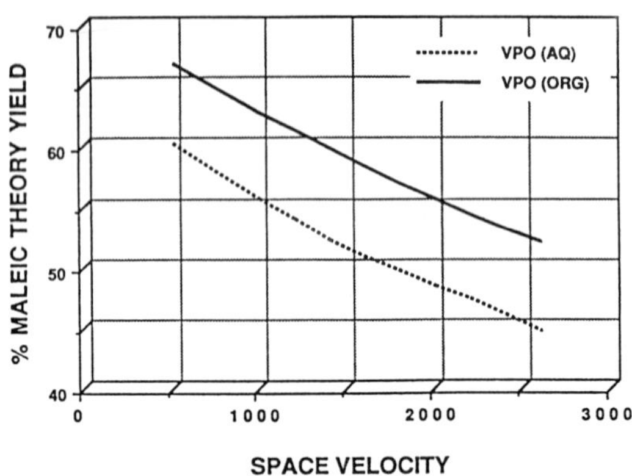

Fig. 1. Catalyst performance for aqueous (dotted line) and organic (solid line) derived vanadyl pyrophosphate catalysts.

and XPS surface atomic phosphorus/vanadium ratios of 1.5 ± 0.3. There are several measurements that distinguish the two catalyst systems. The BET surface areas are 10 and 14 m^2/g for aqueous and organic derived catalyst systems, respectively. The SEM determined morphologies of the $[VOHPO_4]_2 \cdot H_2O$ precursors are large hexagonal plates and thin, rose-like platelets for the aqueous and organic catalysts, respectively. However, these major morphological differences are somewhat diminished in the aged catalysts. SEM images indicate in both systems significant fracturing of the platelets and formation of rectangular and rodlike crystal habits 0.1 - 1.0 microns in size, with the organic system having more rods than the aqueous counterpart and generally smaller in size by a factor of two. The previously published XRD patterns (Fig. 2a-b) of the resulting aqueous and organic derived catalysts (6) differ in two respects: (1) the overall intensities of the primary peaks in the powder patterns are greater in intensity for the aqueous derived catalyst; and (2) a single reflection at 3.87Å ($2q_{CuKa} = 22.90^{o}$) is significantly broadened in the organic derived catalyst. The latter difference has been attributed to layer stacking disorder in vanadyl pyrophosphate (10,12,14). The exact form of the orientational disorder contributing to this key, distinguishing feature has not previously been reported.

The solid-state structure of vanadyl pyrophosphate has previously been reported (15), but inconsistencies in these studies have generated doubt concerning the accuracy of the crystallographic model. Because, as the previous discussion indicates, the structural nuances of vanadyl pyrophosphate are of such great importance to catalysis, a re-examination of this structural determination seemed prudent. Thus, single crystals were obtained via a new route: crystal growth from a near-solidified melt of the microcrystalline catalyst. The structural linkage between the synthesized crystals

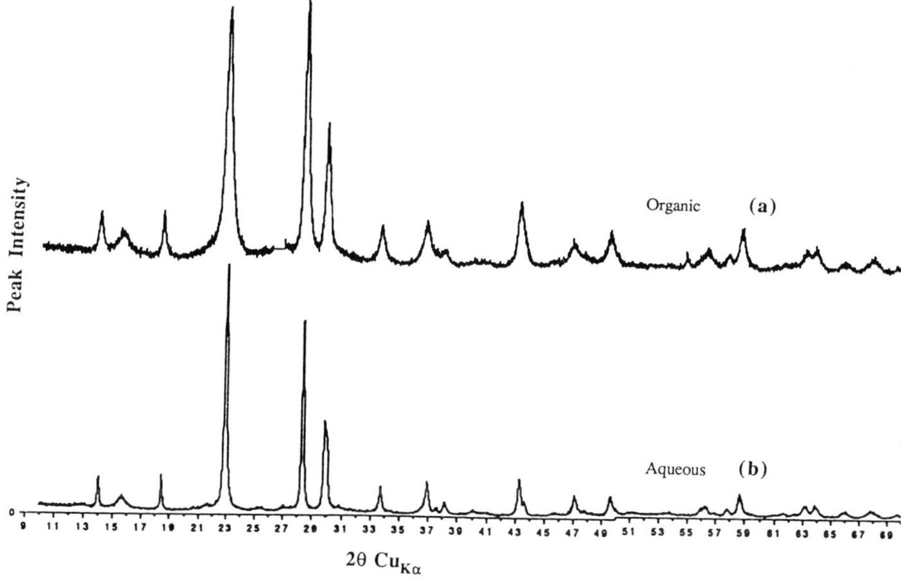

Fig. 2. Observed X-ray powder patterns for microcrystalline catalysts prepared in (a) aqueous and (b) organic media.

and the microcrystalline catalyst samples was firmly established by comparing crystal and catalyst vibrational spectra using laser Raman and FTIR microprobe techniques. With the new crystals, we have sought to redetermine the crystal structure in order to clarify the origin of apparent crystal defects. In addition, we have made an attempt to assemble zeroth-order models of the surface topology parallel to (1,0,0) which would accommodate the non-stoichometric surface phosphorus. Differences between bulk and surface P/V stoichometry for microcrystalline vanadyl pyrophosphate catalysts have been reported previously (16), as well as the decrease of reaction selectivity associated with surface phosphorus loss during the butane reaction (17). It is clear that any structural model for the catalyst must account for the disposition of excess surface phosphorus, and its role in selectively enhancement.

The Crystal Structure of Vanadyl Pyrophosphate, Revisited

Simultaneously in 1978, Linde and Gorbunova (15a), and separately, Middlemiss (15b), reported the X-ray structure determination of vanadyl pyrophosphate. Unfortunately, the results of both studies possessed serious flaws in their crystallographic models. Middlemiss attempted to phase the structure via Patterson synthesis ("heavy-atom" techniques), which resulted in the refinement of a projection of the structure in the non-centrosymmetric space group P_{bc2_1} (18a). Linde utilized statistical phasing methods to solve the structure, which resulted in a model with identical connectivity as that described by Middlemiss, and lower residuals: R_1=0.089, and R_w=0.091. However, the Linde model contained several unusual bonding interactions and puzzling pseudo-symmetry. The bonding interactions in question involved the vanadyl moiety (V=O), which in the fully refined model gave two chemically inequivalent bonds: two in the range of 1.54(2)Å, and the remaining two interactions of approximately 1.72(2)Å. Furthermore, the structure was reported in a brief format with little experimental information.

Diffraction quality crystals of vanadyl pyrophosphate, used to redetermine the crystal structure, have been obtained from near-solidified melts derived from microcrystalline catalyst samples. Recrystallization experiments utilized samples taken from a fixed-bed reactor after more than 5000 hours in the butane oxidation reaction. Surprisingly, the single crystals harvested from these melts, which index as vanadyl pyrophosphate in diffraction experiments, are variable in color, ranging from emerald-green to gray, and from yellow-brown to red-brown. Single crystals described by Linde et al. and Middlemiss were reportedly emerald-green in color. Color variations have been noted previously in the preparation of microcrystalline catalyst materials (19).

Emerald-green crystals of vanadyl pyrophosphate are orthorhombic (20), with $a = 7.710(2)$Å, $b = 9.569(2)$Å, $c= 16.548(3)$Å, $V= 1220.9(8)Å^3$, $r_{calc}= 3.359(2)\ g/cm^3$. Red-brown crystals exhibit lattice parameters which are slightly dilated relative to their emerald-green counterparts: $a=7.746(2)$Å, $b= 9.606(2)$Å, $c=16.598(3)$Å, $V= 1235.0(8)Å^3$, $r_{calc}= 3.320(2)\ g/cm^3$. Lattice parameters cited by the previous authors are nearly identical to the emerald-green specimens studied here. Aside from the color variation, the intensities and peak widths for numerous reflections collected from the single crystals show marked differences between the materials. These differences are similar to those reported for the X-ray powder diffraction patterns for catalysts prepared from aqueous or organic media. Significant differences also exist in the Raman spectra for emerald-green and red-brown crystals. For example, emerald-green and gray crystals exhibit a strong sharp doublet centered at 922 cm^{-1}, which is diminished to a broad weak singlet at 928 cm^{-1} for the yellow-brown and red-brown materials (21). We believe that the structural differences apparent in these single crystals are likely those which have been identified with the microcrystalline catalysts and relate to the ordering of the metal atoms within the structure.

The diffraction data taken from ten single crystals of vanadyl pyrophosphate have been extensively studied. Due to the complexity of this crystal structure, only a terse discussion of our results will be presented here. Our primary interest was to verify that the solution of the crystal structure reported by Linde was correct, and secondly, to determine the cause of the poor refinement results.

The atomic coordinates reported by Linde indicated strong pseudo-symmetry, especially apparent for the heavy atoms (22), suggestive of higher space group symmetry than that chosen. However, no higher symmetry description of the lattice could be found. The space group extinctions are rigorously consistent only with the choice of P_{ca2_1} or the centrosymmetric counterpart, P_{cam} (18b). Structural solutions for emerald-green and red-brown crystals (and for the data published by Middlemiss) can be found in non-centrosymmetric P_{ca2_1} consistent with structure reported by Linde. While not fully indicative of a correct solution, these twenty-six atom models refine to conventional residuals in the range of R_1=0.089-0.096 and R_w=0.093-0.099 for the crystals studied. A perspective plot of one layer of the structure projected on the bc-plane is illustrated in Fig. 3. We have also been able to find numerous solutions to the structure, consistent with the connectivity of vanadyl pyrophosphate, in the centrosymmetric space groups P_{cam} and P_{caa} (18c). Efforts to refine these centrosymmetric structures have failed to yield models which converge at residuals less than R_1=0.15.

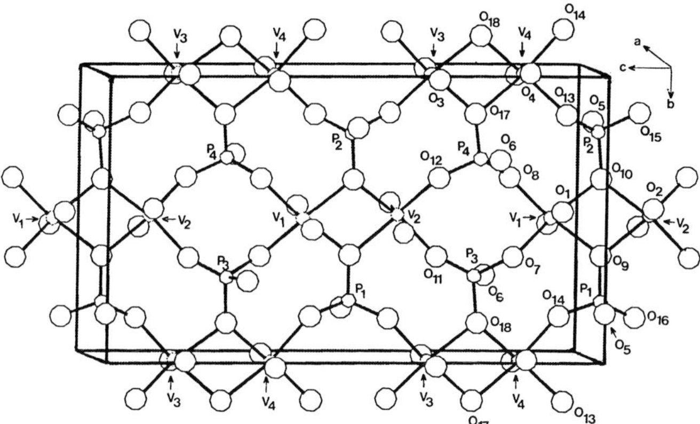

Fig. 3. Projection of the continuous structure of vanadyl pyrophosphate onto the crystallographic bc-plane.

Difference Fouriers computed from the twenty-six atom models in P_{ca2_1} for both emerald-green and red-brown crystals indicate residual electron density consistent with disorder of the vanadium sites. The disordered positions for the metal atoms are oriented approximately 0.65Å across the basal plane of the distorted octahedral vanadium coordination sphere. This type of disorder is common for crystal structures which possess square-pyramidally distorted octahedral metal centers and is the cause of the previously reported diffraction streak effects noted in electron diffraction studies of the microcrystalline catalysts (23). For this structure the disorder represents a columnar re-orientation of the vanadyl bonds, reversing the direction of the entire column along the a-axis. It should be noted that for all crystals studied, there is no such disorder indicated for the phosphorus atoms.

In order to better understand the symmetry and structure of the crystallographic model, consider the schematic representation of a small fragment of the continuous solid reported by Linde, depicted in Fig. 4.

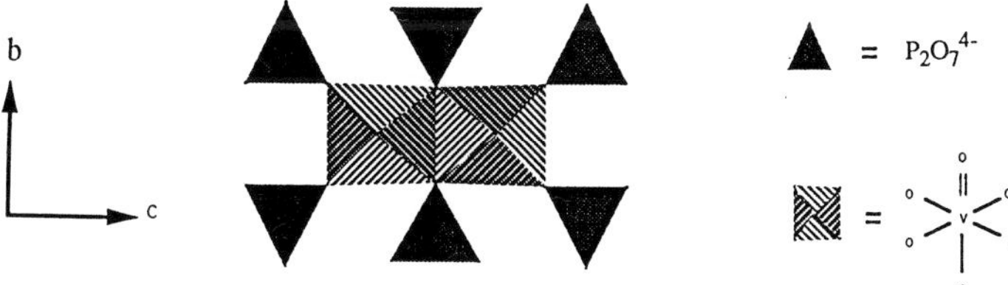

Fig. 4. Schematic representation of the vanadium centered dimeric unit which comprises the layered structure of vanadyl pyrophosphate.

Neglecting the exact direction of the vanadyl and pyrophosphate groups, this two-dimensional fragment is representative of the unit of structure comprising the close-packed layers of vanadyl pyrophosphate. The d^1 vanadium centers possess pyramidally distorted octahedral coordination. The vanadium and the vanadyl oxygen, and a concomitantly weak interlayer oxygen interaction trans to V=O (24). The close-packed layer structure is comprised of this dimeric unit in which adjacent vanadium centered octahedra share a common polyhedral edge. Each dimer is surrounded by six apex-shared pyrophosphate groups. The pyrophosphate groups form interlayer bonds via the pyrophosphate oxygen (P-O-P), are oriented perpendicular to the plane of the paper in Fig. 4, and bridge to adjacent layers above or below the plane. If, for instance, the direction of these six pyrophosphates alternate their orientation relative to the close-packed plane, ie. oriented up-down-up-down-up-down traversing the perimeter of the dimer, then a center of symmetry could be defined, and a centrosymmetric structure would result. For emerald-green and red-brown single crystals, the six pyrophosphate groups are oriented non-centrosymmetrically with respect to the dimer (25). The structure described by Linde in P_{ca2_1} possesses an orientation for the six pyrophosphate groups of up-up-down-down-up-down (or the converse). No model of the structure of vanadyl pyrophosphate possessing this symmetry can be constructed with less than four independent phosphorus atoms contained in the asymmetric unit of the cell. This condition forces the use of the non-centrosymmetric space group, and as will be shown below, presents an intriguing structure at the termination of the crystal parallel to the (1,0,0) surface.

Accounting for the disorder of the metal atoms in the crystallographic refinement improves the results earlier reported by Linde, but the site disorder exhibited is not simple nor statistical. The dimeric vanadium polyhedra within the crystal form a chain-like structure parallel to the c-axis, with the two independent chains lying at approximately $y = 0$ and $y = 1/2$, as illustrated in Fig. 3. Interestingly, only two of the four independent vanadium sites disorder for emerald-green crystals, while all four sites disorder for their red-brown counterparts (26). In emerald-green crystals, those vanadium atom sites which lie in a chain along the c-axis at $y=1/2$, disorder with approximate 3:1 site occupation for the two possible positions above or below the basal plane. More massive disorder of all four vanadium sites is found for the red-brown crystals. Preliminary counter-weighted isotropic least-squares refinement of the disordered model leads to convergence at $R_1=0.0344$ and $R_w=0.0355$ for a typical green crystal, and $R_1=0.0540$ and $R_w=0.0560$ for a typical red-brown crystal. The aberrant bonding interactions reported by Linde are not present in the fully refined disordered models. For the two typical refinements noted above, the four independent vanadyl bonds average 1.604(20)Å and 1.621(13)Å in emerald-green and red-brown crystals, respectively. At this time we do not have an exact explanation of the cause of this pattern of disorder. However, considering the results in the case of the emerald-green crystal, a possible explanation of the disorder would involve the co-crystallization of two polytypes of vanadyl pyrophosphate whose structures differ in the relative orientation of adjacent dimer chains lying along $y=0$ and $y=1/2$ as illustrated in Fig. 5.

There is strong evidence apparent in the single crystal step scans which indicate that the patterns of disorder of the metal atom sites in green and brown crystals follow the differences in peak intensities and peak widths in X-ray powder patterns for samples prepared via differing synthetic routes. In order to determine the correlation between the single crystals and the catalyst powder patterns we have generated simulated XRD's (Fig. 6), based on the convolution of individually measured intensities from the single crystal studies (27). These patterns indicate excellent correspondence with experimental patterns of catalyst samples. Convolutions generated from data derived from emerald-

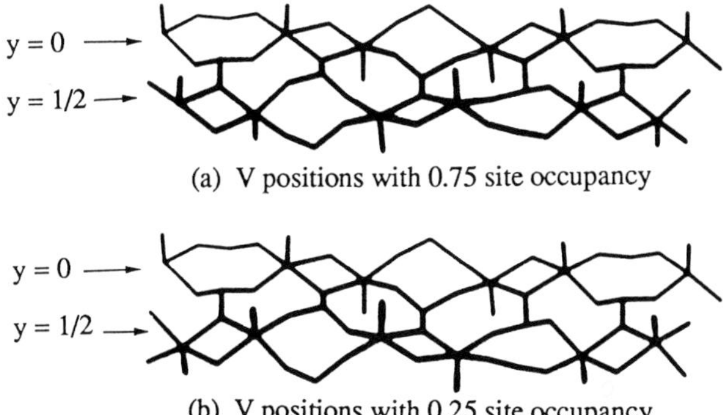

Fig. 5. Models of vanadyl pyrophosphate illustrating the disordered chain structure along the crystallographic c-axis.

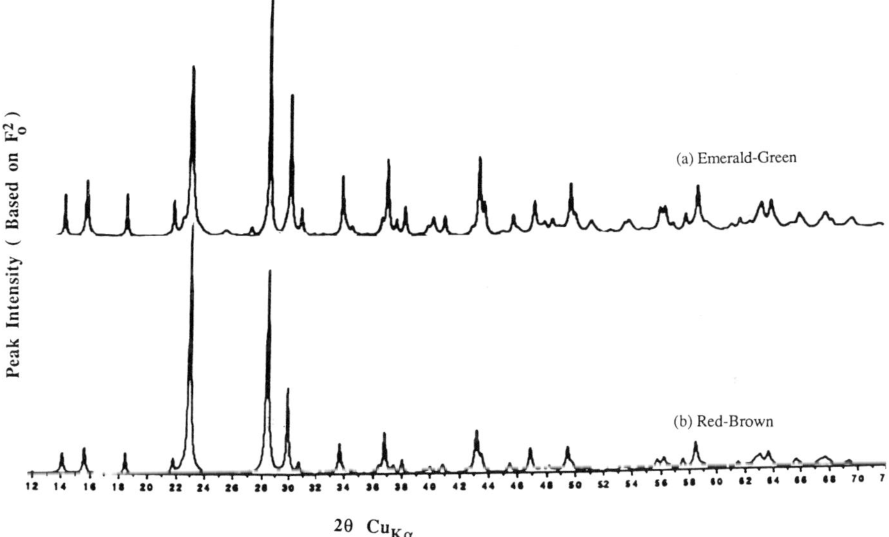

Fig. 6. Calculated powder patterns based on the convolution of peak intensity and observed peak widths taken from single crystal step scans for (a) emerald green and (b) red-brown crystals.

green crystals are representative of experimental patterns of vanadyl pyrophosphate synthesized from organic alcohol intercalated precursors, while those generated from step scans taken from red-brown crystals resemble the patterns observed for materials generated from aqueous preparation. In summary, these results provide support for the idea that the differences in the XRD patterns between aqueous and organic derived catalysts arises from structural disordering, and this disordering is associated with the orientation of the vanadyl columns in the structure.

Models of Surface Topology Parallel to (1,0,0) and Site Isolation.

Obvious questions relate to what significance, if any, these structural attributes might have on the catalytic behavior of vanadyl pyrophosphate. We believe that both effects, namely, the asymmetric orientation of the pyrophosphate groups around the vanadium dimer, and the columnar disordering of the vanadyl moieties can have a profound structural effect on the surface topology parallel to the (1,0,0) surface.

In order to premise our models of surface topology, it is instructive to consider the simple topotactic reaction which transforms the orthophosphate precursors into the pyrophosphate phase, and the consequences of the topotaxy at the (1,0,0) surface of an isolated single crystal (28). It is important to recognize that for each intact layer of either the orthophosphate or pyrophosphate phase, the atomic P/V ratio is 1.0. In the case of the orthophosphate hemihydrate precursor, two equivalents of water are released as a result of topotaxy in the generation of the pyrophosphate phase: loss of the water of solvation and one equivalent from the dehydration of two adjacent interlayer orthophosphate groups in the formation of the pyrophosphate (P-O-P) bond. In the ordered structure of the orthophosphate precursor, half of the hydroxyl protons of the $HOPO_3^{2-}$ moieties are oriented above or below the close-packed plane. Dehydration results in half of the pyrophosphate bonds being formed in bridging positions to a layer above the plane, and half to a layer below (29). However, at the surface of this hypothetical isolated crystal, the dehydration to form the pyrophosphate bond can proceed with the formation of only one half of an equivalent: only those which will bridge between the surface and the first sub-surface layer. The surface atomic P/V ratio of this material will be identical with that of the bulk (P:V=1.0), and the surface layer will be chemically representative of a mixed orthophosphate/pyrophosphate. If the topotaxy is accomplished in excess phosphorus, as is generally the case for the material found to have the highest selectivity, the remaining surface orthophosphate can be transformed into pyrophosphate. This material would possess a surface atomic P/V ratio of 1.5, in agreement with experimental observation. We believe that these arguments are rational and chemically intuitive, and should be valid regardless of the exact nature of the topology of the (1,0,0) surface (i.e., flat or stair-step) since they are premised on the stoichometry of the compound and the topotaxy which relates the structure of the precursor to the product. Surface relaxation effects and surface reconstruction would be expected to be minor considerations due to the fact that the protonated phosphate moieties at the surface can retain full valance around each oxygen and phosphorus atom (30). The five coordinate vanadium atoms which terminate in vanadyl columns oriented into the crystal can easily solvate or chemisorb a labile sixth ligand.

When considering static models of termination of the crystal structure parallel to (1,0,0), in which all surface terminating phosphorus groups are represented as pyrophosphate moieties, the most intriguing feature relates to the manner in which the pyrophosphate groups orient about the vanadium dimer. In particular, the direction of two pairs of two adjacent groups orient together, ie. up-up-down-down-up-down, traversing the perimeter of the dimer. Fig. 7 illustrates a model of the structure parallel to (1,0,0) and its accessible van der Waal surface (31). This model possesses a surface layer

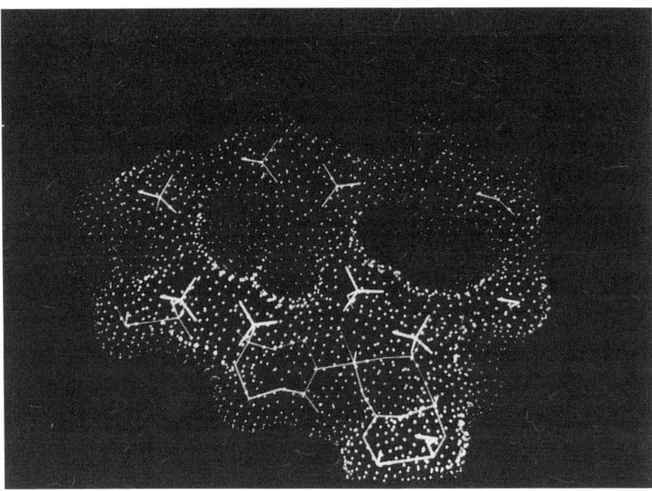

Fig. 7. A model of surface termination parallel to (1,0,0) for vanadyl pyrophosphate. The dot surface is used to illustrates the accessible van der Wall surface (Connolly surface) using a probe sphere radium of 2.2Å.

exhibiting an atomic P/V ratio of 1.5 and clearly indicates a surface cavity created by the "vacancy" of two adjacent pyrophosphate groups. This cavity is roughly elliptical in shape and the internal parimeter of the ellipse is bordered by several sets of surface terminating vanadyl groups. The vanadyl groups (or conversely, open sixth coordination sites) that are positioned at the ends of the elliptical cavity are severely hindered by two adjacent overshadowing pyrophosphate groups while other vanadyl moieties more central to the cavity are accessible.

The importance of the directional character of the vanadyl columns, and the symmetry of the structure can be appreciated when considering models such as that shown in Fig. 7. If there exist stable polytypes of vanadyl pyrophosphate (or alternatively, statistically disordered structures) which relate differing chain symmetry within the crystal, then the number, accessibility, and symmetry of unhindered vanadium coordination sites and vanadyl groups will be different for the different polytypes. We are currently exploring the structural consequences of differing eutaxy in models of vanadyl pyrophosphate in order to gain some understanding of the magnitude of the total energy differences. These calculations are being performed using an *ab-initio* self-consistent-field Hartree Fock formalism which fully treats crystallographic translation and symmetry (CRYSTAL, Pisani et al.) (32).

Conclusions

Our structural study indicates that the vanadyl pyrophosphate compound can crystallize with varying degrees of disorder of the vanadium positions. The best description of the disordering is variability in the directional orientation of the vanadyl columns running perpendicular to the (1,0,0) surface. This columnar disorder becomes very important when considering the non-stoichometric P/V surfaces parallel to (1,0,0). Terminating the surface in pyrophosphate groups places the reactive vanadium centers in cavities with varying degrees of steric hindrance by surface pyrophosphate moieties. The degree of vanadium center hindrance in the cavity is influenced significantly by the orientation and symmetry of the vanadyl columns within the structure. The analysis of the surface topology reveals an isolation of vanadium centers as surface clusters of up to four accessible vanadium centers per cavity. This model provides a means for active site isolation, an important general property for selective oxidation catalysts. First described by Grasselli (33), the site isolation principle requires that active oxygen be distributed in an arrangement that provides for limitation of numbers of active oxygen in various isolated locations so as to restrict overoxidation. Our proposed surface model encompasses both of these key features. Furthermore, it is clear from our surface model that loss of phosphorus through rupture of surface pyrophosphate groups will enlarge the surface cavities and thus expose larger expanses of accessible active oxygen leading to selectivity loss. As to the differences between organic and aqueous based catalyst preparations, this work leads to the new concept that the two catalyst systems may have surfaces parallel to (1,0,0) that differ subtly in surface topology. The distribution of accessible reactive sites associated with the surface cavities are different in the two systems because of the columnar disorder differences. We suggest this difference can have a significant effect on the number of surface sites well suited for oxidation of butane to maleic anhydride.

REFERENCES

1. Operated by the Battelle Memorial Institute for the United States Department of Energy under contract DE-AC06-76RLO-1830.

2. J.R. Ebner, V. Franchetti, G. Centi and F. Trifiro, Chem. Rev., 88 (1988) 55.

3. J.R. Ebner and J.T. Gleves, in: A.E. Martell and D.T. Sawyer (Ed.), Oxygen Complexes and Oxygen Activation by Transition Metals, Plenum Press, New York , 1988, p. 273.

4. G. Centi, F. Trifiro, G. Busca, J.R. Ebner, and J.T. Gleves, in: M.J. Philips and M. Ternan (Ed.), Proc. 9th Int. Congr. Catal., The Chemical Institute of Canada, Ottawa, 1988, p. 1538.

5. M.A. Pepera, J.L. Callahan, M.J. Desmond, E.C. Millberger, P.R. Blum and M.J. Bremer, J. Am. Chem. Soc., 107 (1985) 4883.

6. G. Centi, F. Trifiro, G. Busca, J.R. Ebner, J.T. Gleaves, Faraday Discuss. Chem. Soc., 87 (1989) 215.

7. J. Ziolkowski, E. Bordes, P. Coutine, J. Catalysis, 122 (1990) 126.

8. G. Busca, F. Cavana, G. Centi, and F. Trifiro, J. Catal., 90 (1986) 400.

9. E. Bordes, in: Petroleum Division Preprints of the Symposium: Hydrocarbon Oxidation, 194th American Chemical Society Meeting, New Orleans, 1987, p. 792.

10. H.S. Horowitz, C.M. Blackstone, A.W. Sleight and G. Tenfer, Appl. Catal., 38 (1988) 193.

11. R.A. Schneider, U.S. Patent 4043943 (1977).

12. F. Cavana, G. Centi and F. Trifiro, J. Chem. Soc., Chem. Commun., (1985) 492.

13. J.W. Johnson, D.C. Johnston, A.J. Jacobson and J.F. Brody, J. Am. Chem. Soc., 106 (1984) 8123.

14. D.C. Johnston and J.W. Johnson, J. Chem. Soc., Chem. Commun., (1985) 1720.

15. (a) Linde, S.A.; Gorbunova, E., Dolk. Akad. Nauk, SSSR (English Trans), 245 (1979), 584; (b) Middlemiss, N.E., doctoral dissertation, Department of Chemistry, McMaster University, Hamilton, Ontario, Canada, (1978).

16. Z. Zazhigalor, V. Belousov, G. Komashko, A. Pyatnitskaya, Y. Komashko, Y. Merkureva, A. Poznyakevich, J. Stoch and J. Haber, in: M.J. Philips and M. Ternan (Ed.), Proc. 9th Int. Congr. Catal., The Chemical Institute of Canada, Ottawa, 1988, p. 1538.

17. J. Haas, C. Plog, W. Maunz, K. Mittag, K. Gollmer, B. Klopries, ibid, p. 1632.

18. (a) The space group P_{bc2_1} is a non-standard setting of P_{ca2_1}-C_{2v} (No.29), in: Theo Hahn, (Ed.), International Tables for Crystallography, Volume A, Reidel, Dordrecht, Holland (1983), p 216, (b) non-standard setting of P_{bcm}-D_{2h} (No.57), ibid, p. 277, (c) pseudo a-glide perpendicular to c is particularly strong for the red-brown materials: Pcaa is a non-standard setting of P_{cca}-D_{2h} (No. 54), ibid, p. 270.

19. Johnson, J.W.; Johnston, D.C.; Jacobson; A.J. Broady, J.F., J. Amer. Chem. Soc., 106 (1984) 8123.

20. Note the use of the standard space group setting of P_{ca2_1}-C_{2v} (No.29).

21. J.J. Freeman, Internal Monsanto Company Report.

22. The pseudo-symmetry is reflective of the site symmetry associated with the positions of the octahedral and tetrahedral intersticies of this particular type of oxide close packing.

23. E. Bordes, P. Courtine, J. Catal., 57 (1979) 236.

24. Vanadyl bond lengths have values of approximately 1.60Å, while those trans to the vanadyl oxygen are approximately 2.25-2.35Å.

25. These exists no mirror or center of inversion which will commute the six groups. A 2-fold parallel to the c-axis will commute the pyrophosphate groups but will disorder the metal atoms.

26. All heavy atom sites have been refined in cycles of full-matrix least-squares using counter-weighted data and recent versions of SHELX86 (G.M. Sheldrick in: G.M. Sheldrick, C. Kruger and R. Goddard (Eds.), Crystallographic Computing 3, Oxford University Press (1985), pp. 175-189). All disordered pairs have been constrained to have a total occupancy of 1.0 and identical isotropic thermal factors. No disorder is noted for the phosphorus atoms within the fully refined structure.

27. Steps scans from the single crystal studies were used to compute widths-at-half-height for each reflection. Peak intensity was computed from the observed structure factors whose 2q values were placed on a wavelength scale for Cu radiation, and back-corrected for absorption (in order to simulate the powder patterns observed using Cu radiation). Convolutions of the set of peaks having $10.0^o \lesssim 2qCu_k \lesssim 80.0^o$ utilized Lorentzian lineshapes. Experimental powder patterns have not yet been performed due to the exceedingly small amount of single crystal material obtained.

28. E. Bordes, P. Courtine, J.W. Johnson, J. Solid State Chem., 55 (1984), 270.

29. Careful consideration of the crystal structures of the orthophosphate hemihydrate indicate that some reorganization of the phosphate moieties must occur in the process of topotaxy. Powder patterns of all microcrystalline samples of the catalyst materials indicate an order of magnitude broadening of odd-odd-odd reflections. This parity group can be shown to be very sensitive to positional disordering of the phosphorus atoms within this structure. No such broadening of this parity group is seen for the single crystals.

30. Major reconstructive processes of surfaces are generally thought to be driven by surface bond rehybridization which accommodates dangling bond states caused by loss of full valance.

31. M.L. Connolly, J. Mol. Graphics, 4 (1986), 3.

32. R. Dovasi, C. Pisani, C. Roetti, M. Causa, and V.R. Saunders, Quantum Chemistry Program Exchange, Publication 577, University of Indiana.

33. R. Grasselli, in: J. Nonnelle, B. Delmon, E. Derouane (Eds.), Surface Properties and Catalysis by Non-Metals, Elsevier, Amsterdam (1983), p. 273.

R.K. Grasselli and A.W. Sleight (Editors), *Structure-Activity and Selectivity Relationships in Heterogeneous Catalysis*
© 1991 Elsevier Science Publishers B.V., Amsterdam

STRUCTURE-ACTIVITY RELATIONSHIPS IN THE OXIDATION OF ALKYLAROMATICS OVER METAL OXIDES

A. ANDERSSON[1], S. HANSEN[2] and M. SANATI[1]

[1]Department of Chemical Technology, Chemical Center, University of Lund, P.O. Box 124, S-221 00 Lund (Sweden)

[2]Department of Inorganic Chemistry 2, Chemical Center, University of Lund, P.O. Box 124, S-221 00 Lund (Sweden)

SUMMARY

In order to accomplish oxidation of alkylaromatic compounds, both cations, which serve as adsorption centers, and oxygen species of suitable bond strength are needed. When compared to partial oxidation, relatively weakly bonded oxygen species are involved in combustion. On the basis of these criteria, using a simple model of the active ensemble, a relationship between reaction rate and bond strength is derived for partial and total oxidation. Its applicability is demonstrated using data for the oxidation of toluene over a large number of binary metal oxides. Some characteristic features following from the model are discussed, considering catalytic results on the structure sensitivity of oxidation and ammoxidation reactions over crystalline V_2O_5 and MoO_3. Furthermore, kinetic results on the oxidation of toluene to benzaldehyde and carbon oxides in presence of ammonia, which serves as an electron donor to the catalyst surface, demonstrate that the oxygen species taking part in partial oxidation and combustion are nucleophilic and electrophilic in character, respectively.

INTRODUCTION

A fundamental understanding of the structure-activity relationships observed in heterogeneous catalytic oxidation, is of basic importance for the development of new catalytic materials and for the decision about possible actions to be taken in order to improve the performance of existing catalysts. Various models or ideas have been presented in the literature, aiming at giving a description of oxidation catalysis which is of general significance.

According to Ziółkowski [1,2], the reaction path depends on the number and configuration of the active oxygen atoms in the vicinity of the adsorbed hydrocarbon molecule. The activity of a surface oxygen atom is assumed to depend on the sum of the bond-strength values to adjacent cations. In a dynamic approach to selectivity in partial oxidation [3], which is a further development of the concept of site isolation [4], the overall kinetic pattern of an oxidation reaction is expressed by a scheme having a common intermediate for the formation of useful and waste products. Reaction temperature and partial pressures of oxygen and hydrocarbon are important factors deciding the selectivity for formation of useful products. A drawback of both of these models is that it is the number of active oxygen species surrounding the adsorbed hydrocarbon molecule that decides whether partial or complete oxidation occurs. The character of the active oxygen species, i.e., whether it is an electrophilic or a nucleophilic reagent, is not considered. Also, the assumption made in the dynamic approach, that useful and waste products are formed from a common intermediate, is not always true [5-8]. In a coordination chemical approach to heterogeneous oxidation [9], reaction mechanisms are discussed in terms of interactions between electron pair donors and electron pair acceptors. According to this model, the mechanism for formation of carbon oxides is believed to involve oxygen species that have higher base strength compared with those participating in partial oxidation. However, such a conclusion is in conflict with the general idea on oxidation catalysis, as it has been formulated by Haber [10]. Namely, that electrophilic oxygen species attack the organic molecule in the region of its highest electron density, leading to degradation and, eventually, formation of carbon oxides. In partial oxidation, on the other hand, a nucleophilic addition of O^{2-} species occurs after an initial activation of the hydrocarbon molecule.

The present work presents a model of the active site, which distinguishes between the character of various oxygen species. A relationship between reaction rate and bond-strength is derived for partial and total oxidation. Its applicability for oxidation and ammoxidation is demonstrated. Furthermore, it is shown that both crystallographic aspects and dynamic processes can be accounted for.

MODEL

A simplified drawing of an oxide surface is shown in Figure 1. In addition to bridging oxygen species, whose number is determined by the bulk structure, a unit surface area consisting of **n** ca-

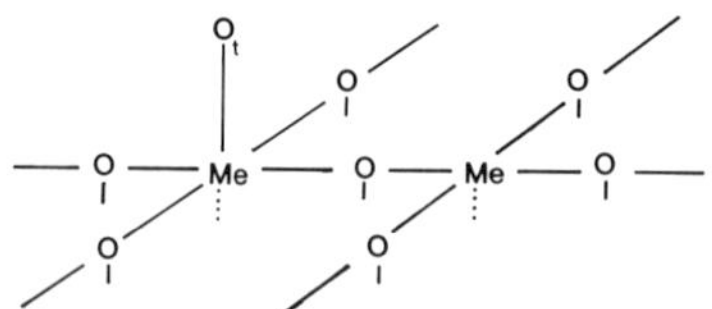

Fig. 1. A schematic illustration of a metal oxide surface.

tions can have between zero and **n** terminal, single-coordinated, oxygen species (O_t). For a surface to be active in the oxidation of alkylaromatics and olefins, it must possess two primary functions. These are to act as a source of active oxygen species and to expose cations, which serve as adsorption centers for the hydrocarbon. In this regard, it is of interest to know whether it is the bridging, the terminal, or both types of oxygen species that react. Let us assume that the terminal oxygen species have a decisive influence on the reaction path in catalytic oxidation. This is not unreasonable, because it seems unlikely that a hydrocarbon molecule approaching a surface should not interact with projecting oxygen species. Also, bridging species are usually more coordinated compared with terminal species and their removal would be energetically unfavourable. On the condition that site reoxidation is facile, a very simple site for the accomplishment of catalytic oxidation can be considered to consist of one cation and one terminal oxygen species, only. The number of such sites per unit surface area cannot exceed **n**/2, and there will be none when all terminal oxygen positions are either empty or completely occupied. The number of active sites as a function of oxygen coverage is in Figure 2.

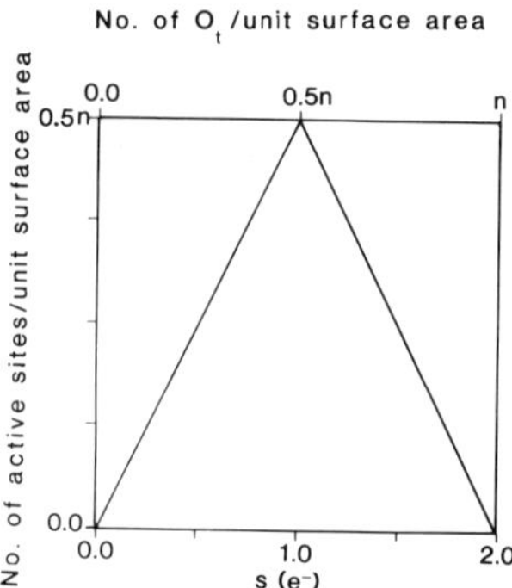

Fig. 2. Number of active sites per unit surface area as a function of either the number of terminal oxygen species (O_t) per unit surface area or the bond-strength value for O_t.

Assuming that surface relaxation phenomena can be neglected, it follows that the more undercoordinated a surface oxygen species is, compared to if it were in a corresponding bulk position, the weaker it is bonded to the surface. As a measure of the undercoordination, bond-strength values can be used [1,2,11,12]. The sum of the bond-strength values of all metal-oxygen bonds around an oxide ion in the structure is always close to 2, indicating a valence of -2. For a terminal oxygen species, the bond-strength value, noted **s**, will be in the range 0 - 2 depending upon the metal-oxygen distance. This suggests that the coverage of terminal oxygen positions, and consequently also the number of active sites, will depend on the bond-strength value. In Figure 2 such a dependence is shown, assuming that it is linear and that all terminal positions are occupied when **s** = 2 and empty when **s** = 0. Of course, the **s**-value required for having maximum number of active sites is dependent on the partial pressures of reactants and the reaction temperature. Additionally, the total number of active sites per unit surface area is also dependent on the number of cations and oxygen species that are required for a specific reaction.

Experimental evidence exists for electrophilic oxygen species, e.g., O_2^- and O^-, to be involved in the combustion of alkylaromatics and olefins by attacking the hydrocarbon in its region of highest electron density, i.e., the aromatic ring or the double bond. Such an attack leads to degradation followed by the formation of carbon oxides [10,12,13]. The reactivity of a site acting as an electrophilic reagent will depend on the **s**-value, or the formal valence, of the O_t species. The turnover number (TON) will be high at small **s**-values due to strong undercoordination of O_t, and it will gradually approach zero with increase in bond strength. It can be expected that O_t has no electrophilic character when **s** ≥ 1.5. In Figure 3 the reactivity of a combustion site is given as a solid line. However, it is likely that the reoxidation of the site is rate-determining at low **s**-values. The TON for site reoxidation can be expected to gradually approach the TON for the electrophilic attack, which is also indicated in Figure 3.

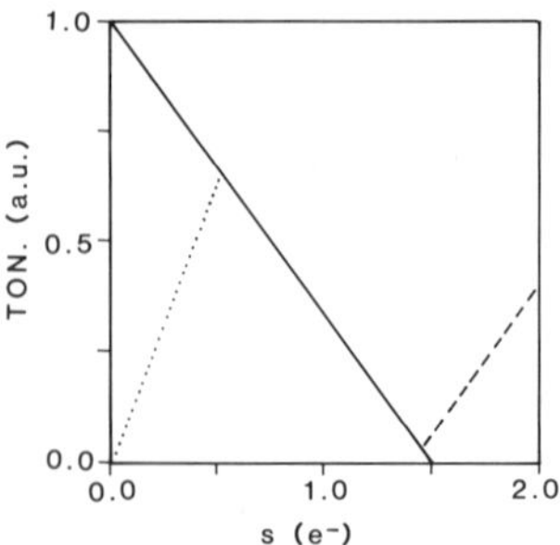

Fig. 3. Turnover number (TON) versus bond-strength. Solid line: TON for electrophilic attack. Dotted line: TON for reoxidation of a combustion site. Dashed line: TON for selective oxidation.

It is generally accepted that the mechanism for selective oxidation of alkylaromatics and olefins proceeds through an initial hydrogen abstraction from the $-CH_3$ group, which is followed by an insertion of oxygen and a second hydrogen abstraction [5,10,13-16]. When the hydrocarbon is adsorbed at a cation, there is a charge transfer from its electron-rich part to the catalyst surface, giving the $-CH_3$ group a positive charge. Thus, the species involved in hydrogen abstraction and oxygen insertion must be able to act as a nucleophilic reagent. Such a species can be a Me=O (Me: metal) group, which has a region of high electron density. It follows, that a site for selective oxidation should have a terminal oxygen species with a relatively high **s**-value. Irrespective of whether hydrogen abstraction, oxygen insertion or site reoxidation is rate-limiting, the TON of a selective site should increase with increase in bond-strength. The TON is given in Figure 3 as a dashed line assuming that the site has no nucleophilic character when $\mathbf{s} \leq 1.4$. In principle, the adsorption of the hydrocarbon can be rate-limiting at high **s**-values, however, this is usually not observed to be the case [5,10,13,14,16-18].

The rate of oxidation depends on the number of active sites and the activity per site (TON). Reaction rates for selective and unselective reactions, obtained by multiplying the dependencies shown in Figures 2 and 3, are given in Figure 4. The main conclusion that can be drawn from this figure is that depending upon the **s**-value, the rate for combustion varies from low to very high values, while the variation in rate for selective oxidation is comparatively smaller. To illustrate the

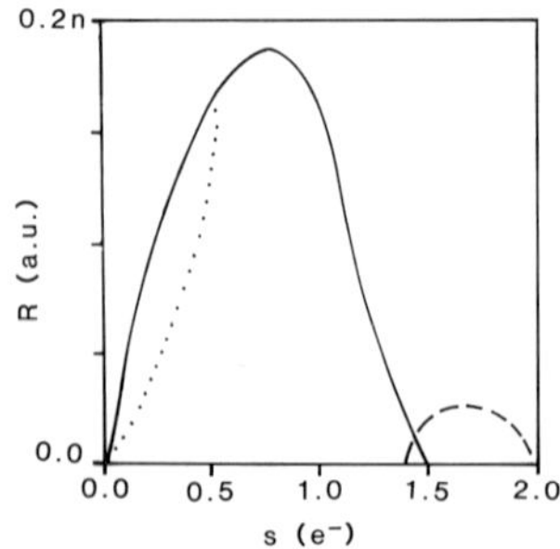

Fig. 4. Reaction rate as a function of bond-strength. Solid line: combustion, electrophilic attack is rate-limiting. Dotted line: combustion, reoxidation is rate-limiting. Dashed line: selective oxidation.

validity of this behaviour, data collected by Germain and Laugier [19] for the oxidation of toluene over 19 binary oxides can be used. This large number of oxides represent a considerable variation in the degree of undercoordination, and consequently also s-value, of terminal oxygen species. Figure 5 shows the variation of selectivity for formation of carbon oxides when plotted against the combustion rate. It is seen that the data fall within the limits given by the two theoretical curves, which

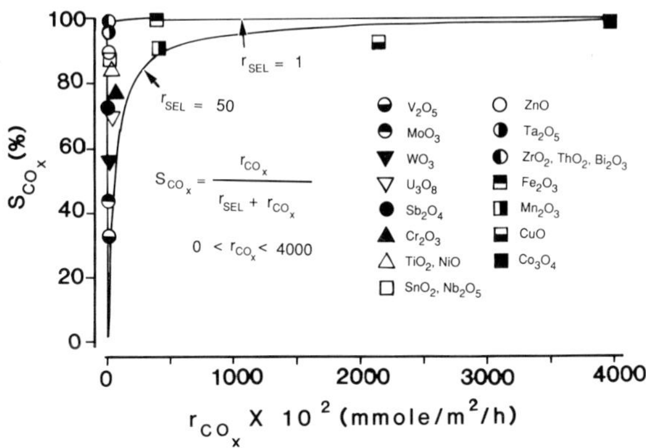

Fig. 5. Oxidation of toluene over 19 binary oxides. The selectivity for formation of carbon oxides is plotted versus the rate for combustion. Reaction temperature, 400°C. Mole ratio air/toluene = 75/1. Data from ref. [19].

have been calculated assuming that the rate for selective oxidation is a constant equal to 1 and 50 mmole·10^{-2}/h/m^2, respectively. Since the rate for combustion varies in the range 0 - 4000 mmole·10^{-2}/h/m^2, it is clearly seen that the variation in rate for combustion is much larger than it is for selective oxidation.

Another conclusion that can be drawn from Figure 5 is that V_2O_5, MoO_3, WO_3 and U_3O_8 are more selective for partial oxidation in comparison with other oxides. If the structures of the oxides are compared, it will be found that cations of the former oxides are in a comparatively more distorted coordination, with at least one relatively short metal-oxygen bond. Consequently, their surfaces can be expected to project groups of the Me=O type, which can act as nucleophilic reagents in selective oxidation. This feature supports the model. From Figure 5, it also follows that the low selectivity of many oxides towards partial oxidation is not always due to their low activity for selective reaction, but due to the fact that they are highly active for combustion. Thus, in order to improve catalyst selectivity, considerable attention should be paid, not only to the mechanism of selective oxidation, but perhaps even more to the combustion mechanism. The fact that several oxides are active, though not selective, for partial oxidation infers that when there is a depletion of terminal oxygen species, due to either reaction or a low s-value, bridging oxygen species may have a decisive influence on the reaction path. Some of these oxygens, with nucleophilic character, can participate in selective reaction routes [20].

STRUCTURE-SENSITIVITY

Even though the simple model derived can be used to explain the general selectivity pattern shown considering data for a large number of oxides, it is a fact that the model was derived assuming a homogeneous surface with only one type of cation and one type of O_t species. However, real oxide surfaces can consist of crystallographically unequal cations as well as unequal O_t species. Furthermore, a real crystal exposes several faces with differing Miller indices. To show the general validity of the principles of the simplified model, some examples of catalytic oxidation over well-defined V_2O_5 and MoO_3 crystals will be treated. Crystal data: V_2O_5, orthorhombic, a = 11.51, b = 4.37, c = 3.56 Å [21]; and MoO_3, orthorhombic, a = 3.96, b = 13.86, c = 3.70 Å [22].

Catalytic data obtained in a study of the ammoxidation of 3-picoline over two V_2O_5 preparations are given in Table 1 [23]. One of the samples had been prepared by melting of V_2O_5, followed

TABLE 1
Effect of the morphology of V_2O_5 in the ammoxidation of 3-picoline[a].

Sample	Selectivity (%)		
	Nitrile	Waste	I(101)/I(010)
Crushed	61	39	0.11
Decomposed	18	82	1.00

[a] $327^{\circ}C$. Mole ratio 3-picoline/NH_3/O_2 = 1/4/4.25.

by cooling and crushing of the solid material. According to X-ray diffraction analysis, this sample showed a preferential exposure of (010) faces. The morphological factor I(101)/I(010), defined as the intensity ratio of the reflections indexed, was low. A much higher factor, indicating a predominance for planes being perpendicular to the (010) plane, was obtained when V_2O_5 was prepared by decomposition of NH_4VO_3 in a stream of air. If the selectivities of the two samples for the formation of nicotinonitrile are compared, it can be concluded that nitrile is formed at the (010) plane, while degradation occurs at planes perpendicular to (010). Also, in a study of the oxidation of *o*-xylene over V_2O_5 samples with different morphologies, it was found that phthalic anhydride is formed at the (010) plane [7].

In the ammoxidation of toluene over a series of samples of MoO_3 which differed with respect to their morphology, the correlations given in Table 2 were obtained [12]. The rates given are initial rates, measured in the absence of molecular oxygen. The data clearly show the (010) plane to be almost inactive. Benzonitrile is formed both at the (100) plane and at the terminations in the [001]-direction. Carbon oxides are almost exclusively formed at [001]-terminations.

TABLE 2
Initial reaction rates[a], mole/h/m^2, over MoO_3 faces.

Faces	C_6H_5CN *10^3	CO_2 *10^4	CO *10^5
(100)	19.1	0.2	1.4
(010)	0.7	1.4	2.0
(001)[b]	96.2	104.0	19.9

[a]452°C. Mole ratio toluene/NH_3 = 1/3.37.
[b](001) and (*h*0*l*).

In Table 3 are the bond-strength values for the O_t species that are present at various faces of MoO_3 and V_2O_5 [12,23]. Relating the catalytic results described to the s-values of O_t species, some general features are revealed. One is that partial (amm)oxidation takes place at planes exposing both cations and O_t species of the double-bonded type, i.e., at V_2O_5(010), MoO_3(100) and MoO_3(001) faces. At MoO_3(100), the O_t positions with s = 0.46 can be expected to be almost vacant, especially, when the ammoxidation is carried out in the absence of molecular oxygen. On the other hand, at

TABLE 3
Bond-strength values (e^-) of O_t.

Plane	MoO_3	V_2O_5
(100)	0.46; 1.50; 2.06	0.48; 1.04
(010)	2.06	0.04; 1.93
(001)	0.78; 2.06	0.76

MoO_3(001) the positions having s = 0.78 are probably partly occupied. Another feature seen is that the formation of carbon oxides is related to the existence of planes having undercoordinated O_t species with moderate s-values, i.e., V_2O_5(100), V_2O_5(001) and MoO_3(001) faces. The features pointed to clearly support the idea that in selective ammoxidation routes, double bonded oxygen species react with activated ammonia to give imido species, =NH, which can be inserted in the activated hydrocarbon [5,16,24]. A comparison of the catalytic behaviours of the (010) faces of V_2O_5 and MoO_3 shows that on V_2O_5 selective (amm)oxidation occurs, while on MoO_3 these faces are almost inactive. If the s-values of the O_t species are considered, Table 3, it is seen that on V_2O_5(010) both cations and double-bonded oxygen species are present. On MoO_3(010), on the other hand, an almost complete coverage of O_t positions can be expected. Clearly, there is a relationship between activity and coverage of terminal oxygen positions, which is indeed a support for the choice of active site that has been made in the model presented.

A stopped-flow experiment was carried out in order to find additional support for the conclusion that on MoO_3, nitrile and carbon oxides are mainly formed at the same crystal plane [25].

Table 4 shows the rates measured under steady-state conditions in the presence of molecular oxygen. Included are also the rates measured immediately after having stopped the supply of gaseous oxygen. A comparison of the data shows that when the reactant stream was depleted of oxygen, the rate for nitrile formation increased by a factor of more than two, while the rates for formation of carbon oxides decreased by a factor two. Such a behaviour is the one to expect, only if nitrile and carbon oxides are competitively formed at the same crystal plane. A depletion of the feed of gaseous oxygen results in a decrease of the rate for combustion, since the concentration of weakly bonded, undercoordinated, O_t species with s = 0.78 is reduced. Also, an initial increase of the rate for nitrile formation occurs due to the fact that the concentration of adsorption sites (naked cations) increases, the concentration of double-bonded oxygen species is almost unchanged and that the competitive degradation is reduced.

TABLE 4

Rates, mole/h/m^2, and selectivities, %, for the ammoxidation of toluene over MoO_3 at 725K.

	C_6H_5CN		CO_2		CO	
	Rate	Sel.	Rate	Sel.	Rate	Sel.
With O_2 [a]	$1.35*10^{-3}$	68.8	$4.74*10^{-4}$	24.1	$5.95*10^{-5}$	3.0
Without O_2 [b]	$3.03*10^{-3}$	92.7	$2.19*10^{-4}$	6.7	$2.09*10^{-5}$	0.6

[a] Mole ratio toluene/NH_3/O_2 = 1/3.4/11.3.
[b] Mole ratio toluene/NH_3 = 1/3.4.

The conclusions drawn about the role of various MoO_3 planes in (amm)oxidation of alkylaromatics, if relevant, should also be applicable to the oxidation of olefins. This is due to the fact that similar mechanisms operate in both cases. Table 5 gives the result of an attempt [12] to find a structure-activity relationship using the data, which had been obtained by Volta and coworkers for the oxidations of propene [26] and isobutene [27] over samples of well-defined MoO_3 crystals. It is seen that an excellent agreement between experimental and theoretical values are obtained assuming the [001]-terminations to be the most active faces for both combustion and selective oxidation. This correlation, even though it is in disagreement with the correlations given by the investigators [26-28], strongly supports the validity of the general views on oxidation catalysis as they are expressed in the present paper. In this regard it is worth mentioning that the habit of MoO_3 crystals, which have grown without constraints due to the environment, e.g., by sublimation, can be described as flattened on (010) and elongated along [001] [12]. Consequently, the rates of crystal growth in the various directions decrease in the order, [001] > [100] > [010]. This reactivity sequence is exactly the same as the one observed in the catalytic measurements, cf. Table 2. Thus, it can be concluded that there is a relationship between crystal habit and catalytic activity. Also, it has been observed that Bi^{3+} ions are deposited selectively on the (100) and (001) planes of MoO_3, but not on the (010) plane [29]. This finding is additional support for the high reactivity of the former faces.

TABLE 5

Selectivity ratios as measured in the oxidations of propene and isobutene over MoO_3.

	$(S_A/S_C)^a$		$(S_{MA}/S_C)^b$	
Sample[c]	Experi-mental[c]	Theo-retical[d]	Experi-mental[e]	Theo-retical[f]
420-6	0.39	0.39	0.67	0.69
420-61	0.59	0.59	0.65	0.69
471-6	0.95	0.98	0.78	0.69
496-6	0.94	0.92	0.67	0.69
496-61	1.44	1.44	0.69	0.69

[a]Selectivity ratio acrolein/CO_2.
[b]Selectivity ratio methacrolein/CO_2.
[c]Notations and data from ref. [26].
[d]Ratio calculated assuming that the relative rates over (100), (010), (101), (001) are 0.138, 0, 1.438, 2.157 for acrolein, and 0, 0.201, 1.000, 1.000 for CO_2.
[e]Experimental data from ref. [27].
[f]Ratio calculated assuming that the relative rates over (100), (010), (101), (001) are 0, 0, 0.688, 0.688 for methacrolein, and 0, 0, 1.000, 1.000 for CO_2.

DYNAMIC FACTORS

The theoretical rates given in Figure 4 are normalized rates in the sense that they have been calculated assuming that reaction occurs, i.e., there is a hydrocarbon molecule adsorbed, at each site. Thus, the effective rate (r) can be expressed as a function of the normalized rate (R) and the fraction of sites (Θ) which is covered with hydrocarbon molecules. It follows, that

$$r - \Theta * R \quad (1)$$

The coverage, Θ, using the Langmuir isotherm can be expressed as follows:

$$\Theta = KP / (1 + KP) \quad (2)$$

where P is the partial pressure of the hydrocarbon and K is the equilibrium constant.

As was concluded from Figure 4, the rate of oxidation is dependent on the strength of the metal-oxygen bond, which can be expressed in terms of the s-value. Of course, the bond-strength is highly affected by the electron distribution at the surface, which is influenced by the adsorption of electron donors like ammonia and hydrocarbons with electron-rich regions, e.g., olefins and aromatics. To illustrate how dynamic factors can be accounted for using the model, some results will be described on the ammoxidation of toluene over a TiO_2(B)-supported vanadium oxide catalyst. The vanadium content of the catalyst used was 4.2 wt. % expressed as V_2O_5, which corresponds to 3.5 theoretical V_2O_5 layers. Catalyst preparation method and experimental procedures were the same as described elsewhere [30]. Experiments were carried out varying the partial pressure of toluene (P_T), while keeping the partial pressures of oxygen (P_O) and ammonia (P_A) constant. The partial pressure of toluene was varied at two constant (one low and one high) values for the pressure of ammonia. The results are given in Figures 6 and 7.

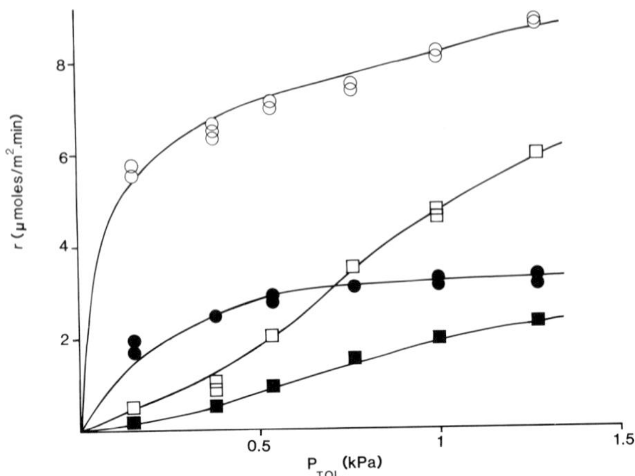

Fig. 6. Rates for formation of CO_2 (circles) and benzaldehyde (squares) as a function of the partial pressure of toluene at ●, ■ 370°C; and ○, □ 400°C. P_O = 11.4 kPa and P_A = 0.14 kPa.

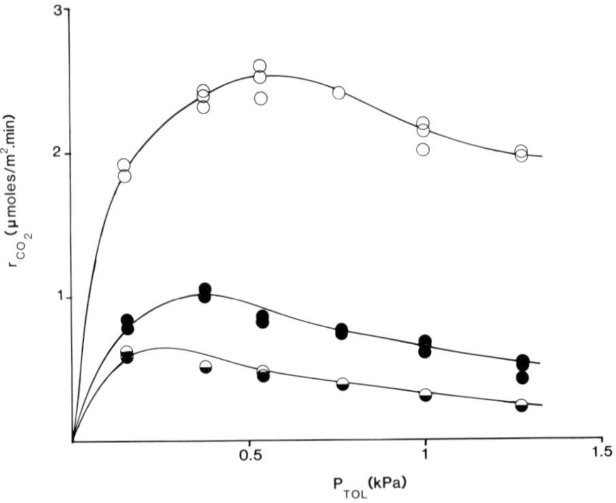

Fig. 7. Effect of the partial pressure of toluene on the rate for formation of CO_2 at ◒ 350°C; ● 370°C; and ○ 400°C. P_O = 11.4 kPa and P_A = 2.85 kPa.

The variation of the rate for formation of CO_2 (r_{CO2}) given in Figure 6, can be shown to be of the form

$$r_{CO2} = k_1P_T / (1 + k_2P_T) \tag{3}$$

where k_1 and k_2 are constants, which possibly further depend on the partial pressures of oxygen and ammonia. Depending on the detailed reaction mechanism, they can include combinations of rate constants and adsorption constants. Equation (3) can alternatively be expressed as

$$r_{CO2} = \Theta_T * C_1 \quad (4)$$

where Θ_T is the fraction of sites that is covered with toluene, and C_1 is a constant. If eqns. (1) and (4) are compared, it follows that in this case the normalized rate, R, is constant. Considering Figure 4, it can be concluded that while varying the partial pressure of toluene, the s-value of O_t species must vary in a region around the rate maximum for unselective reaction. In such a case, R can be considered to be almost constant, which is in agreement with the experimental observation. On the other hand, when the partial pressure of toluene is varied at a high partial pressure of ammonia, the s-value will vary in a higher region. According to Figure 4, a decrease in R with the partial pressure of toluene can be expected. Figure 7 shows the rate dependence on the partial pressure of toluene for formation of CO_2, at a high and constant pressure of ammonia. Analysis shows that the dependence is of the form

$$r_{CO2} = k_3P_T / (1 + k_4P_T + k_5P_T^2) \quad (5)$$

where k_3, k_4, and k_5 are constants of similar type to k_1 and k_2. This dependence can be rearranged to

$$r_{CO2} = \Theta_T * C_2 / (1 + C_3P_T) \quad (6)$$

where C_2 and C_3 are constants. A comparison with eqn. (1) shows that in this case R is a function, which decreases with increase of the partial pressure of toluene, and consequently with increase in bond-strength. Thus, the results on the formation of CO_2 clearly show the validity of the general behaviour given in Figure 4. For the first time, direct kinetic evidence has been presented that electrophilic oxygen species are involved in the degradation and combustion of hydrocarbons.

The dependence on the partial pressure of toluene that was observed for the formation of benzaldehyde, at a low pressure of ammonia, is also included in Figure 6. No aldehyde was formed at the alternate high partial pressure of ammonia that was also studied. In addition to aldehyde, benzonitrile was formed. However, any discussion of nitrile formation is not necessary for the current purpose. Considering Figure 6, it can be concluded that the rate expression for aldehyde formation (r_{CHO}) can be given as

$$r_{CHO} = (k_6P_T + k_7P_T^2) / (1 + k_8P_T + k_9P_T^2) \quad (7)$$

where k_6, k_7, k_8, and k_9 are constants of the same type as those in eqns. (3) and (5). Furthermore, it can be shown that $k_6 << k_7$. Equation (7) can be expressed as follows:

$$r_{CHO} = \Theta_T * (C_4 + C_5P_T) / (1 + C_6P_T) \quad (8)$$

where C_4, C_5, and C_6 are constants. From a comparison of eqns. (1) and (8), it follows that in this case R is a function, which increases with increase of the partial pressure of toluene, also, $C_4 << C_5$ since $k_6 << k_7$. If the expression derived for R is considered in relation to the effect to be expected from the adsorption of an electron donor, which is shown in Figure 4 as a dashed curve, it can be concluded that there is agreement. A low value of C_4 indicates that R is low when the electron induced effect of toluene adsorption is not accounted for. Possibly, the s-value is in the range 1.4 - 1.5. When the partial pressure of toluene is increased, and consequently also the s-value, the rate according to the curve in Figure 4 should increase. This behaviour is in perfect agreement with the dependence observed. Thus, reaction kinetic evidence has been given that oxygen species participate as nucleophilic reagents in the selective oxidation mechanism.

If the dependencies shown in Figures 6 and 7 are discussed in more common kinetic terms, it must be concluded that benzaldehyde is predominantly formed at a site accommodating two toluene molecules, while a site with only one toluene species has a comparatively much lower activity. Concerning the formation of CO_2, it must be concluded that site deactivation occurs as a result of adsorption of a second toluene molecule at the active site. Deactivation is, however, observed only when the partial pressure of ammonia is high. Indeed, such explanations to the observed dependencies seem rather unlikely in comparison with those offered by the dynamic approach here given.

CONCLUSIONS

For the oxidation of alkylaromatics and olefins, cations serving as adsorption centers for the hydrocarbon, and oxygen species of a certain character are needed. Use of ammonia as an electron donor to the catalyst surface has for the first time made it possible to obtain reaction kinetic evidence that oxygen species of electrophilic character are involved in combustion, while the oxygen species participating in partial oxidation are nucleophilic reagents.

A relationship between the bond-length of metal-oxygen bonds and selectivity has been observed. For a crystal face to be selective for partial oxidation, it should exhibit both long and short terminal metal-oxygen bonds, corresponding to oxygen vacancies and double-bonded oxygen species, respectively. However, it cannot be excluded that bridging oxygen species can also participate in partial oxidation, since several oxides lacking very short metal-oxygen bonds are active, though not selective, for partial oxidation. The decisive role of undercoordinated terminal oxygen species in unselective reaction has clearly been observed.

It has been demonstrated that useful information can be obtained by considering the catalyst surface as an extension of the bulk structure. If the coordinations of a surface oxygen species and a corresponding bulk species are compared, roughly, three different types of surface positions for oxygen can be distinguished as follows.

i) Positions for strongly undercoordinated oxygen species, which can be considered to be vacant, acting as adsorption centers for the hydrocarbon.

ii) Positions for moderately undercoordinated oxygen species, which are partly filled positions. These are occupied by electrophilic oxygen species, which participate in the degradation and the combustion of hydrocarbons.

iii) Positions occupied by strongly coordinated oxygen species, which participate as nucleophilic reagents in the partial oxidation of hydrocarbons.

To make possible a direct comparison of the coordinations of different oxygen species, bond-strength values calculated from bond-strength-bond-length relationships can be used.

It has been shown that in kinetic expressions, reactant induced electron redistributions at the surface can be accounted for by the use of a simple model of the active site, combined with consideration of how the bond-strength of the metal-oxygen bond is affected by the adsorption of reactant.

ACKNOWLEDGMENT

Financial support from the National Swedish Board for Technical Development (STU), the National Energy Administration (STEV) and the Swedish Natural Science Research Council (NFR) is gratefully acknowledged.

REFERENCES

1 J. Ziółkowski, *J. Catal.*, 80 (1983) 263.
2 J. Ziółkowski, *J. Catal.*, 81 (1983) 311.
3 F. Cavani, G. Centi, F. Trifirò and R.K. Grasselli, *Catal. Today*, 3 (1988) 185.
4 J.L. Callahan and R.K. Grasselli, *AIChE J.*, 9 (1963) 755.
5 J.C. Otamiri and A. Andersson, *Catal. Today*, 3 (1988) 211.
6 J.C. Otamiri and A. Andersson, *Catal. Today*, 3 (1988) 223.
7 M. Gasior and T. Machej, *J. Catal.*, 83 (1983) 472.
8 R.Y. Saleh and I.E. Wachs, *Appl. Catal.*, 31 (1987) 87.
9 H. Noller and H. Vinek, *J. Mol. Catal.*, 51 (1989) 285.
10 J. Haber, in R.K. Grasselli and J.F. Brazdil (Eds.), *Solid State Chemistry in Catalysis*, ACS Symposium Series, Vol. 279, American Chemical Society, Washington, D.C., 1985, pp. 3-21.
11 I.D. Brown and K.K. Wu, *Acta Crystallogr. Sect. B*, 32 (1976) 1957.
12 A. Andersson and S. Hansen, *J. Catal.*, 114 (1988) 332.
13 J. Haber, in J.P. Bonnelle, B. Delmon and E. Derouane (Eds.), *Surface Properties and Catalysis by Non-Metals*, NATO ASI Series, Ser. C, No. 105, Reidel, Dordrecht, 1983, pp. 1-45.
14 C.R. Adams and T.J. Jennings, *J. Catal.*, 2 (1963) 63.
15 W.M.H. Sachtler and N.H. de Boer, *Proc. 3rd Int. Congr. on Catalysis, Amsterdam, 1964*, p. 252.
16 R.K. Grasselli and J.D. Burrington, in D.D. Eley, H. Pines and P.B. Weisz (Eds.), *Advances in Catalysis*, Vol. 30, Academic Press, New York, 1981, pp. 133-163.
17 C. Cavani, G. Centi and F. Trifirò, *Ind. Eng. Chem. Prod. Res. Dev.*, 22 (1983) 570.
18 P. Cavalli, F. Cavani, I. Manenti, F. Trifirò and M. El-Sawi, *Ind. Eng. Chem. Res.*, 26 (1987) 804.
19 J.-E. Germain and R. Laugier, *Bull. Soc. Chim. France*, 2 (1972) 541.
20 J. Haber and E.M. Serwicka, *React. Kinet. Catal. Lett.*, 35 (1987) 369.
21 H.G. Bachmann, F.R. Ahmed and W.H. Barnes, *Z. Kristallogr.*, 115 (1961) 110.
22 L. Kihlborg, *Ark. Kemi*, 21 (1963) 357.
23 A. Andersson, J.-O. Bovin and P. Walter, *J. Catal.*, 98 (1986) 204.
24 J. Haber, *Proc. 8th Int. Congr. on Catalysis, Berlin(West), July 2-6, 1984*, Verlag Chemie, Weinheim, 1984, Vol. V, pp. 85-111.
25 A. Andersson and S. Hansen, *Catal. Lett.*, 1 (1988) 377.
26 J.C. Volta, M. Forrisier, F. Theobald and T.P. Pham, *Faraday Discuss. Chem. Soc.*, 72 (1981) 225.
27 J.C. Volta, J.M. Tatibouet, Ch. Phichitkul and J.E. Germain, *Proc. 8th Int. Congr. on Catalysis, Berlin(West), July 2-6, 1984*, Verlag Chemie, Weinheim, 1984, Vol. IV, pp. 451-461.
28 J.C. Volta and J.M. Tatibouet, *J. Catal.*, 93 (1985) 467.
29 K. Brückman, J. Haber and T. Wiltowski, *J. Catal.*, 106 (1987) 188.
30 M. Sanati and A. Andersson, *J. Mol. Catal.*, 59 (1990) 233.

R.K. Grasselli and A.W. Sleight (Editors), *Structure-Activity and Selectivity Relationships in Heterogeneous Catalysis*
© 1991 Elsevier Science Publishers B.V., Amsterdam

Catalyst Oxide Support Oxide Interaction to Prepare Multifunctional Oxidation Catalysts

Yoshihiko Moro-oka, De-Hua He, and Wataru Ueda

Research Laboratory of Resources Utilization, Tokyo Institute of Technology, Nagatsuta-cho 4259, Midori-ku, Yokohama, 227 Japan

Working mechanism of multicomponent bismuth molybdate catalyst system was investigated by surface and structural analysis of the catalysts and activity tests and ^{18}O tracer experiments in the oxidation of propylene to acrolein. The results obtained with model catalysts, $Bi_2Mo_3O_{12}$ supported on cobalt molybdate with or without iron molybdate clearly show the importance of the strong interaction between the catalyst oxide and support oxide for the preparation of active catalyst systems. A concept of multifunctional system including the transportation of active oxygen species through the bulk diffusion of oxide ion and different kinds of active site for the activation of dioxygen as well as the consecutive dehydrogenation and oxygenation of hydrocarbon molecule is presented for the design of the effective oxidation catalyst.

Introduction

In the recent 20 years, remarkable developments in the heterogeneous oxidation catalysts have been achieved by blending of the main active components with several kinds of metal oxide additives. Typical examples are shown in the multicomponent bismuth molybdate catalysts which are known as the most active and selective catalysts for the allylic oxidation of lower olefin[1,2]. They have been widely used in the industrial oxidations. However, owing to their complicated compositions and structures[3-10], little has been reported for the working mechanism and role of each component in the catalyst systems.

Multicomponent bismuth molybdate catalysts claimed in the recent patents are composed of a number of elements in addition to molybdenum and bismuth. The general formula of them may be written as follows;

$Mo\text{-}Bi\text{-}M^{II}\text{-}M^{III}\text{-}M^{I}\text{-}X\text{-}Y\text{-}O$	M^{II}	: Co, Ni, Mn, Mg, (Pb),...
Fundamental structure	M^{III}	: Fe, Cr, Al,... especially Fe.
	M^{I}	: Na, K, Cs, Tl,...
	X	: Sb, W, V, Nb, Te,...
	Y	: P, B,...

The first four elements are essential and consist of a fundamental structure of the catalysts systems, and the remains are usually added for enhancement of the catalyst life and mechanical strength and minor improvement of the catalytic activity and selectivity. Since the role of alkali metal and other two additives are further complicated, the investigation of our group has been directed only to the fundamental structure of the catalysts.

Our recent studies have been undertaken to clarify the working mechanism of multicomponent metal oxide catalysts systems, $Mo\text{-}Bi\text{-}M^{II}\text{-}M^{III}\text{-}O$ using $^{18}O_2$ tracer technique[11,12,13]. It was concluded that the real active component of $Mo\text{-}Bi\text{-}M^{II}\text{-}M^{III}\text{-}O$ systems is bismuth molybdate, especially its α-phase and transition metal molybdates, $M^{II}MoO_4$, $M^{III}{}_2(MoO_4)_3$ and their solid solutions, serve as a kind of support for the active catalyst component. It was also suggested that a rapid bulk migration of oxide ion through lattice vacancies of the oxide support as well as bismuth molybdate plays an important role to increase the catalytic activity of the multicomponent oxide systems.

Bismuth molybdates supported on a cobalt molybdate or cobalt-iron molybdate were prepared as model catalysts and their catalytic behaviors were investigated in the oxidation of propylene to acrolein in order to confirm the above concept[14]. We show here clear evidences that the catalytic activity of bismuth molybdate depends deeply on the type of the support metal molybdates and a strong interaction between bismuth molybdate and support oxide is essential for the active multicomponent bismuth molybdate catalysts.

Strong interaction of metallic catalysts with oxide support, SMSI, was already proposed and has been extensively investigated[15]. However, SMSI usually brings about only negative effects on the catalytic activity. Different from the case of metallic catalyst, it is demonstrated in this article that the strong interaction between catalyst oxide and support oxide is the most important concept to prepare effective industrial oxidation catalysts.

Experimental

Catalyst Preparation and Characterization

Two kinds of tri- and tetra-component bismuth molybdate catalysts were prepared by different methods. A series of $Mo_{12}Bi_1Co_8M_3{}^{II\text{ or }III}O_x$ catalyst containing Fe^{3+}, Al^{3+}, or Ni^{2+} as the fourth component and $Mo_{12}Bi_1Co_{11}O_x$ were prepared from the corresponding metal nitrate solutions and molybdic acid by the coprecipitation method reported by Wolfs and Batist[3]. BET surface area and composite oxides detected by XRD analysis are listed in Table 1. Surface composition of each catalyst was also examined by XPS. It was confirmed that molybdenum and bismuth were concentrated in the surface of the catalyst particle as first pointed

Table 1 Characterization of Tri- And Tetra-component Bismuth Molybdate Catalysts

Catalyst	Phase detected by XRD	Surface area (m^2/g)	Oxidation of propylene(450°C) Rate (10^{-5}mol/min·g)	Selectivity (%)
$Mo_3Bi_2O_{12}$	$Bi_2(MoO_4)_3$	1.8	17.5	93
$Mo_{12}Bi_1Co_{11}O_x$	β-$CoMoO_4$, $Bi_2(MoO_4)_3$	3.8	9.5	97
$Mo_{12}Bi_1Co_8Ni_3O_x$	β-$CoMoO_4$, $Bi_2(MoO_4)_3$ $NiMoO_4$	6.5	4.6	97
$Mo_{12}Bi_1Co_8Fe_3O_x$	β-$CoMoO_4$, $Bi_2(MoO_4)_3$ $Fe_2(MoO_4)_3$, $FeMoO_4$	7.1	218.7	96
$Mo_{12}Bi_1Co_8Al_3O_x$	β-$CoMoO_4$, $Bi_2(MoO_4)_3$ $Al_2(MoO_4)_3$	8.5	43.4	95

out by Wolfs et al.[3,4]. Although reduced amounts of cobalt and iron were also detected in the surface, those concentrations were far lower compared to molybdenum and bismuth. It may be concluded that most part of the surface of the tri- and tetra-component systems is covered by thin layer of α-phase of bismuth molybdate and the core of the catalyst particle is composed of $M^{II}MoO_4$ and $M^{III}{}_2(MoO_4)_3$.

Another series of multicomponent bismuth molybdate catalysts were prepared by loading of $Bi_2(MoO_4)_3$ on $CoMoO_4$ or $Co_{11/12}Fe_{1/12}MoO_x$ support. $CoMoO_4$ was prepared according to the literature[3] and has a surface area of 14.3 m^2/g. $Co_{11/12}Fe_{1/12}MoO_x$(11.5 m^2/g) was prepared by the solid state reaction between cobalt oxalate and ferrous oxalate at 520 °C for 40 h. Complete solid solution between $CoMoO_4$ and $FeMoO_4$ was formed and no XRD peaks except those of $CoMoO_4$ were detected after the calcination. It was also found by ESR measurement that a part of Fe^{2+} was oxidized to Fe^{3+} during the calcination.

Supported $Bi_2Mo_3O_{12}$ catalysts were prepared by the impregnation method. Support oxides were first impregnated with aqueous ammonium molybdate solution and then triphenyl bismuth in acetone. After drying, samples were calcined at 500 °C for 20 h. Thus, a series of $Bi_2Mo_3O_{12}$ catalysts supported on the $CoMoO_4$ or $Co_{11/12}Fe_{1/12}MoO_x$ were prepared by changing the loading amount of $Bi_2Mo_3O_{12}$ from 0 to 0.3 in molar ratio of $Bi_2Mo_3O_{12}$ against the support molybdates. The supported catalysts were characterized by various physicochemical measurements. It was confirmed that no new composite oxides were formed between the catalyst and support oxides. It was found by TEM and X-ray emission measurements that bismuth molybdate was first located on the surface of the support oxide particles forming the thin layer and then small crystals of $Bi_2Mo_3O_{12}$ grew on the surface. The surface area of the catalysts decreased with increasing loading amount of $Bi_2Mo_3O_{12}$ until it reached the surface area of pure $Bi_2Mo_3O_{12}$, ca. 2.0 m^2/g.

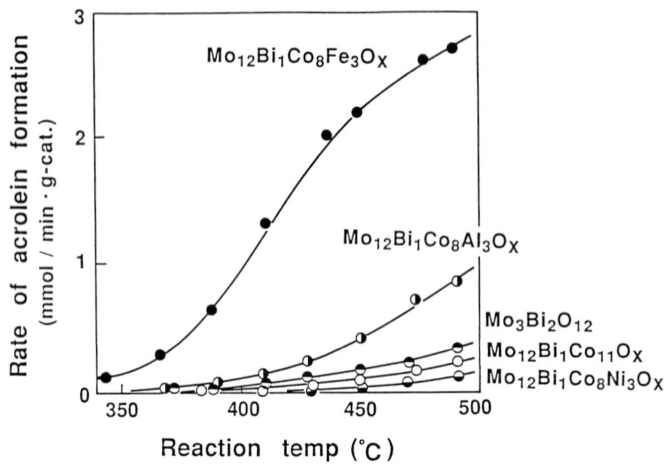

Fig. 1 Catalytic activity of the multicomponent bismuth molybdates in the oxidation of propylene.

Reaction Procedure

Activity of each catalyst was tested in the oxidation of propylene to acrolein using a conventional flow reactor under an atmospheric pressure (propylene; 16%, oxygen; 16%, nitrogen; balance). $^{18}O_2$ tracer experiments were carried out using a closed circulating system under a reduced pressure (propylene; 9.3 kPa, $^{18}O_2$; 9.3 kPa). The reactant and product gases were analyzed by gas chromatography and mass spectrometry.

RESULTS AND DISCUSSION

Catalytic Activity of Multicomponent Bismuth Molybdate

Catalytic activities to form acrolein for typical tri- and tetra-component bismuth molybdate catalysts are plotted against the reaction temperature in Fig. 1. Selectivity to acrolein of every catalyst tested was quite high as partly described in Table 1 but catalytic activity differs remarkably depending on the type of the catalysts. Summarizing the results in this work and reported previously[12,13], the following trends are prominent.

1. In terms of the specific activity, pure α-phase of bismuth molybdate, $Bi_2(MoO_4)_3$ is moderately active. However, owing to its low surface area (1.8 m^2/g), the activity per unit weight of the catalyst is quite low.
2. Although the addition of the third element M^{II} (M^{II} is Co^{2+}, Ni^{2+}, Mn^{2+} or Mg^{2+}. Ionic radii of them are smaller than 0.80 Å) to the pure bismuth molybdate

increases the surface area of the catalyst systems (3-7 m2/g), but the specific activity of the tri-component system, Mo-Bi-M^{II}-O never exceeds that of pure bismuth molybdate.

3. Replacement of a part of third component, M^{II}, in the tri-component system by a trivalent metal cation, M^{III}, increases the specific activity as well as the surface area of the catalyst system. Although chromiun and aluminum are effective to some extent, iron increases the specific activity remarkably with keeping excellent selectivity to acrolein.
4. On the contrary, replacement of M^{II} by another divalent cation, M'^{II} such as Ni^{2+}, is ineffective for improving the specific activity of Mo-Bi-M^{II}-O system.

In conclusion, a part of the increasing activity of Mo-Bi-M^{II}-M^{III}-O system compared to pure bismuth molybdate comes from increase of the surface area but most part arises from increase of the specific activity.

Catalytic Behaviors of Model Catalysts

Catalytic activities of the supported bismuth molybdate catalysts were also examined in the oxidation of propylene to acrolein with variation of the loading amount of $Bi_2Mo_3O_{12}$. The results are shown in Figs. 2 and 3. The catalytic activity to form acrolein of each support molybdate is quite low at the conditions adopted in this investigation. As shown in both Figs. 2 and 3, the observed catalytic activity of $Bi_2Mo_3O_{12}/Co_{11/12}Fe_{1/12}MoO_x$ increases steeply with increasing loading amount of $Bi_2Mo_3O_{12}$. When the loading amount of $Bi_2Mo_3O_{12}$ reaches the value of 0.005 in the molar ratio of the catalyst to the support (this corresponds to the monolayer coverage of $Bi_2Mo_3O_{12}$ on the $Co_{11/12}Fe_{1/12}MoO_x$ surface), its specific activity is

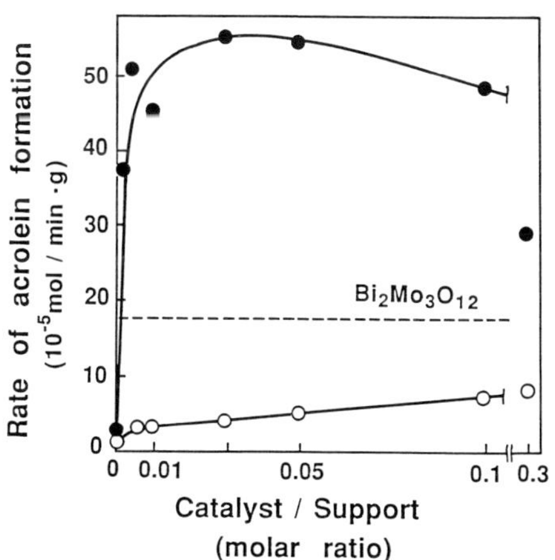

Fig. 2 Catalytic activities per unit weight of the supported bismuth molybdate, (O) $Bi_2Mo_3O_{12}/CoMoO_4$ and (●) $Bi_2Mo_3O_{12}/Co_{11/12}Fe_{1/12}MoO_x$(450 °C).

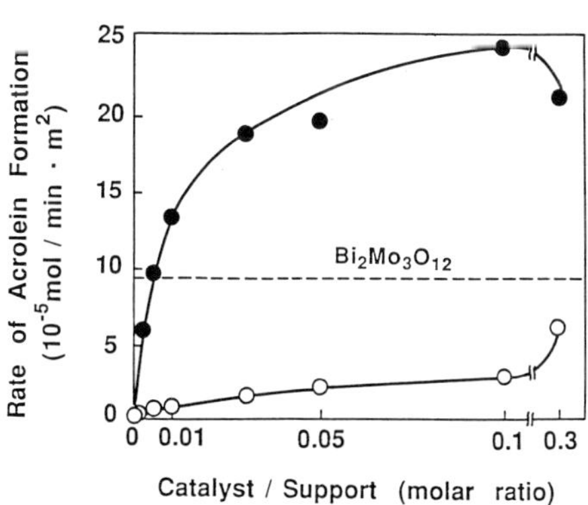

Fig. 3 Catalytic activities per unit surface of the supported bismuth molybdate, (O) $Bi_2Mo_3O_{12}/CoMoO_4$ and (●) $Bi_2Mo_3O_{12}/Co_{11/12}Fe_{1/12}MoO_x$(450 °C).

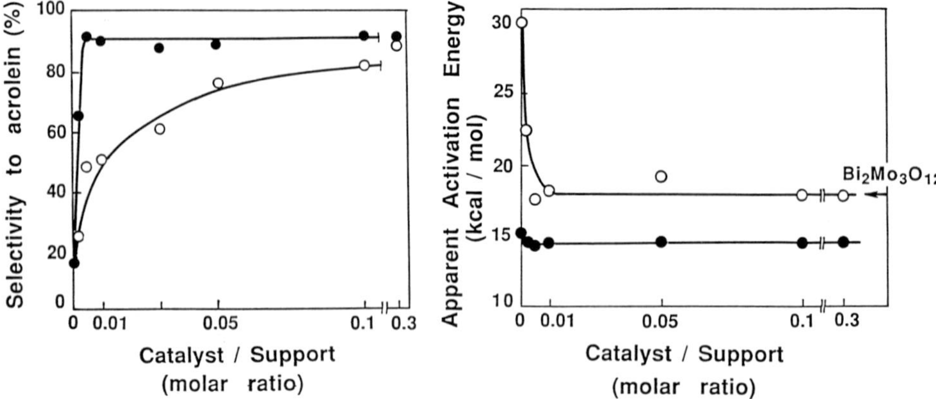

Fig. 4 The selectivity to acrolein in the oxidation on the supported catalysts, (○) $Bi_2Mo_3O_{12}/CoMoO_4$, and (●) $Bi_2Mo_3O_{12}/Co_{11/12}Fe_{1/12}MoO_x$.

Fig. 5 The apparent activation energy of the reaction on the supported catalysts, (○) $Bi_2Mo_3O_{12}/CoMoO_4$ and (●) $Bi_2MO_3O_{12}/Co_{11/12}Fe_{1/12}MoO_x$.

almost comparable to that of the pure $Bi_2Mo_3O_{12}$. Since the support has larger surface area compared to the pure $Bi_2Mo_3O_{12}$, the activity per unit weight exceeds far beyond that of the pure $Bi_2Mo_3O_{12}$. The specific activity of the $Bi_2Mo_3O_{12}/Co_{11/12}Fe_{1/12}MoO_x$ increases further with increasing the loading amount of $Bi_2Mo_3O_{12}$ and reaches a much higher value than that of the pure $Bi_2Mo_3O_{12}$. On the other hand, increase of the activity of the $Bi_2Mo_3O_{12}/CoMoO_4$ system with the loading amount of $Bi_2Mo_3O_{12}$ is quite low. It never exceeds that of the pure $Bi_2Mo_3O_{12}$. It is noteworthy that this tendency just corresponds to the results obtained in the activity test for the tri- and tetra-component bismuth molybdate catalyst(Fig. 1).

The prominent difference in the catalytic activity depending on the type of the support oxide reflects the selectivity to acrolein and the apparent activation energy of the reaction as shown in Figs. 4 and 5, respectively. The apparent activation energy of the $Bi_2Mo_3O_{12}/CoMoO_4$ system varies rapidly with the loading amount of the catalyst and coincides with that of the pure $Bi_2Mo_3O_{12}$. The apparent activation energy of $Bi_2Mo_3O_{12}/Co_{11/12}Fe_{1/12}MoO_x$ system shows a minor change with the loading amount of the catalyst and is lower than that of the $Bi_2Mo_3O_{12}$ by 3 kcal/mol.

These results clearly show that the oxidation of propylene proceeds on pure $Bi_2Mo_3O_{12}$ in the $Bi_2Mo_3O_{12}/CoMoO_4$ system and $CoMoO_4$ serves as a simple support for the active component. On the other hand, it is evident that $Bi_2Mo_3O_{12}$ supported on the $Co_{11/12}Fe_{1/12}MoO_x$ is modified considerably and this suggests the existence of some strong interaction between $Bi_2Mo_3O_{12}$ and the support oxide.

Tracer Experiment 1. Comparison of active oxygen species of tri- and tetra-component bismuth molybdate catalysts

Incorporation of the lattice oxide ions into the oxidized products was examined in the oxidation of propylene to acrolein using 99.1% $^{18}O_2$ on various tri- and tetra-component bismuth molybdate catalysts. Slow heterogeneous exchange reactions between gaseous oxygen or the oxidized products with catalysts were observed but they were not so rapid to exert any serious effect on the evaluation of the participation of the lattice oxide ions in the oxidation of propylene. Typical results are shown in Fig. 6 where ^{18}O concentrations of acrolein produced on Mo-Bi-Co^{2+}-O and Mo-Bi-Co^{2+}-Fe^{3+}-O are plotted against the total amount of oxygen consumed in the oxidation. In the case of tri-component system, $Mo_{12}Bi_1Co_{11}O_x$, incorporation of $^{16}O_{lattice}$ at the initial stage terminates within a relatively short time and ^{18}O concentration in the products reaches the same value with that of gaseous oxygen. All tri-component system (except Mo-Bi-Pb-O[12]) gave the same results and it is clear that only a part of the lattice oxide ions in the catalyst system are active for the reaction. Comparison between the total amount of ^{16}O incorporated into the oxidized products and $^{18}O^{2-}$ in each composite oxide of the catalyst gives a conclusion that lattice oxide ions only in the bismuth molybdate phase are active and participate in the reaction in the tri-component systems, Mo-Bi-M^{II}-O(M^{II} is Co, Ni, Mn, or Mg)[12].

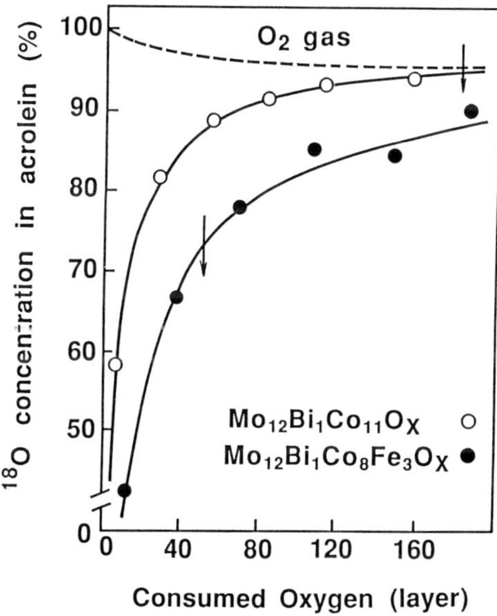

Fig. 6 ^{18}O concentration of acrolein as a function of the consumed oxygen(450°C). The arrow mark shows the point where consumption of $^{16}O^{2-}_{lattice}$ reaches to the same amount of oxide ions in the $Bi_2(MoO_4)_3$.

On the other hand, incorporation of ^{16}O into the oxidized products from the catalyst does not terminate within the reaction time on the tetra-component systems, Mo-Bi-M^{II}-M^{III}-O. As shown in Fig. 6, the ^{18}O concentration in acrolein obtained on the tri-component system reaches the same value as that of gaseous oxygen at the point indicated by the arrow mark which implies consumption of all $^{16}O^{2-}_{lattice}$ in the bismuth molybdate phase. On the contrary, incorporation of $^{16}O^{2-}_{lattice}$ still continues beyond this point on the tetra-component system including both divalent and trivalent metal cations. It is evident that the lattice oxide ions not only in the bismuth molybdate phase but also in other transition metal molybdates are active and can participate in the reaction. This phenomenon was also observed on other tetra-component systems, $Mo_{12}Bi_1Co_8Cr_3O_x$ and $Mo_{12}Bi_1Co_8Al_3O_x$[16]. However, replacement of a part of cobalt cation in the $Mo_{12}Bi_1Co_{11}O_x$ by another divalent cation, Ni^{2+}, did not improve the degree of the incorporation of lattice oxide ions at all and the lattice oxide ions only in the bismuth molybdate phase is active on the $Mo_{12}Bi_1Co_8Ni_3O_x$ catalyst.

Tracer Experiment 2. Oxidation of propylene over ^{18}O labeled model catalyst

In order to clarify the role of each element in the multicomponent bismuth molybdate catalyst, another tracer experiment was carried out using an $Bi_2Mo_3O_{12}$ catalyst supported on $Co_{11/12}Fe_{1/12}MoO_x$ whose lattice oxide ion was labeled by ^{18}O. The ^{18}O-labeled catalyst was prepared by the same method except that a part of the lattice oxide ion in the support oxide, $Co_{11/12}Fe_{1/12}MoO_x$ was replaced with ^{18}O by $^{18}O_2$ reoxidation after H_2 reduction. It was confirmed that the lattice oxide ions both in $Bi_2Mo_3O_{12}$ and $Co_{11/12}Fe_{1/12}MoO_x$ did not exchange in each other during the catalyst preparation. The oxidation of propylene was carried out with $^{16}O_2$ and propylene on the labeled catalyst, $Bi_2Mo_3{}^{16}O_{12}$/ $Co_{11/12}Fe_{1/12}Mo^{18}O_x$ and ^{18}O concentration in acrolein was followed with the reaction time. The results are shown in Fig. 7. The ^{18}O concentration in acrolein was quite low at the initial stage. It increased with the reaction time, reached a maximum and then decreased. On the basis of the result, it is suggested that the incorporation of lattice oxide

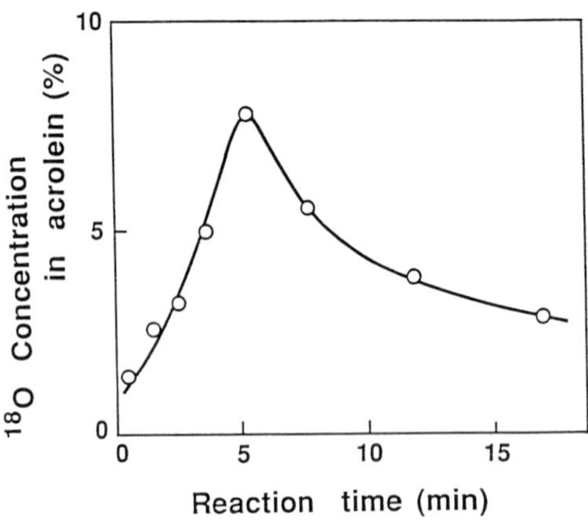

Fig.7 ^{18}O concentration of acrolein in the oxidation on ^{18}O-labeled catalyst, $Bi_2Mo_3{}^{16}O_{12}$/ $Co_{11/12}Fe_{1/12}Mo^{18}O_x$.

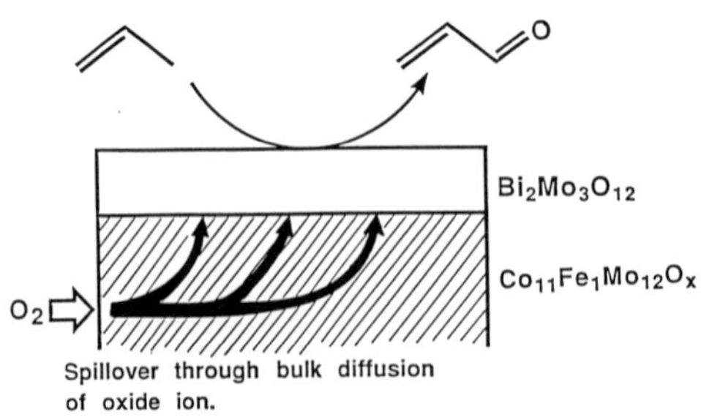

Fig. 8 Scheme for bulk diffusion of lattice oxide ion in multicomponent bismuth molybdate catalyst.

ion into the reaction product does not come from simple oxygen exchange reaction between surface active oxygen with bulk oxide ion. The result may be more reasonably interpreted by the assumption that activation of molecular oxygen and oxidation of propylene take place on the different active sites and oxygen species activated by the support oxide involving iron and/or cobalt cation spillover through bulk diffusion to the reaction site of bismuth molybdate (Fig. 8). This assumption may also explain the reason why multicomponent bismuth molybdate is active only when it includes both divalent and trivalent cations as the third and fourth component. Divalent cation and trivalent cation can easily exchange in each other to make lattice vacancies when both cations have almost the same ionic radii. It has been well known that bulk migration of oxide ion is extensively accelerated in the presence of those vacancies. $Bi_2Mo_3O_{12}$ itself has a scheelite type structure including many lattice defects. Thus, it is reasonably understood that lattice oxide ions not only in the bismuth molybdate phase but also in $M^{II}MoO_4$ and $M^{III}_2(MoO_4)_3$ phases are active and can participate in the reaction in the tetra-component catalyst system, Mo-Bi-M^{II}-M^{III}-O. Rapid supplying of active oxygen species to the reaction site will increase the number of the reaction site and change the nature of it. Clear correspondence between the catalytic activity and the bulk migration of oxide ion was reported in several other oxide systems[11,17-20].

Conclusion

It is demonstrated that the catalytic activity of the multicomponent bismuth molybdate is clearly associated with the degree of the participation of lattice oxide ions not only in bismuth molybdate but also in other transition metal molybdates.

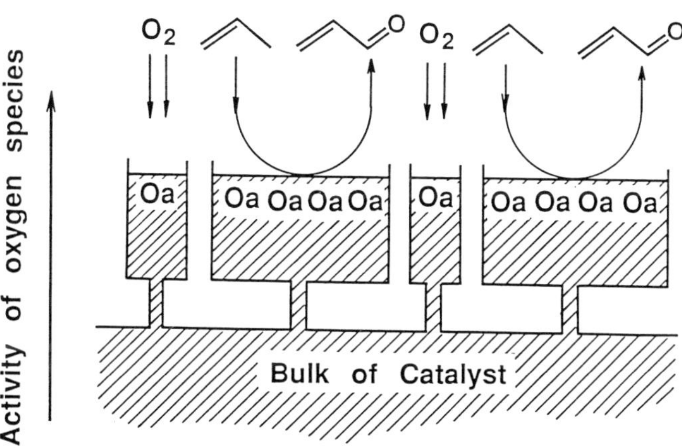

Fig. 9 A water tank model for multicomponent metal oxide catalysts.

Metal molybdates of the third and fourth components, $M^{II}MoO_4$ and $M^{III}_2(MoO_4)_3$, serve as the support for the active component, $Bi_2Mo_3O_{12}$ but strong interaction between the catalyst oxide and the support oxides seems to be essential for the preparation of effective catalyst system. Bulk migration of oxide ion through lattice vacancies plays an important role to increase the catalytic activity. It is suggested that the active catalyst system should have multifunctional nature to promote activation of oxygen, migration of oxygen species and consecutive dehydrogenation and oxygenation of propylene molecule. Collaboration of different kinds of active site and different functions of the catalyst enhance the catalytic activity significantly. The concept is briefly illustrated in Fig. 9 using a water tank model. The concept will be applicable not only for the oxidation of lower olefin but also to many kinds of oxidation using multicomponent metal oxide catalysts.

References

1 Grasselli, R. K., and Burrington, J. D., Ind. Eng. Chem. Prod. Res. Dev. **23**, 393(1984).
2 Grasselli, R. K., Burrington, J. D., and Brazdil, J. F., Faraday Discuss. Chem. Soc. **72**, 203(1982).
3 Wolfs, M. W. J., and Batist, Ph. A., J Catal., **32**, 25(1974).
4 Matsuura, I., and Wolfs, M. W. J., J. Catal., **37**, 174(1975).
5 Matsuura, I., Proc. 7th Intern. Congr. Catal. Tokyo, 1980 (T. Seiyama and K. Tanabe, Eds.), Kodansha, Tokyo/Elsevier, Amsterdam, Part B, p.1099(1981).
6 Prasada Rao, T. S. R., and Menon, P. G., J. Catal., **51**, 64(1978).
7 Ooij, W. J. V., and Muizebelt, W. J., Proc. Intern. Vac. Congr. & 3rd Intern. Conf. Solid Surf., Vienna, p.839(1977).
8 Umemura, S., Ohdan, K., and Asada, H., 5th Soviet-Japan Catal. Seminar, p.60 (1979).
9 Krylov, O. V., Kinet. Katal., **25**, 955(1984).
10 Prasada Rao, T. S. R., and Krishnamurthy, K. R., J. Catal., **95**, 209(1985).
11 Moro-oka, Y., Ueda, W., Tanaka, S., and Ikawa, T., Proc. 7th Intern Congr. Catal., Tokyo, 1980 (T. Seiyama and K. Tanabe Eds.), Kodansha, Tokyo/Elsevier, Amsterdam Part B, p.1086(1981).
12 Ueda, W., Moro-oka, Y., and Ikawa, T., J. Catal., **70**, 409(1981).
13 Ueda, W., Moro-oka, Y., Ikawa, T., and Matsuura, Y., Chem. Lett., 1365(1982).
14 He, D-H., Ueda, W., and Moro-oka, Y., to be published.
15 Tauster, S. J., Fung, S. C., and Garten, R. L., J. Am. Chem. Soc., **100**, 170(1978).
16 He, D-H., Ueda, W., and Moro-oka, Y., to be published.
17 Keulks, G. W., Krenzke, L. D., and Notterman, T. M., Adv. Catal., Academic Press, New York, **27**, 183(1978).
18 Ueda, W., Asakawa, K., Chen, C. L., Moro-oka, Y., and Ikawa, T., J. Catal., **101**, 360(1986).
19 Ueda, W., Chen, C. L., Asakawa, K., Moro-oka, Y., and Ikawa, T., J. Catal., **101**, 369(1986).
20 Kim, Y-C., Ueda, W., and Moro-oka, Y., Studies in Surf. Sci. Catal., **55**, 491(1990).

R.K. Grasselli and A.W. Sleight (Editors), *Structure-Activity and Selectivity Relationships in Heterogeneous Catalysis*

© 1991 Elsevier Science Publishers B.V., Amsterdam

STRUCTURAL-SENSITIVITY IN PROPYLENE MILD OXIDATION ON NEW [100] ORIENTED MoO_3 CATALYSTS

M. ABON, B. MINGOT, J. MASSARDIER and J.C. VOLTA

Institut de Recherches sur la Catalyse, C.N.R.S.
2, Avenue Albert Einstein
69626 Villeurbanne Cédex, France

ABSTRACT

New unsupported MoO_3 catalysts have been prepared and characterized with respect to the orientation and the shape of the crystallites. A preferential [100] orientation has been infered by diffraction techniques, the (100) surface planes being actually truncated to (120) faces whose relative area depends on the temperature of preparation. These (120) faces appear to be the active ones in propylene mild oxidation to acrolein, whereas the basal (010) faces lead to complete oxidation. The structural sensitivity would be in line with the peculiar atomic arrangement of the (120) faces, as discussed in relation with Redox and acidic properties.

INTRODUCTION

Propylene oxidation has been previously studied by Volta et al (1-3) on MoO_3 supported on graphite, prepared by oxyhydrolysis of a graphite -$MoCl_5$ intercalation compound. The same reaction was also investigated by Tatibouet (4) and Abou-Akar (5) on large vapor-grown MoO_3 crystallites. Both research groups concluded that the active crystal face in acrolein formation is the side (100) face. Haber et al (6,7) also agreed with an initial activation of propylene to allylic species on the (100) side faces but claimed that the further steps, including oxygen insertion, occur on the basal (010) face. On the basis of a theoretical bond-strength model, Ziolkowski (8) considered that (101) and (001) faces are responsible for the acrolein formation, the complete oxidation occuring on the (100) face whereas the (010) face would be inactive. In a recent experimental work by Oyama (9), the structure sensitivity character of this reaction on MoO_3 has been questioned.

We have then reexamined the propylene oxidation reaction on MoO_3 in order to ascertain the contribution of the various crystal faces to the formation of the different products. For that purpose, a new preparation of unsupported [100] oriented MoO_3 catalysts was developed.

Their catalytic properties have been compared to those of vapor-grown MoO_3 crsytallites exposing mainly the basal (010) faces which is the natural cleavage plane of lamellar α-MoO_3. Redox and acidic properties of these MoO_3 catalysts have been also investigated in relation with their reactivity in propylene mild oxidation.

1. PREPARATION AND CHARACTERIZATION OF [100] ORIENTED MoO_3 CATALYSTS

As already reported (10), these catalysts have been prepared by complete oxidation of thin molybdenum foils (25μm in thickness) under an oxygen flow at different temperatures (from 500 to 680°C). A pure α-MoO_3 phase has been obtained as checked by XRD, Raman and XPS spectroscopies. In so far as the characterization has been previously described (10), we just recall here the main points with some additional precisions relevant to the present study.

First, a comparison of the XRD patterns of the oxidized sheets (Fig. 1a, 1b, 1c, and 1d) with the usual powder pattern of α-MoO_3 (Fig. 1e) and with the pattern relative to vapor-grown α-MoO_3 crystallites (Fig. 1f) clearly shows a preferential [100] orientation (the (200) diffraction peak is by far the most intense in patterns a, b, c and d). However this preferential [100] axis normal to the surface of the oxidized sheet is less and less favored when the temperature of oxidation increases : at 536°C (Fig. 1a), the XRD pattern exhibits nearly the single (200) diffraction peak whereas the intensity of additional (110), (210), (320) and (301) peaks increases at higher oxidation temperatures (see Fig. 1b, 1c and 1d). This peculiar [100] orientation has been also confirmed at a single crystallite scale by Electron Microdiffraction and High Resolution Electron Microscopy (10, 11). For vapor-grown MoO_3 crystallites, the XRD pattern is quite different (Fig. 1f) and shows nearly the only (OkO) peaks (with even k values).

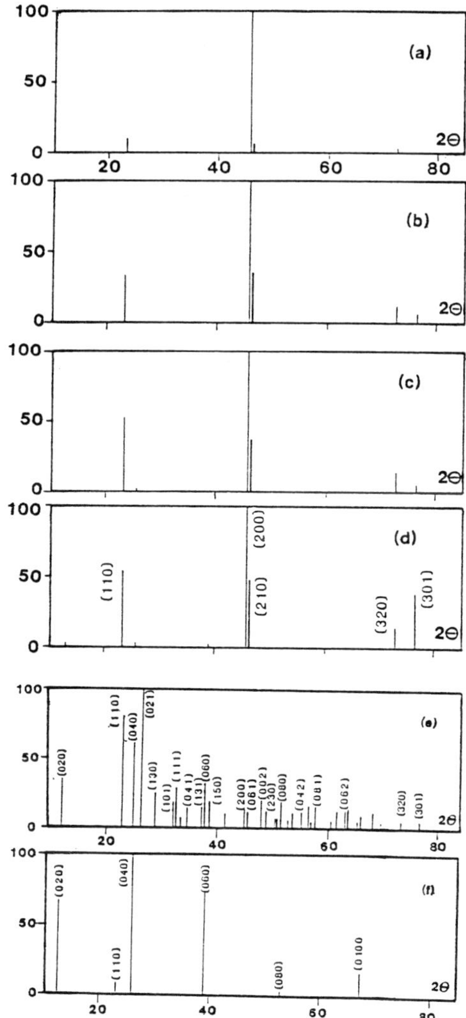

Fig. 1 Reflexion XRD patterns of :
- α-MoO_3 prepared by oxidation of a molybdenum sheet at 536°C (a), 590°C (b), 650°C (c), 680°C (d).
- α-MoO_3 powder (ASTM files), (e).
- vapour-grown α-MoO_3 crystallites prepared by sublimation under an oxygen stream (f).

A detailed morphologic study of the crystallites shape has been performed by Scanning Electron Microscopy (SEM). On vapour-grown crystals, the SEM observations have confirmed that the main exposed faces are the natural (010) cleavage faces, as already known. Within the oxidized sheets, it was observed that the (100) surface planes are actually truncated to (1kO) exposed faces, mainly (120). A schematic drawing of the shape of MoO_3 crystallites prepared by oxidation of a Mo sheet and by vapor-growth process is given in Fig. 2.

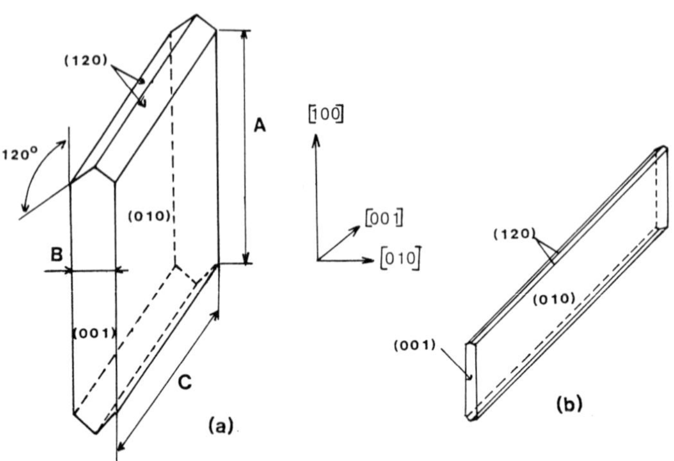

Fig. 2 Schematic drawing of MoO_3 crystallites prepared : (a) by oxidation of a molybdenum sheet at 650°C (magnification : x 10^4) ; (b) by sublimation under oxygen of a MoO_3 powder and subsequent vapor-growth (magnification : x 50).

A model based on the bulk MoO_3 structure shows that (120) faces may be considered as stepped surfaces composed of (100) terraces (section of the double chain of octahedra) and normal (010) steps, as shown in Fig. 3.

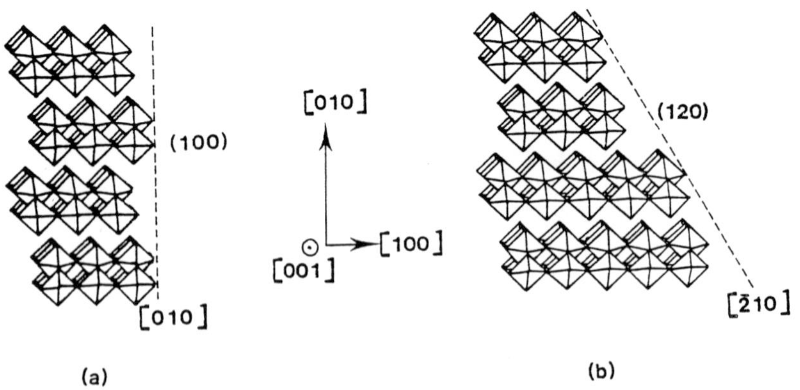

Fig. 3 Cross section view of α-MoO_3 (100) and (120) planes (projection of the lattice on the (001) plane).

However crystallites within the oxidized sheet keep roughly their usual shape depicted in Fig.2, with developed (010) faces normal to the surface of the sheet and still accessible to gaseous reactants. A statistical analysis of SEM pictures showed that the size of [100] oriented crystallites and the relative area of the (120) faces increase with the temperature of oxidation, as shown in table 1.

TABLE 1
Mean dimensions and relative area of the exposed faces of MoO_3 crystallites as a function of the oxidation temperature of the Mo sheet. Similar data relative to vapor-grown cristallites are given for comparison. A, B, and C respectively refer to the dimensions of the crystallites in the [100], [010] and [001] axis, as represented in Fig. 2.

T°C	A(μm)	B(μm)	C(μm)	B/C	B/A	%area (120)	%area (010)	%area (001)
536	1.6	0.2	0.9	0.22	0.13	10.5	72.8	16.7
570	2.9	0.4	1.6	0.25	0.14	11.2	70.5	18.3
590	4.0	0.6	2.0	0.30	0.15	11.7	67.3	21.0
620	4.6	0.9	3.2	0.28	0.20	14.8	65.7	19.5
650	5.0	1.1	4.0	0.28	0.22	16.4	64.7	18.9
660	6.0	1.6	5.8	0.28	0.27	19.2	62.3	18.5
670	5.8	1.7	5.7	0.30	0.29	20.4	60.2	19.4
680	8.0	2.6	7.8	0.33	0.33	21.6	57.5	20.9
Vapor Grown MoO_3	350	25	1000	0.025	0.071	7,4	90.3	2.3

More precisely, data in Table 1 indicate that the relative area of surface (120) planes is roughly enhanced by a factor 2 when the oxidation temperature increases from 536 to 680°C, at the expense of the % area of the (010) faces in so far as the contribution of the apical (001) faces remains nearly constant. This evolution would be a consequence of an anisotropic growth of the crystallites as a result of steric hindrance mainly in the usual growth direction along the [001] axis (denoted by the dimension C). Vapor grown MoO_3 crystallites are much larger and look quite thinner in the [010] axis (denoted by the dimension B) : $B/C \approx 0.025$ instead about 0.22 to 0.30 on [100] oriented crystallites which therefore exhibit more developed side (100) and apical (001) exposed faces.

The complementary SEM analysis of the shape and geometry of the MoO_3 crystallites also demonstrated that neither the nature of the exposed surface planes nor the relative area of these planes can be infered on the sole basis of diffraction techniques which only indicate that the crystallites within the oxidized sheet have a preferential [100] orientation normal to the surface of the sheet.

2. CATALYTIC PROPERTIES IN PROPYLENE OXIDATION

Experiments have been performed in a differential microreactor in the temperature range 350-410°C. These temperatures are low enough to prevent any significant contribution of the non-catalytic homogeneous reaction. The partial pressure of gases was propylene/oxygen/nitrogen : 100/100/560 and the gas flow was kept constant ($0.28\ cm^3.s^{-1}$). The reaction was followed at low conversion (less than 2%) by gas chromatography. It has been checked that no mass or heat transfers limitation was occuring. Some deactivation was observed in the course of the runs and then measurements were performed after a 24hours stabilization period. Besides acrolein and carbon dioxide, the formation of ethanal and propanal was observed as minor products, with traces of acetone.

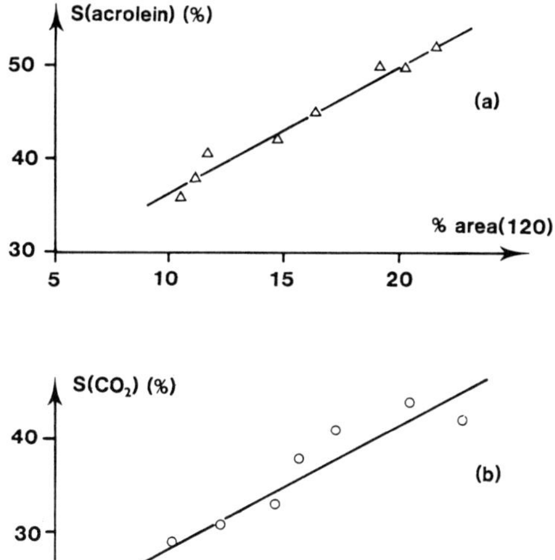

Fig. 4 Propylene oxidation on [100] oriented MoO_3 catalysts at 400°C (a) Selectivity in acrolein vs the % area of (120) faces; (b) Selectivity in CO_2 vs the % area of (010) faces.

The selectivity for ethanal and propanal was respectively about 9 and 4% (with respect to the sum of the products), with no significant changes with the % area of (120) or (010) faces. Results for a reaction temperature of 400°C are summarized in Fig 4 which compares the evolution of the selectivity for acrolein and carbon dioxide as a function of the % area of exposed (120) and (010) faces. A linear correlation is clearly evidenced in Fig. 4 between the selectivity for acrolein and the % area of (120) faces and between the selectivity for carbon dioxide and the % area of (010) faces. The measurements are only slightly dependent on the temperature since the apparent activation energy for acrolein formation is not much higher than the corresponding value for CO_2 formation (respectively 120 and 105 $kJ.mole^{-1}$). Comparative results on a vapour-grown MoO_3 crystallites are in fairly good agreement with the structure-sensitive character evidenced in Fig 4 : selectivity for CO_2 is much higher than for acrolein in agreement with the high % area of the (010) faces with respect to (120) faces (respectively about 90 and 7% as shown in Table 1). A more detailed analysis of the experimental data including a mathematical treatment already used by Volta and Tatibouet (1) confirmed that CO_2 is formed nearly exclusively on the basal (010) faces whereas acrolein is formed on the side faces with a small contribution of about 10% of the (001) faces.

DISCUSSION AND CONCLUSIONS

The catalytic oxidation of propylene to acrolein is known to obey a Mars and Van Krevelen Redox mechanism. The initial step believed to be rate-determining would correspond to the adsorption and first dehydrogenation of propylene to allylic species on an acidic surface site. The peculiar activity of the (120) faces for acrolein formation should then be related to a conjunction of suitable acidic and redox properties allowing dehydrogenation and also oxygen insertion (12).

The usually proposed atomic model of the surface structure of the MoO_3 (010) and (100) faces is shown if Fig. 5 (recall that (120) faces can be viewed as composed of the addition of (100) and (010) faces, as shown in Fig. 3).

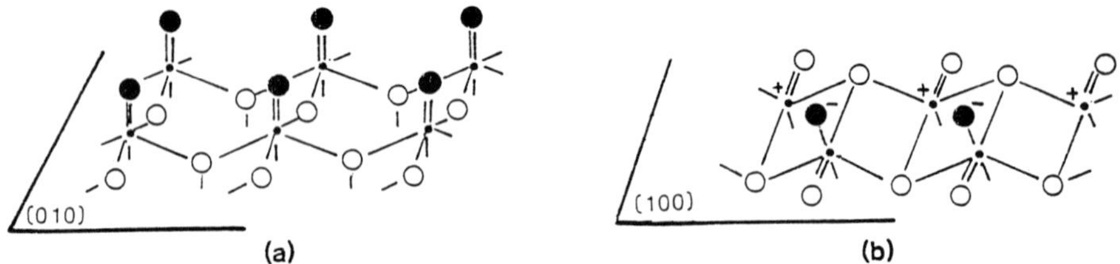

Fig. 5 Ideal atomic surface arrangement of the (010) (a) and (100) (b) MoO_3 faces as derived from the bulk structure. Black dots are surface oxygens pointing outwards. On the (100) face, one over two of these oxygens (coordinated to two Mo cations in the bulk) is missing to comply with surface electroneutrality.

First it has been observed by XPS analysis (10, 13) that upon progressive heating under vacuum, some partial reduction from Mo^{VI} to Mo^{V} state is occuring, the extent of reduction being more pronounced (roughly by a factor 3) on the (010) faces than on (120) faces as shown in Fig. 6. Upon heating at 400°C under a low propylene pressure (0.4 Torr), the reduction to the Mo^{V} state is more important than under vacuum.

Propylene, then appears to be a more effective reducing agent than vacuum, as expected. The fact that (010) faces appear to be more easily reduced than (120) faces (that is actually mainly (100) terraces considering the geometry of the XPS analysis) is a more surprising observation in so far as singly coordinated surface oxygens (Mo=O groups) are frequently assumed to be strongly bonded. However the reduction of surface Mo=O groups has been also found easier than expected by Chung (14) or Ueno and Bennett (15). Anyway twofold and threefold coordinated surface oxygens are also present on the (010) face (see Fig. 5) and further studies are required to know what kind of oxygen is the more easily lost. In this respect, it may be added that Haber (7) claimed that two-fold coordinated oxygens on (010) MoO_3 faces are involved in the oxygen insertion step in allylic species leading to acrolein.

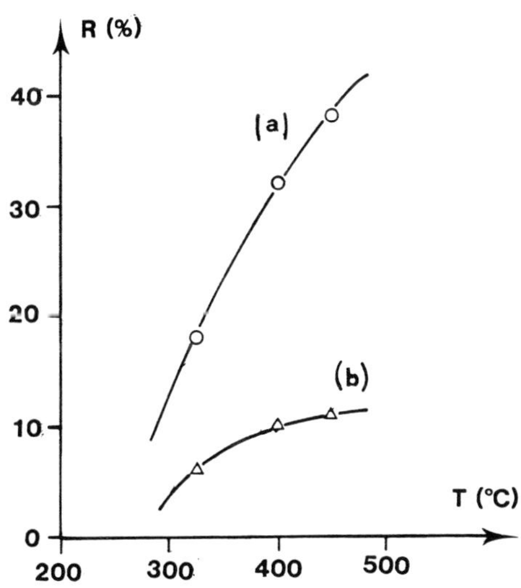

Fig. 6 XPS determination of the relative amount of Mo^{V} as a function of the heating temperature under vacuum of : (a) MoO_3 (010) faces ; (b) MoO_3 (120) faces ; $R=I(Mo^{V})/[(I(Mo^{V})+I(Mo^{VI})]$.

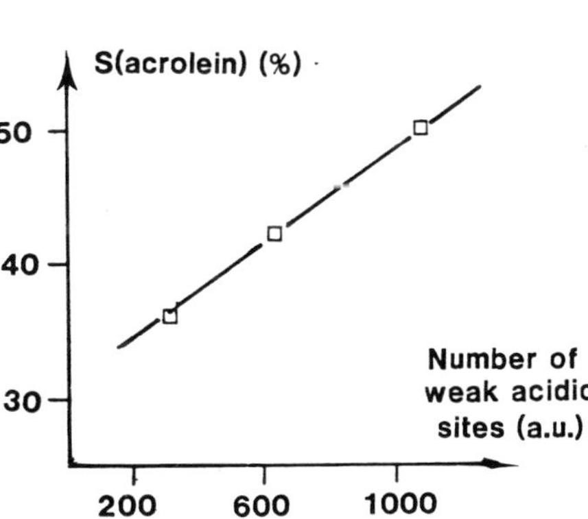

Fig. 7 Selectivity for acrolein as a function of the number of weak acid sites (as measured by TPD of pyridine) on [100] oriented MoO_3 catalysts.

With respect to acidic properties as studied by TPD of probe basic molecules (ammonia, pyridine), it has been mainly concluded (11, 13) that weak or medium acidity evidenced on (120) faces (likely on the (100) terraces) must be involved in the acrolein formation. On [100] oriented MoO_3 samples, the selectivity for acrolein is well correlated with the number of these weak acid sites, as shown in Fig. 7.

The peculiar selectivity of the (120) faces for acrolein formation as evidenced in the present study could be then related to the local structure of such faces, with the complementary role of :

- the (100) terraces with weak acidic sites (coordinatively unsaturated surface cations acting as Lewis acid sites responsible of the propylene activation to allylic species)
- the (010) steps more easily reduced and supplying then the lattice oxygen involved in acrolein formation.

REFERENCES

1 J.C. VOLTA and J.M. TATIBOUET, J. Catal. 93 (1985) 467.
2 J.C. VOLTA, J.M. TATIBOUET, C. PHICHITKUL and J.E. GERMAIN, Proc. 8th. Intern. Congress Catalysis, Dechema, Ed., Berlin 1984, 451.
3 J.C. VEDRINE, G. COUDURIER, M. FORISSIER and J.C. VOLTA, Catalysis Today, 1, (1987) 261.
4 J.C. VOLTA, W. DESQUESNES, B. MORAWECK and J.M. TATIBOUET, Proc. 7th. Inter. Congress on Catalysis (Tokyo 1980), Elsevier, Amsterdam, 1981, p. 1398.
5 A. ABOU-AKAR, Thesis, Lyon (1987).
6 K. BRÜCKMANN, R. GRABOWSKI, J. HABER, A. MAZURKIEWICZ, J. SLOCZYNSKI and T. WILTKOWSKI, J. Catal. 104, (1987) 71.
7 J. HABER in "Structure and Reactivity of Surfaces", C. Morterra et al. Ed., Elsevier, 1989 p. 447.
8 J. ZIOLKOWSKI, J. Catal. 80, (1983) 263.
9 S.T. OYAMA, Bull. Chem. Soc. Japan, 61 (1988) 2588.
10 B. MINGOT, N. FLOQUET, O. BERTRAND, M. TREILLEUX, J.J. HEIZMANN, J. MASSARDIER and M. ABON, J. Catal. 118 (1989) 424.
11 B. MINGOT, Thesis, (1989).
12 A. GUERREDO-RUIZ, J. MASSARDIER, D. DUPREZ, M. ABON and J.C. VOLTA, Proc. 9th Inter. Congress Catalysis, M.J. Philip and M. Ternan Ed., Calgary (1988) p.1601.
13 M. ABON, B. MINGOT, J. MASSARDIER and J.C VOLTA, in "New Developments in Selective Oxidation", G. Centi and F. Trifiro' Ed., 1990, p.747, Elsevier (Amsterdam).
14 J.S. CHUNG, Ph. D. Thesis, Connecticut University, USA.
15 A. UENO and C.O. BENNETT, Bull. Chem. Soc. Japan 52 (1979) 2551.

R.K. Grasselli and A.W. Sleight (Editors), *Structure-Activity and Selectivity Relationships in Heterogeneous Catalysis*
© 1991 Elsevier Science Publishers B.V., Amsterdam

A COMPARISON BETWEEN LOW AND HIGH TEMPERATURE $Bi_2O_3.MoO_3$ PHASES FOR 1--BUTENE REACTIONS

M. FARINHA PORTELA, CARLA PINHEIRO, CRISTINA DIAS and MARIA JOÃO PIRES

GRECAT - Grupo de Estudos de Catálise, Technical University of Lisbon, Instituto Superior Técnico, Aven. Rovisco Pais, 1096 Lisboa Codex (Portugal)

SUMMARY

Low (γ) and high (γ') temperature $Bi_2O_3.MoO_3$ phase have the same stoichiometry but markedly different structures. The catalytic activities of these phases for the 1-butene reactions in the presence of oxygen were studied with the aim of relating them to the structural differences. Low temperature phase is very active and selective. High temperature modification is much less active and leads to considerable degradation. TPD results allow to infer that the differences between the phases lead to formation of different active sites, involving oxygen vacancies on Mo^{6+} cations for the γ-phase and oxygen vacancies on Bi^{3+} cations for the γ'-phase. Furthermore an oxygen species not found on γ'-phase was detected on γ-phase.

INTRODUCTION

γ-$Bi_2O_3.MoO_3$ phase has a koechlinite structure with alternating sheets of Bi_2O_2 layers and layers consisting of Mo^{6+} ions in octahedral surrounding. The octahedra share corners in the sheets and their apices are directed toward the Bi_2O_2 layers.

γ'-$Bi_2O_3.MoO_3$ modification is formed by heating γ-phase at a temperature higher than 550°C. Transformation is complete at 640°C. The Bi_2O_2 layers and the Mo positions remain unchanged. Only the O^{2-} ions in the MoO_4 layers are rearranged in such a way that such layers consist of MoO_4 tetrahedra having no anions in common.

Taking into account such structural differences it was found interesting to compare the catalytic activity of such solids, having the same stoichiometry, for the 1-butene in the presence of oxygen.

The differences in structure between the two phases must induce differences in the nature and amounts of active oxygen species on the catalysts. In this way a study of temperature programmed desorption of oxygen on both solids was undertaken.

EXPERIMENTAL

The unsupported γ-$Bi_2O_3.MoO_3$ pure phase was prepared by reproducible coprecipitation technique (1). The precursor underwent a final calcination temperature of 450°C (8h) under air. The γ'-phase was prepared submitting the γ-phase at a 650°C (4h) calcination.

X-ray diffraction and the infra-red and Raman spectroscopies did not show traces of impurities. Analysis by XPS confirmed the right Bi/Mo atomic ratio.

The BET surface area was 1.0 m^2/g for the γ-phase and 0.4 m^2/g for the γ'.

To study oxygen removal under temperature programmed desorption (TPD) conditions:

- chromatograms of oxygen were recorded in the range 303-763 K at a heating rate of 10°C/min in a helium stream (1 mL/s).
- the losses of oxygen - measured by a Cahn RG electrobalance, with 1 μg sensibility in the same temperature range at the same heating rate in helium and air streams (1 mL/s) - were recorded.

For the catalytic tests a classical continuous apparatus was used with Pyrex reactor. The operating pressure was near atmospheric one and the results reported here are for olefin partial pressure 0.5-2 kPa and oxygen pressure 20 kPa. Investigated temperature range was 280-350°C for the γ-phase. Very lower reactivity of γ'-phase imposed a temperature increase to the range of 350-440°C. The experiments were carried out so as to obtain differential conversions, in order to eliminate the effect of reaction products. The catalysts showed a stable activity.

RESULTS

TPD chromatograms of both phases exhibit desorption peaks around 473, 593 and 673 K, but the size of such peaks is very different for the two modifications.

Furthermore a fourth peak is visible in the γ-phase chromatogram at 733 K (Fig. 1).

The colour of the samples after TPD experiments becames gray and in the XRD spectrum of the γ' sample is visible the formation of metallic bismuth, evidencing a reduction. It was observed that, after temperature programmed heating of the γ-phase in hellium up

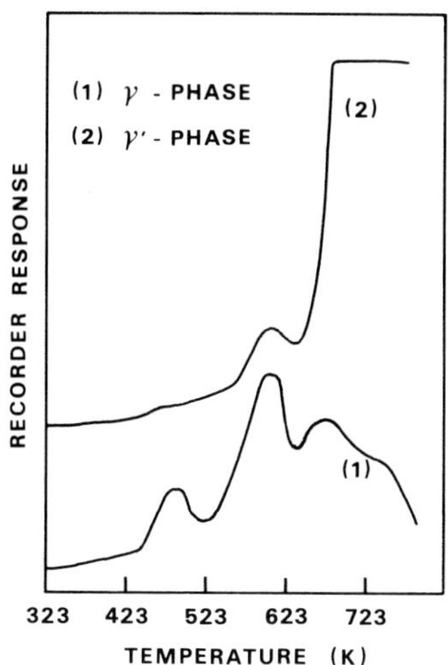

Fig.1 - TPD chromatograms of oxygen from γ and γ' phases.

to 633 K in the electrobalance, the catalyst recovered the lost weight after subsequent identical treatment in dried air. As mentioned the catalysts were treated with dried air for 2 hours before the experiments. Such evidences point out that the peaks observed in the chromatograms correspond to removed oxygen. Fig.2 shows the thermograms recorded with the electrobalance in hellium and air streams.

Considering that the maximum or minimum points of the chromatograms correspond to inflection points of the thermograms in helium, it is seen that both types of information match. The amounts of oxygen (as percentage of sample weight) related to the chromatogram peaks, measured from the thermograms, are presented in Table 1.

The total losses of weight recording during temperature programmed heating in helium up to 763 K are 2.5×10^{-2} % wt for the γ-phase and 1.1×10^{-2} % wt for γ'.

The comparison of thermograms in helium and in air for both phases shows that they match for γ'-phase up to 700 K, after which some aditional removal takes place in helium. For the γ-phase it is noticeable that the thermogram in air, also up to 700 K, is the thermogram in helium with an uniform displacement of ordinates of 0.8×10^{-2} % wt. After 700 K the difference becomes higher the higher the temperature. There is a loss of weight of 0.8×10^{-2} % wt in he-

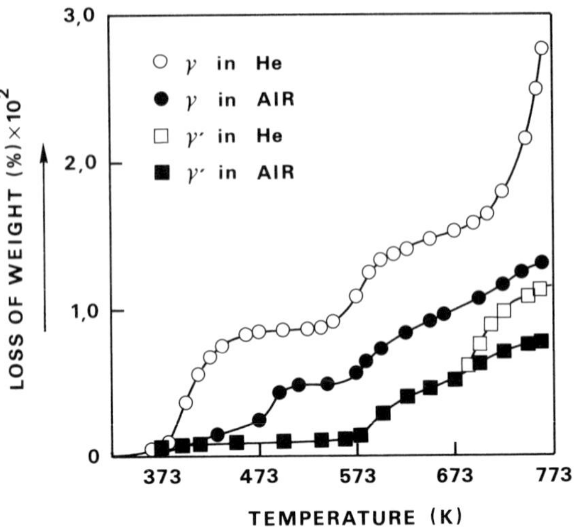

Fig.2 - Thermograms

TABLE 1
Oxygen removed corresponding to the TPD peaks

Peak Temperature	O_2 removed (% wt)	
	γ phase	γ' phase
473	0.10×10^{-2}	0.1×10^{-3}
593	0.60×10^{-2}	0.30×10^{-2}
673	0.20×10^{-2}	0.70×10^{-2}

lium up to 430 K practically absent in the thermogram in air.

Such initial removal of oxygen from the γ-phase in helium is not visible on the TPD chromatograms which means that the concentration of oxygen is constant in such circunstances. It corresponds to an oxygen species the amount of which is kept constant by the catalyst in air up at least 700 K. When heating γ-phase in air up to 623 K and then replacing air by helium an amount of 0.8×10^{-2} % of oxygen is quickly removed. Such oxygen is quickly reinserted when helium is subsequently replaced by air (Fig.3).

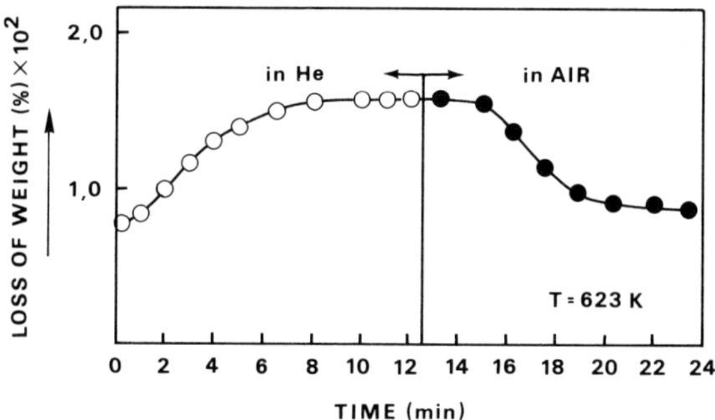

Fig.3 - Changes of weight at 623 K of γ-phase heated under air up to 623 K, when put under He, and after under air again.

With γ'-phase when air was replaced by helium at 673 K no change in weight was recorded and occured the same when subsequently helium was replaced by air, which is in agreement with matching of thermograms recorded in helium and air for this phase.

It is noteworthy that the comparison of thermograms in air and in helium for both phases indicates that the processes, related with the TPD chromatograms peaks at 473, 593 and 673 K, take place both in air and helium.

The results of catalytic activity tests, on both modifications presented in Figs. 4 and 5 evidence that the rates of formation of butadiene, 2-butenes and CO_2 have apparent first order dependency on olefin for the low partial pressures used (Figs. 4 and 5).

The computed rate constants based on surface area are presented in Tables 2 and 3.

TABLE 2

Rate constants for γ modification

Temperature K	Butadiene	cis-2 butene	trans-2 butene	CO_2
		(g mol h^{-1} m^{-2} Pa^{-1} x 10^{-8})		
553	8.9	6.4	3.6	-
573	18.3	10.0	5.4	1.5
593	35	8.4	4.4	5.5
623	68	10.8	6.0	9.2

TABLE 3
Rate constants for γ' modification

Temperature K	Butadiene	cis-2-butene	CO_2
	($g\ mol\ h^{-1}\ m^{-2}\ Pa^{-1}\ x\ 10^{-8}$)		
623	0.55	0.58	0.70
673	1.80	0.78	2.1
693	2.8	0.65	2.7
713	4.8	0.68	3.7

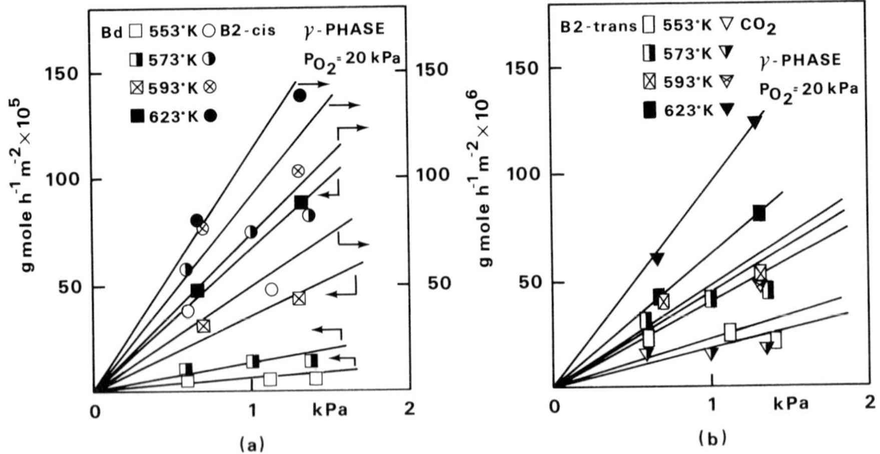

Fig. 4 - Effect of 1-butene partial pressure on products formation rates - - γ-PHASE

It is seen that the activity of the γ'-phase is considerably lower than the activity of the γ-modification. So, at 623 K, in the investigated range of low pressures of olefin, activity of γ'-phase for oxidative dehydrogenation is about 120 times lower on surface basis. For isomerization to cis-2-butene is about 20 times lower, for degradation to CO_2 12 times lower and for trans-2-butene isomerization activity of γ'-phase was found null.

The oxidative dehydrogenation and degradation pathways obey to Arrhenius law and calculated activation energies are presented in Table 4.

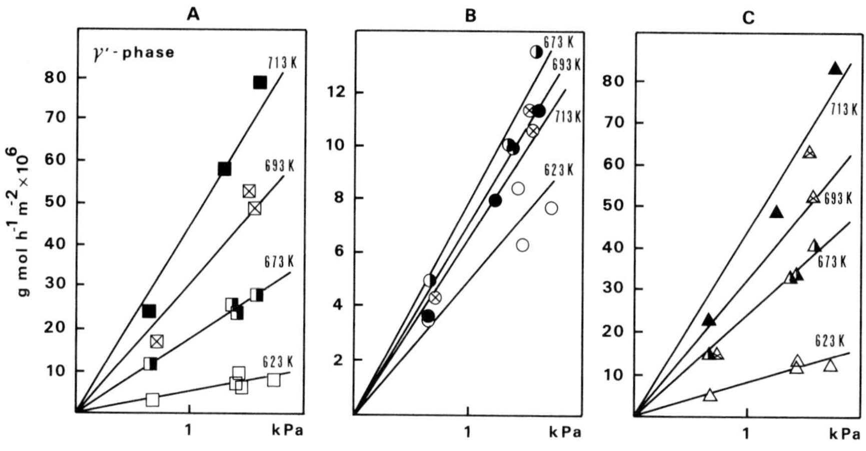

Fig.5 - Effect of 1-butene partial pressure on products formation rates: (A) butadiene, (B) cis-2-butene, (C) CO_2 (P_{O_2} = 21 kPa)

TABLE 4

Apparent activation energies of oxidative dehydrogenation and degradation

$Bi_2O_3.MoO_3$	Butadiene	CO_2
	($kJ\ mol^{-1}$)	
γ	77	98
γ'	88	65

TABLE 5

Selectivities for γ-modification

Temperature K	Selectivities (%)			
	Butadiene	cis-2 butene	trans-2 butene	CO_2
553	47	34	19.0	0
573	52	28	15.3	4.3
593	66	15.8	8.3	10.3
623	72	11.5	6.4	9.8

TABLE 6

Selectivities for γ'-modification

Temperature K	Selectivities (%)		
	Butadiene	cis-2-butene	CO_2
623	30	32	38
673	39	16.7	45
693	46	10.6	44
713	52	7.4	40

In Tables 5 and 6 are shown the selectivities found for the reaction products. It is visible that γ'-phase besides being considerably less active it degrades considerably more. Furthermore with the γ-phase CO_2 selectivity increases with temperature as was expected of the higher activation energy of degradation. It is surprising the absence of trans-2-butene formation with γ'-phase.

DISCUSSION AND CONCLUSIONS

Our results evidence that the losses of oxygen recorded in the TPD chromatograms at around 473, 593 and 673 K take place in air and helium for both phases.

Dadyburjor and Ruckenstein (2) admitted that for bismuth molybdate the energy barrier for loss of an intermediate layer O^{2-} ion as O_2 is less than that for a Mo^{6+} - bound O^{2-} ion as O_2, which is less than that for a Bi^{3+} bound O^{2-} ion as O_2. On this basis the chromatogram peaks observed for both phases would be assignable

473 K to oxygen from intermediate layer O^{2-} ions

593 K to oxygen from Mo^{6+} - bound O^{2-} ions

673 K to oxygen from Bi^{3+} - bound O^{2-} ions

It was seen that the XRD patterns obtained for both catalysts after treatment under hellium up to 763 K show the characteristic peaks of pure phases but in the γ'-phase spectrum is visible the formation of metallic bismuth, which is in agreement with the big peak recorded at about 673 K for such phase.

The highest proportions of removed oxygen related with TPD peaks are oxygen proceeding from Mo^{6+} - bound O^{2-} ions, for γ-phase and Bi^{3+} - bound O^{-2} ions for γ'.

In this way oxygen vacancies are formed on γ- and γ'-phases when heated under air at the temperatures used in catalysis. Such vacancies are predominantly over Mo^{6+} cations in case of γ-phase chiefly within the temperature range of 553-623 K. On the other hand they are predominantly over Bi^{3+} cations in the case of γ'-phase for temperatures over 623 K. These are precisely the temperature ranges within which we were induced to investigated catalytic activity for the two phases to have measurable activity. This leads to suggest that the different structures of γ- and γ'-phases induce the formation of two types of active sites. One, very active, found predominantly on γ-phase, involving oxygen vacancies on Mo^{6+} cations and leading mainly to selective oxidation and isomerization to cis-2- and trans-2-butene. The other, considerably less active, found predominantly on γ'-modification, involving oxygen vacancies on Bi^{3+} cations and leading to considerable degradation, without isomerization to trans-2-butene.

It seems that difference in the structure of active sites is not the sole reason for the observed difference in reactivity between the two phases. The oxygen species yielded by γ-phase and kept under air up to high temperatures, easily removable and reinsertable would contribute to facilitate the reoxidation of the sites reduced by reaction. Such species was not found in the γ'-modification.

REFERENCES

(1) Pires, M.J.; Portela, M.F.; Oliveira, M.; Saraiva, A.; Miranda, T. In Proceedings of the 7th Iberoamerican Symposium on Catalysis, La Plata (Argentina), 1980, p. 109.

(2) Dadyburjor, D.B. and Ruckenstein, E., J. Catal. 63, 383-388, 1980.

R.K. Grasselli and A.W. Sleight (Editors), *Structure-Activity and Selectivity Relationships in Heterogeneous Catalysis*
1991 Elsevier Science Publishers B.V., Amsterdam

SURFACE- AND BULK-TYPE CATALYSIS OF HETEROPOLYMOLYBDATES. IMPORTANCE OF THE CONCEPT IN THE STRUCTURE-ACTIVITY RELATIONSHIPS FOR CATALYST DESIGN

Makoto Misono,[1] Noritaka Mizuno,[2] Hiro-o Mori, Kwan Y. Lee, Jinbao Jiao,[3] and Toshio Okuhara

Department of Synthetic Chemistry, Faculty of Engineering, The University of Tokyo, Bunkyo-ku, Tokyo 113, Japan

Abstract

Oxidations of acetaldehyde, H_2 and CO, and oxidative dehydrogenation of cyclohexene over several 12-heteropolymolybdates and molybdovanadates (supported and unsupported) have been studied, in relation to the structure-activity relationships. First, the concept of the surface- and bulk-type catalysis are described with experimental evidence. Then, the experimental results concerning the effects of redox properties and acidity of the catalysts on the catalytic activity for the two types of reactions are given. The catalytic activity for oxidation of acetaldehyde (a surface-type catalysis) is closely related to the surface redox property measured by the rate of reduction of catalyst by CO, while the catalytic activity for bulk-type catalysis (e.g., oxidative dehydrogenation of cyclohexene) is correlated with the bulk redox property measured by the reduction by H_2. It was found that the rates of acetaldehyde oxidation over the acid form dispersed on Cs salt showed a very similar pattern to those for acidic Cs salts, as regards the H/Cs ratio.

INTRODUCTION

We have reported for the heterogeneous catalysis of heteropoly compounds that under certain conditions the reactant molecules are absorbed into the catalyst bulk and react there. We called this novel bulk-type behavior "pseudoliquid phase" (1, 2).

Later, we noticed that there was another kind of bulk-type catalysis for catalytic oxidation at high temperatures. It was found that stoichiometric (noncatalytic) oxidation by and catalytic oxidation over 12-heteropolymolybdates could be classified into two groups based on the dependency of the rate on the surface area (2, 3). One is the surface-type catalysis, where the rate is proportional to the surface area of catalyst as

[1] To whom correspondence should be addressed.
[2] Pesent address: Catalysis Research Center, Hokkaido University, Sapporo 060, Japan
[3] University of Science and Technology of China, Hefei, Anhui, China

in the ordinary solid catalysts. In the other group, the rate is little dependent on the specific surface area, but is proportional to the weight or the volume of catalyst. We called the catalysis of this group "bulk-type (II) catalysis." While the bulk-type (I) is the catalysis in the pseudoliquid phase of the catalyst bulk (2), in the case of bulk-type (II) the reactants usually remain on the surface. Owing to the rapid diffusion of protons, the whole catalyst bulk participates in the redox cycle of oxidation catalysis, even though the reactant molecules are transformed into products on the surface. The classification to two types depends on the relative rate of diffusion to the rate of reaction and on the size of catalyst particle, so there are reactions intermediate between the two extremes.

The followings are the reactions that have already been found to belong either of the two groups in the case of 12-molybdophosphoric acid. These examples clearly indicate that the concept of bulk- and surface-type catalysis is important to understand and design the catalytic oxidations over heteropoly compounds, since the structure(property)-activity relationships are quite different between the two types. This difference as well as the inhomogeneity of the catalyst composition are the main reasons why the relationships reported so forth are not consistent (2).

Surface-type: Oxidation of CO, acetaldehyde, and methacrolein.

Bulk-type(II): Oxidation of H_2, oxidative dehydrogenation of cyclohexene and isobutyric acid.

In this work, it has been attempted to elucidate the relationships between the catalytic activity and the structure-related properties of catalysts such as redox and acidic properties for both bulk- and surface-type reactions, based on the reported and newly obtained data, and compared the differences between the two types of catalysis.

EXPERIMENTAL

Catalysts. 12-Molybdophosphoric acid ($H_3PMo_{12}O_{40}$, abbreviated as PMo12) commercially obtained was used after purification, as described previously (4-6). Its alkali salts (Na2PMo12, etc.) were carefully prepared as in the literature (5) and $H_{3+x}PMo_{12-x}V_xO_{40}$ (PMo10V2, etc.)'s were obtained commercially.

Reactions. Catalytic oxidations of CO and H_2 were carried out at 350°C in a closed recirculation system (5). Catalytic oxidations of acetaldehyde and cyclohexene were conducted with a flow reactor mainly at 300°C. The standard feeds were acetaldehyde : O_2 : N_2 = 2.3% : 9.6% : balance (total flow rate: 70 cm^3 min^{-1}) and cyclohexene : O_2 : N_2 = 1.2% : 13% : balance (total flow rate: 30 cm^3 min^{-1}).

The rates of reduction of catalyst by CO and H_2 and its reoxidation by O_2 were measured in a closed recirculation system as in the previous works (5, 6). The rate of reduction by CO (denoted by r(CO)) and that by H_2 (denoted by r(H)) in the initial stage of reduction are regarded to express respectively the surface and bulk oxidizing ability of catalysts (5). The acidic property was measured by thermal desorption of absorbed pyridine with the aid of IR (1, 7).

RESULTS AND DISCUSSION

Catalytic oxidation of H_2 and CO.

When catalytic oxidation of H_2 was carried out in the recirculation system over various lots of Na2PMo12's that had the same composition but different specific surface areas (1.0 - 2.9 m^2 g^{-1}), the rates were almost independent of the surface area as shown in Fig. 1a, but approximately proportional to the weight of catalyst. It was confirmed for the same lots of catalysts that the rate of catalytic oxidation of CO was proportional to the surface area as usually observed for solid catalysts (also in Fig. 1a). The dependencies were the same in the case of PMo12, too. Similar dependencies on the surface area corresponding to the two types were also observed for the stoichiometric reduction of catalysts by H_2 and CO (r(H) and r(CO)). Here, 10^{-7} mol g^{-1} s^{-1} = 2.1 - 2.4 electrons $anion^{-1}$ min^{-1}, slightly depending on the molecular weight and the water content. In the case of reduction by H_2, it has been quantitatively demonstrated that the diffusion of proton and/or water in the bulk is very rapid and thus the slow step is the formation of water in the bulk from proton and oxygen of polyanion (4). As the slow step is the reaction which proceeds in the bulk, the rate (r(H)) little depends on the surface area, but on the bulk volume.

Therefore, there are two parallel correlations between the catalytic oxidation and stoichiometric reduction independently for the bulk- and surface-type reactions, as shown in Fig. 2a and b (5). The correlations have been interpreted based on a redox mechanism. Results in Table 1 demonstrate that those reactions proceeded by a redox mechanism (5); the rates of three reactions for each catalyst, that is, catalytic oxidation, noncatalytic reduction and reoxidation by O_2 of catalyst, agreed well when the rates were measured at the stationary oxidation state of catalytic oxidation.

The process of the reoxidation of the catalyst by O_2 (r(O)) which is involved in the redox cycle is also a surface-type reaction and depends on the surface area (5). So, one may wonder why the catalytic oxidation which

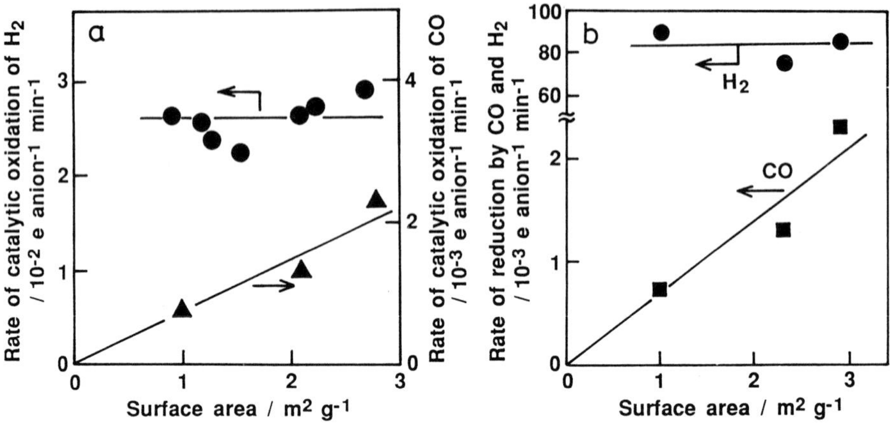

Fig. 1. Dependencies of the catalytic and noncatalytic oxidation of H_2 and CO on the specific surface area of $Na_2HPMo_{12}O_{40}$ at 350°C. (a) Catalytic and (b) noncatalytic reactions.

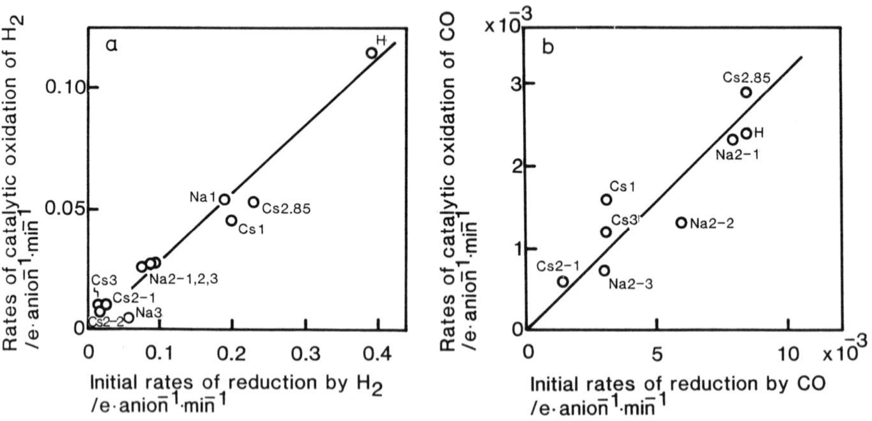

Fig. 2. Correlation between the catalytic oxidation over and stoichiometric reduction of various heteropolymolybdates for H_2 and CO at 350°C. a; H_2, b; CO. (From J. Phys. Chem., 89 (1985) 80).

Table 1 Rates of catalytic oxidation, noncatalytic reduction and reoxidation

	DR	Rate of catalytic oxidation	Rates of noncatalytic reduction	reoxidation
H_2 (300°C)	0.11	7.3×10^{-2}	7.5×10^{-2}	6.8×10^{-2}
CO (350°C)	0.038	2.4×10^{-3}	3.1×10^{-3}	2.6×10^{-3}

[Unit, DR (degree of reduction of catalyst at catalytic conditions): electron anion^{-1}, rate; electron anion^{-1} min^{-1}). Catalyst: PMo12.]

contains a surface-type reaction little depends on the surface area. The reason may be understood by Fig. 3. The r(H) and r(O) are shown as a function of DR (degree of reduction of catalysts) in this figure for two samples of Na2PMo12 having different specific surface areas (1.0 and 2.2 m^2g^{-1}), as well as one PMo12. Two r(H)'s of Na2PMo12 fall on the same curve, but r(O)'s are different (proportional to the surface area). However, since the slope of r(H) is very gentle near the crosspoints, the rate at the crosspoint, which corresponds to the rate of catalytic oxidation in a redox mechanism, is not much dependent on the reoxidation curve. Furthermore, the presence of water vapor in the system accelerates the migration of oxide ion in the bulk in the form of OH^- or H_2O (5), so that the rate of reoxidation under the catalytic conditions (in the presence of water formed by catalytic oxidation) may become less dependent on the surface area than shown in Fig. 3.

Fig. 3 also illustrates the reason why the rates of catalytic oxidation of H_2 at stationary states are in parallel, as shown in Fig. 2a, with r(H)'s that were measured in the initial stage of reduction. Since the r(H) and r(O) curves are more or less parallel with each other, (for example for PMo12 and Na2PMo12 in Fig. 3), the rates at the crosspoints reflect the initial r(H) values in those experiments.

On the other hand, both of the rates (per gram) of catalytic oxidations of H_2 and CO increased linearly with the specific surface area in the case of Cs salts. As expected from the slow diffusion of proton and water in the Cs salts (8, 9), the oxidation of H_2 became like surface-type.

Oxidation of acetaldehyde and oxidative dehydrogenation of cyclohexene. The results for PMo12 and Na2PMo12 indicated that the oxidation of acetaldehyde belongs to the surface-type catalysis (linear increase of the rate with specific surface area) and the dehydrogenation of cyclohexene to the bulk-type (small dependency of the rate on the surface area). The results for PMo12 loaded on SiO_2 confirmed this idea (10). When the amount of PMo12 loaded on support increased, the rate of the former reaction showed saturation due to the increase in the particle size of PMo12, while the rate of the latter reaction

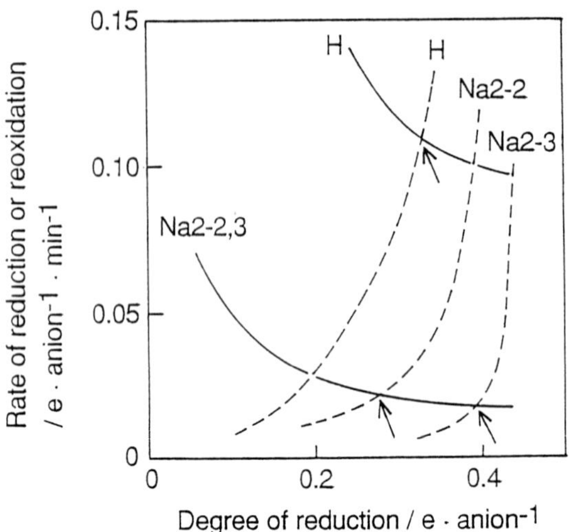

Fig. 3. Rates of reduction by H_2 and reoxidation by O_2 over $Na_2HPMo_{12}O_{40}$ having different surface areas (Na2-2 and -3; 1.0 and 2.2 m^2 g^{-1}) and $H_3PMo_{12}O_{40}$ (H; 1.0 m^2 g^{-1}) at 350°C. Solid line; reduction by H_2, broken line; reoxidation by O_2. Reoxidation was carried out for the catalysts which had been reduced by 0.5 electrons per polyanion. Arrows indicate the crosspoints of the reduction and reoxidation curves for each catalyst.

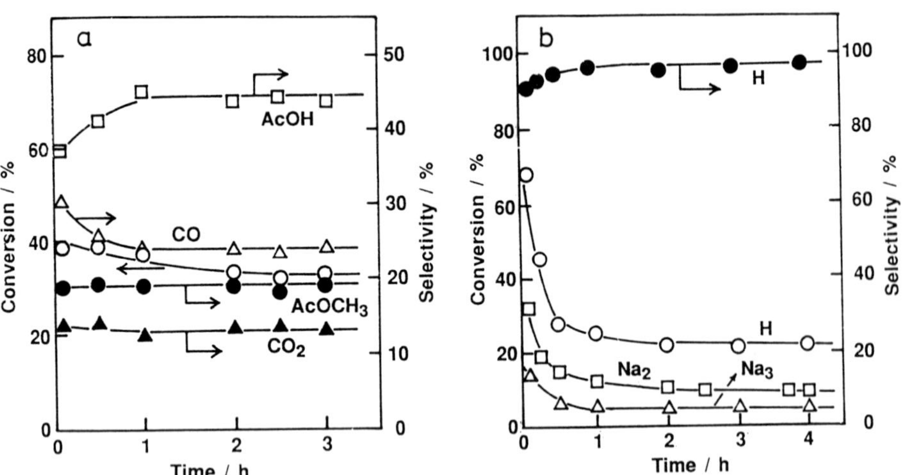

Fig. 4. Variation of conversion and selectivities with time on stream for (a) oxidation of acetaldehyde and (b) oxidative dehydrogenation of cyclohexene over $Na_2HPMo_{12}O_{40}$ at 300°C.

increased linearly with the amount of PMo12 loaded to a higher loading level.

Fig. 4 shows the variations with time on stream of the oxidation reactions of acetaldehyde and cyclohexene which were carried out with a continuous flow reactor. In the case of acetaldehyde, the rate became constant after 2 h with an initial small decrease. Products were acetic acid, methyl acetate, CO and CO_2 as reported previously (10). The activities of the various alkali salts of PMo12 are plotted in Fig. 5a and b as a function of x in $M_xH_{3-x}PMo_{12}O_{40}$ (M = Na or Cs). It may be noted in this figure that the activity (rate) data scattered when they were normalized to the catalyst weights (Fig. 5a), but they showed monotonous decrease with x when normalized to the specific surface areas of the catalysts (Fig. 5b). This indicates that the reaction is of surface-type.

In the case of cyclohexene, the products were mainly benzene (selectivity; > 90%) with small amounts of CO and CO_2. As shown in Fig. 4b, the deactivation in the initial stage was significant, due to the reduction of catalyst as in the case of H_2-O_2 reaction (4). The rate at the stationary state, however, reflects the order in the initial rate, and the rate normalized to catalyst weight decreased monotonously with x, as shown in Fig. 5c. Therefore, this reaction belongs to the bulk-type and the rates may be in parallel with r(H).

In Fig. 6 the relationships between the rates of catalytic oxidations and the oxidizing abilities of catalysts (r(CO) and r(H)) are shown. Good correlations are noted between the rate of catalytic oxidation of acetaldehyde (denoted by r(aldehyde)) and r(CO) (both are surface-type) (Fig. 6a) and between the catalytic oxidation of cyclohexene (denoted by r(hexene)) and r(H) (both are bulk-type)(Fig. 6b). In contrast, there were poor correlations between r(aldehyde) and r(H), and between r(hexene) and r(CO).

As for the side reactions for the oxidation of acetaldehyde, a fair correlation was found between the rate of decomposition of acetic acid (a main side reaction) and the acidity of catalyst. So, to improve the catalytic performance, we attempted to increase the oxidizing ability and decrease the acidity of catalyst by introducing V into a part of Mo (PMo11V and PMo10V2). Significant improvement in the selectivity for acetic acid was obtained, but the catalytic activity deceased in the order of PMo12 > PMo11V > PMo10V2. The r(CO) value (the oxidizing ability of catalyst) increased, as expected, in the opposite order PMo10V2 > PMo11V >> PMo12. This discrepancy is probably because the oxidation state of the catalyst under the catalytic oxidation was much lower for vanadium-substituted molybdophosphates, as suggested by Akimoto et al. (11). In order to fully explain the results, more information on the redox and acidic properties under the working conditions is necessary.

Fig. 7 shows the activity pattern of two series of Cs salts for acetaldehyde oxidation obtained by continuous flow experiments. One series (A) was prepared

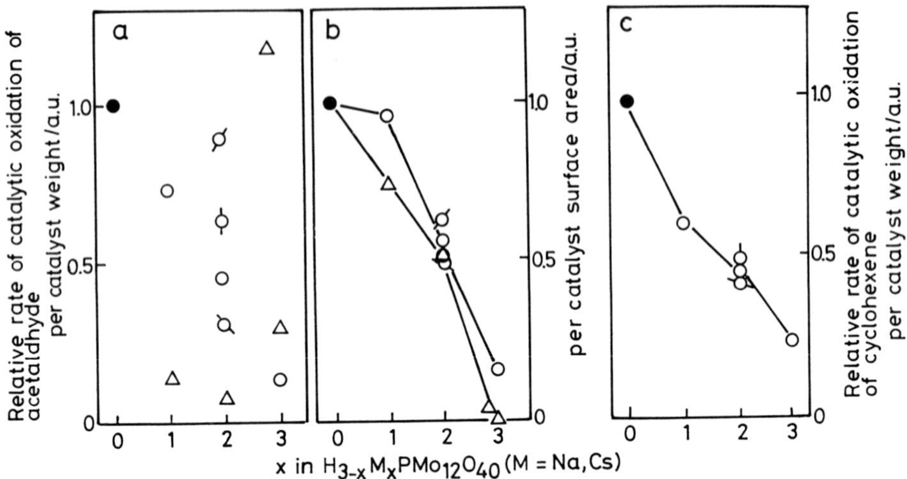

Fig. 5. Relative rates of catalytic oxidations of acetaldehyde (a and b) and cyclohexene (c) over $H_3PMo_{12}O_{40}$ and its alkali salts at 300°C.
○; Na salts, △; Cs salts. Flags attached to marks indicate different lots.

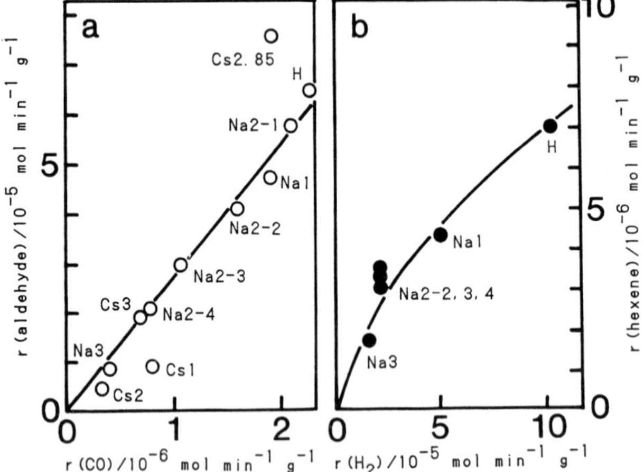

Fig. 6. Correlations between the oxidizing ability (r(CO) and r(H_2)) of catalyst and the rates of catalytic oxidation of (a) acetaldehyde and (b) cyclohexene. Mx denotes $M_xH_{3-x}PMo_{12}O_{40}$. Na2-1,2,3 means different lots of $Na_2HPMo_{12}O_{40}$.

by the ordinary method (5): Measured amount of Cs_2CO_3 aqueous solution was titrated into aqueous solution of $H_3PMo_{12}O_{40}$ and evaporated to dryness. The other series (B) was prepared by impregnation of $Cs_3PMo_{12}O_{40}$ with aqueous solution of $H_3PMo_{12}O_{40}$ and dried. Then they were in air heated at 300°C. Prior to the reaction, they were treated in the reactor for 1 h in the stream of O_2(10%)/N_2 mixture at 300°C. The reaction temperature was 250°C.

It is interesting to note that both series of catalysts showed very similar activity patterns. The highest activity was obtained with Cs2.5PMo12 (series B). It may be further noted that this pattern is somehow in parallel with the specific surface area. The high activities of Cs2.5PMo12's are mainly due to their high surface area. A similar activity pattern as regards the H/Cs ratio has been observed for the acid catalysis of $H_3PW_{12}O_{40}$ (PW12) (12). According to our previous study on acidic Cs salts of PW12 (12), the catalysts were mixtures of free acid (PW12) and Cs salt. For a support material, Cs3PMo12 and Cs3PW12 are suitable, since they have very high surface areas (100 - 200 m^2/g). So, the structure of Cs2.5PMo12, the most active catalyst among the catalysts shown in Fig. 7, is presumably PMo12 highly dispersed as thin films or small particles on the surface of Cs3PMo12 particles. A similar structure has been implied in the review of Ueshima et al.(13) and proposed by Black et al. (14). In the case of Cs salts of PW12 we observed phenomena which indicated the migration of proton and Cs ion by a treatment at 300°C (12, 15). So the treatment at a high temperature may have caused the diffusion of cations and closer contact between PMo12 and Cs3PMo12, e. g., formation of epitaxially grown thin films of PMo12 on Cs3PMo12 or the formation of acidic Cs salts having more nearly uniform composition.

SUMMARY

The heterogeneous catalysis of heteropoly compounds is classified into three types regarding the reaction field; surface-type, bulk-type (I) and bulk-type (II). Surface-type catalysis is the ordinary heterogeneous catalysis on the two-dimensional surface of solid catalyst (including pore walls). Bulk-type (I) is the catalysis in the pseudoliquid phase, where the reactant molecules are absorbed in the bulk lattice and react there. This is usually observed at low temperatures ($< 150°C$). The same phenomena may exist also in the case of liquid-solid heterogeneous catalysis. The bulk-type (II) which is described in the present article, was found for redox-type oxidation catalysis at high temperatures (200 - 350°C). Although the reactant molecules usually remain on the surface, the whole bulk can take part in the redox cycle of catalyst through the diffusion of the redox carriers. The three types of catalysis are schematically illustrated in Fig. 8.

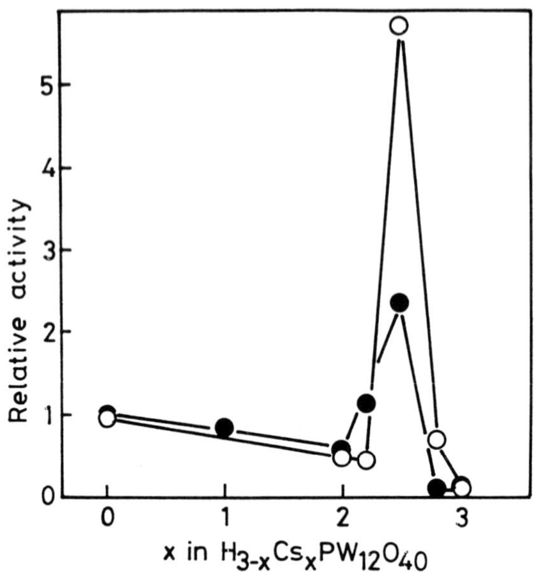

Fig. 7. Catalytic activity for oxidation of acetaldehyde at 250°C over two series of Cs containing 12-molybdophosphates.
●; prepared by titration (series A), ○; prepared by impregnation (series B).

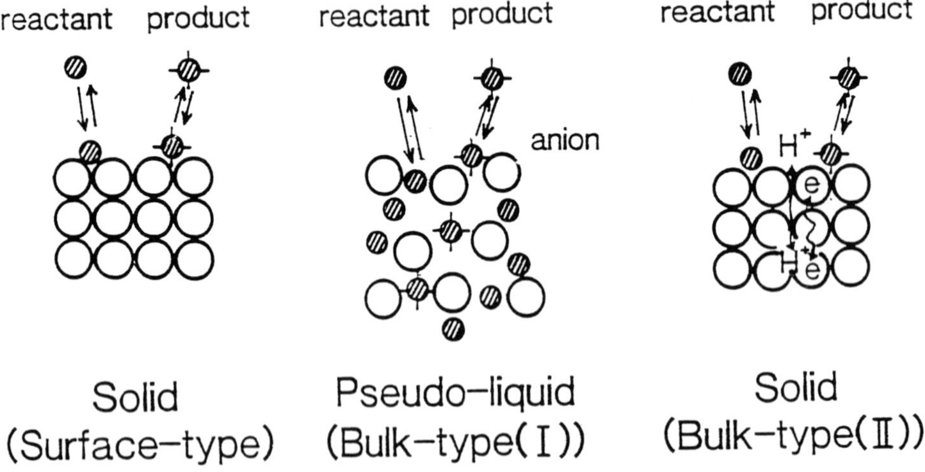

Fig. 8. Schematic model of the three types of catalysis of heteropoly compounds in the solid state. (From J. Catal., 123 (1990) 157).

Since the structure (or property) vs. activity patterns are quite different between the surface- and bulk-type catalysis (Figs. 2 and 6), we believe that the concept presented here is important for the understanding and design of industrial catalysts based on heteropoly compounds. Further, this concept may be also applicable to some of other solid catalysts.

Heteropoly compounds are useful and interesting catalyst materials for both fundamental studies and for practical applications. Although there are already several industrial processes which use heteropoly compounds as catalysts, so forth in all cases except one (oxidation of methacrolein) heteropoly compounds are used as solved in the liquid phase. In order to develop new catalytic processes with solid heteropoly compounds taking full advantage of the merits of those compounds such as controllable acidity and oxidizing ability and well definable structure (2), understanding of the structure-activity relationships in which the concept of the bulk- and surface-catalysis is properly taken into account may be indispensable.

Acknowledgment This study was supported in part by Grant-in-Aid for Scientific Research from the Ministry of Education, Science and Culture of Japan.

REFERENCES

1 M. Misono, K. Sakata, Y. Yoneda, and W. Y. Lee, "Proc. 7th Int. Congr. Catal., Tokyo, 1980," Kodansha(Tokyo) and Elsevier(Amsterdam), 1891, p1047.

2 M. Misono, Catal. Rev. -Sci. Eng., 29 (1987) 269; (addenda and errata) 32 (1988) 339.

3 T. Komaya and M. Misono, Chem. Lett., (1983) 1177.

4 N. Mizuno and M. Misono, J. Phys. Chem., 93 (1989) 3334; ibid., 94 (1990) 890.

5 N. Mizuno, T. Watanabe, and M. Misono, J. Phys. Chem., 89 (1985).

6 M. Misono, N. Mizuno, and T. Komaya, Proc. 8th Intern. Congr. Catal., 1984, Vol. 5, Verlag Chem., 1984, p487.

7 M. Misono, N. Mizuno, K. Katamura, A. Kasai, Y. Konishi, K. Sakata, T. Okuhara, and Y. Yoneda, Bull. Chem. Soc. Jpn., 55 (1982) 400.

8 T. Hibi, K. Takahashi, T. Okuhara, M. Misono, and Y. Yoneda, Appl. Catal., 24 (1986) 69.

9 T. Okuhara, S. Tatematsu, K. Y. Lee, and M. Misono, Bull. Chem. Soc. Jpn., 62 (1989) 717.

10 N. Mizuno, T. Watanabe, H. Mori, and M. Misono, J. Catal., in press (vol. 123 (1990) 157).

11 M. Akimoto, K. Shima, K. Sato, and E. Echigoya, J. Catal., 72 (1981) 83.

12 S. Tatematsu, T. Hibi, T. Okuhara, and M. Misono, Chem. Lett., (1984) 865.

13 M. Ueshima, H. Tsuneki, and N. Shimizu, Hyoumen, 24 (1986) 582.

14 J. B. Black, N. J. Claydon, P. L. Gai, J. D. Scott, E. M. Serwicka, and J. B. Goodenough, J. Catal., 106 (1987) 1.

15 N. Mizuno and M. Misono, Chem. Lett., (1987) 967.

R.K. Grasselli and A.W. Sleight (Editors), *Structure-Activity and Selectivity Relationships in Heterogeneous Catalysis*
1991 Elsevier Science Publishers B.V., Amsterdam

ACTIVE IRON OXO CENTERS FOR THE SELECTIVE CATALYTIC OXIDATION OF ALKANES

James E. Lyons, Paul E. Ellis, Jr., and Vincent A. Durante

Research and Development Division, Sun Refining and Marketing Company
P.O. Box 1135, Marcus Hook, PA 19061

ABSTRACT

Much work has been done in an effort to understand the nature of iron oxo complexes and their roles in the selective catalytic oxidation of alkanes. Iron oxo (ferryl) species (Fe=O) have been proposed to be the active intermediates responsible for both the enzymatic and biomimetic oxidations of alkanes to alcohols, while it is generally accepted that iron(III) μ-oxo species [Fe(III)-O-Fe(III)] are not catalytically active. We have synthesized a number of iron complexes having μ-oxo bridges in several molecular environments including porphyrinato, polyoxometallate, and silicometallate structures and examined the catalytic activity of these compounds for alkane oxidation in both liquid and vapor phase. The activity and selectivity of these catalysts depend upon the molecular environment of the μ-oxo species used as the catalyst precursor. In some instances in situ conversion of μ-oxo to ferryl oxo species may be the key to catalysts capable of direct hydroxylation of alkanes with air or oxygen.

INTRODUCTION

There currently exists no commercial one-step catalytic air oxidation process to convert light alkanes to alcohols. Such a one-step route would represent superior useful technology for the utilization of natural gas and similar refinery-derived light hydrocarbon streams. Natural gas or its components (methane, ethane, propane and the butanes) could not only be a valuable alternative to crude oil but processes for converting these light alkanes to alcohols for use as motor fuels would produce a clean-burning, high octane alternative to conventional gasoline.

As desirable as a process for the direct one-step air-oxidation of light alkanes to alcohols might be, the low reactivity of alkanes coupled with the lack of active catalysts which can accomplish such a selective transformation under mild conditions, has made this a formidable challenge. Oxidations catalyzed by either homogeneous or heterogeneous catalysts are generally too deep. Homogeneous catalysts which operate in the liquid phase via metal-mediated radical pathways give rise to large amounts of carbon-carbon bond cleavage which can be the predominant pathway at the elevated temperatures needed for good rates. For example, cobalt salt catalyzed reactions of isobutane at 135°C produce significant quantities of C_1 and C_3 oxidation products (1), and cobalt acetate catalyzed oxidation of n-butane at 165°C gives acetic acid as the major product (21). Heterogeneous catalysts used to promote vapor phase oxidations can be quite selective but give deep oxidations to α,β-unsaturated carboxylic acids or their cyclic anhydrides in the case of propane or butane (3,4). Selective oxidation of methane to formaldehyde has been accomplished at low conversions using a ferric molybdate catalyst (5). A similar catalyst has been reported to produce methanol from methane at low conversions (6) and some thermally generated gas phase radical reactions are reported to give rather high methanol selectivities from methane or natural gas (7) but greater rates and selectivities are needed for a practical process. To date enzymatic systems are the only relatively selective catalysts for the reaction of alkanes with oxygen to give alcohols.

Cytochrome P-450 is a heme-iron catalyst which promotes air-oxidation of alkanes to alcohols (8) and methane monooxygenase is a non-heme catalyst for the conversion of methane to methanol (9). Both catalyst systems are believed to be capable of generating high oxidation state ferryl intermediates which directly hydroxylate the alkane, Figures 1,2 (8,9). Both enzymatic systems, however, have the requirement of a coreductant (usually NADH) to furnish the electrons and protons which are stoichiometrically consumed in the catalytic cycles, Figures 1,2. This requirement imposes a severe limitation on the commercial use of these or similar systems. The biological systems have another drawback as well. The reductive binding of dioxygen is accomplished by protons and electrons in addition to the iron center which produce a mole of water in so doing. Thus these systems are at best only 50% efficient in oxygen.

In considering how one might design a catalyst which could promote the direct air-oxidation of an alkane, we wondered whether it would be possible to tune the redox potential of an Fe(II) center so that instead of rapidly and irreversibly being converted to the μ-oxo complex, Fe(III)OFe(III), it would instead more efficiently produce and cleave the μ-peroxo species, (Fe(III)OOFe(III)) to form an active ferryl and complete the hypothetical catalytic cycle shown in Figure 3. In this conceptual model, we have reductively bound the dioxygen molecule at both ends, have no stoichiometric requirement for electrons and protons, and utilize the oxygen completely. A catalyst which could operate in this manner might be called suprabiotic rather than biomimetic.

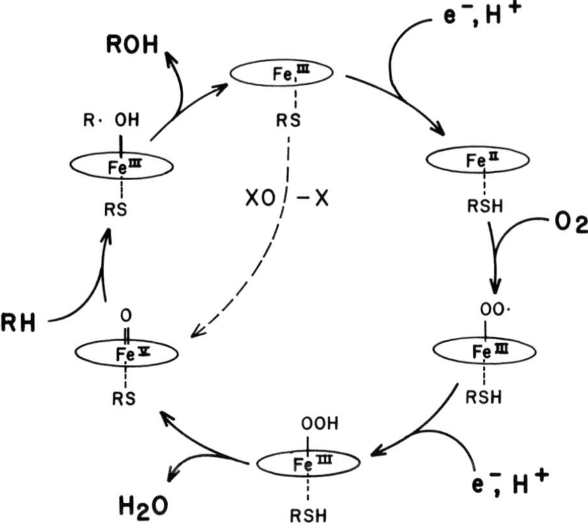

Figure 1. Proposed Mechanism for Alkane Oxidations Catalyzed by Cytochrome P-450 (8)

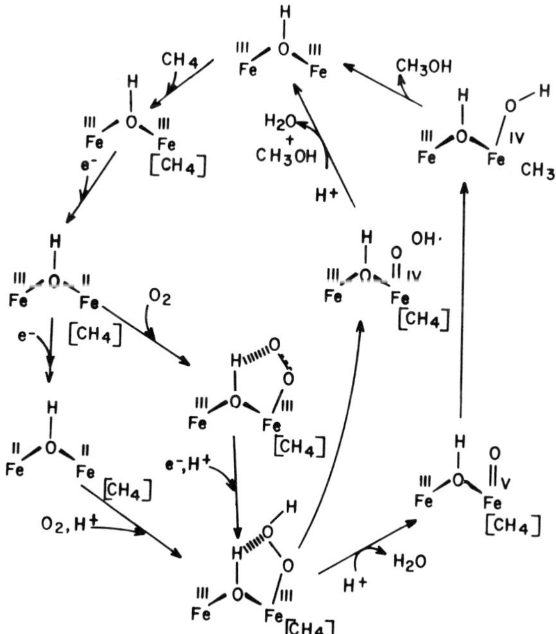

Figure 2. Proposed Mechanism for Alkane Oxidations Catalyzed by Methane Monooxygenase (9)

In this paper we will present the results of the catalytic oxidation of light alkanes in the presence of four catalyst types which were prepared with this catalyst design concept in mind: iron perhaloporphyrin complexes, Keggin structures with iron in the framework, Keggin structures with proximate iron centers, and a synthetic crystalline silicometallate catalyst having both framework and exchange sites containing iron.

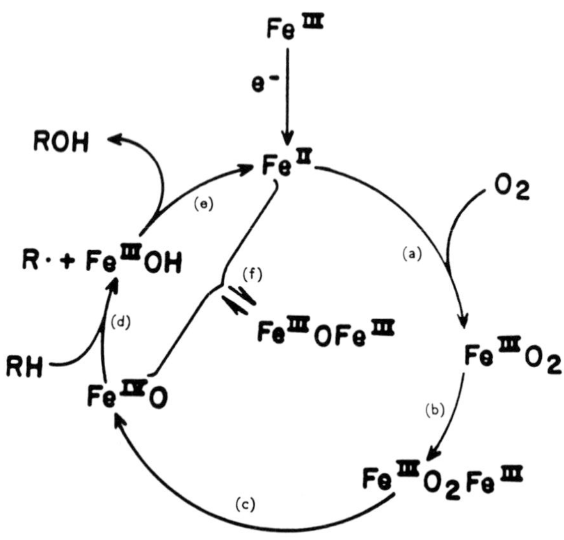

Figure 3. Conceptual Catalytic Cycle for the Direct Oxidation of an Alkane With O_2

EXPERIMENTAL

<u>Iron Haloporphyrin Complexes</u>

Iron(III)porphyrin azide and halide complexes were prepared by metathesis of the corresponding iron(III)porphyrin chloride or μ-oxo complex with the appropriate acid, HX (X = F, Cl, Br, N_3) (10% aqueous solution) in methylene chloride. Iron(III)porphyrin hydroxide complexes were prepared by treatment of the iron(III)porphyrin chloride with aqueous KOH in methylene chloride (10). The μ-oxo complex, $Fe(TPP\beta\text{-}Br_4]_2O$, was prepared using the method of Callot (11). Preparation of $[Fe(TPP)]_2O$ and $[Fe(TPPF_{20})]_2O$ were accomplished by known procedures (12). Preparation of $(TPPCl_8\beta\text{-}Br_4)$ complexes was achieved by bromination of $Zn(TPPCl_8)$ with NBS (13) followed by iron insertion using the $FeCl_2$/DMF method (14). $Fe(TPPF_{20}\beta\text{-}Br_8)Cl$ was prepared by the direct reaction of $Fe(TPPF_{20})$ with 6M Br_2 in carbon tetrachloride at reflux.

Oxidations were carried out in a barricaded laboratory equipped for experimentation at high pressure since many reactions were conducted within the explosion limits. Reactions reported in Table 1 were carried out in 50 cc Fisher-Porter glass aerosol tubes with magnetic stirring. Reactions reported in Tables 2 and 3 were conducted in a glass lined stirred autoclaves with Teflon coated internals (impeller, dip-tube, etc.) Gas and liquid products were analyzed by a combination of standardized gc, ms and gcms analyses.

In-situ absorption spectroscopy was performed using a Guided Wave Model 200 spectrometer equipped with an optical transmission wand probe which is inserted into the bottom of a stirred autoclave, using high pressure fittings and is fixed 1 cm below the impeller blades.

Iron-Containing Polyoxometallates

The complex $K_4(PW_{11}FeO_{39})$, was prepared according to the method of Tourné (15). The heteropolyacids: $H_6(PW_9Fe_3O_{37})$ and $H_7(PW_9Fe_2MO_{37})$ (M = Ni, Zn, Co, Mn) were prepared from the trilacunary complex B-$Na_9[PW_9O_{34}]$ (16) by inserting the corresponding trimetal acetate bridged complexes: $M_3O(CH_3CO_2)_6(H_2O)_3$ (17) into the trilacunary complex, converting the resulting products to the tetrabutylammonium salts and pyrolyzing these complexes to give the free heteropolyacids.

The oxidations shown in Table 4 were run in a glass lined stirred autoclave as described above.

Iron Sodalite

The silicoferrate, iron sodalite, [Fe]SOD, was synthesized by a modification of the method of Szostak and Thomas (18-20). The x-ray diffraction pattern was consistent with the sodalite structure. The crystalline product was bound with sodium silicate solution calcined at up to 555°C in air for one hour. Analysis: Fe, 10.11%: Si, 30.23%. A combination of evidence including crystallinity, measurements by PXRD relative to standard samples, measurement of relative ion exchange capacity, and esr measurements, indicates that approximately 80% of the iron is in the framework of the zeolite while about 20% is in extraframework sites (20).

Methane oxidations were carried out in a previously described (21) glass lined continuous reactor which limited the contact of products with catalyst and which provided an open reaction zone for reaction intermediates which may have been formed. Reaction conditions are given in Table 5.

RESULTS AND DISCUSSION

Alkane Oxidations Catalyzed by Haloporphyrin Complexes

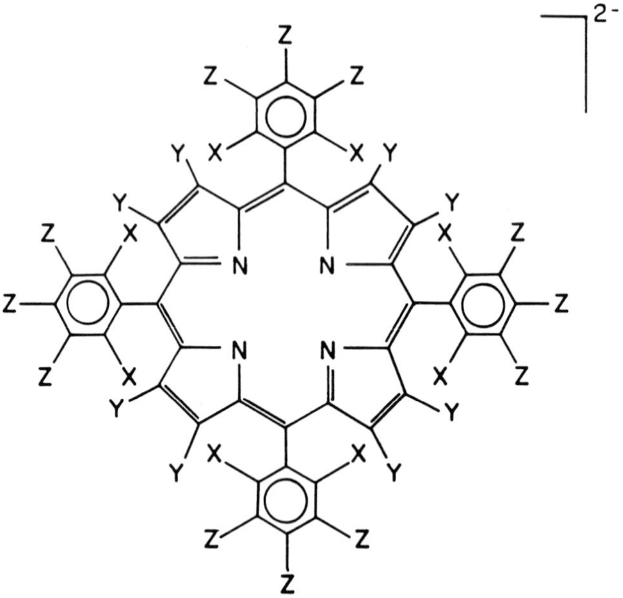

LIGAND	X	Y	Z
(TPP)	8H	8H	12H
(TPPβ-Br$_4$)	8H	4H,4Br	12H
(TPPCl$_8$)	8Cl	8H	12H
(TPPCl$_8\beta$-Br$_4$)	8Cl	4H,4Br	12H
(TPPF$_{20}$)	8F	8H	12F
(TPPF$_{20}\beta$-Br$_8$)	8F	8Br	12F

Figure 4. Halogenated Porphyrin Ligand System For Iron(III) Oxidation Catalysts

We have found that as the twenty-eight hydrogens in a series of iron tetraphenylporphyrin (TPP) complexes are successively replaced by halogen atoms, Figure 4, there is a steady increase in the catalytic activity of the complexes for the reaction of isobutane with molecular oxygen to give <u>tert</u>-butyl alcohol (TBA), Table 1 (22-24). The reduction potential [Fe(III)/Fe(II)] of the complexes also increases with extent of porphyrin ring halogenation (from -0.29 to +0.31 volts as one goes from Fe(TPP) to Fe(TPPF$_{20}\beta$-Br$_8$) complexes, Figure 4). Thus, there is a relationship between the ease of reduction of iron(III) and its ability to function as a catalyst for the selective reaction of an alkane to an alcohol.

It is also of interest to note that not only are the highly halogenated porphyrinatoiron halides, azides, and hydroxides active catalysts, but the corresponding μ-oxo diiron complexes of the haloporphyrins are also catalytically active. This is particularly interesting because μ-oxo complexes of the parent tetraphenylporphyrins are inactive, and μ-oxo dimers are considered to be inactive forms (25) of biomimetic porphyrinatoiron(II) oxidation catalysts since their formation, Figure 3, is often irreversible and the resulting Fe(III)OFe(III) species is incapable of binding dioxygen.

TABLE 1

Effect of Ring Halogenation on the Isobutane Oxidation Activity Of Porphyrinato Iron (III) Complexes[a]

Catalyst	mmoles	O_2 Uptake, mmoles	TON[b]	Selectivity[c] To TBA, %
Fe(TPP)Cl	0.025	0.0	0	-
Fe(TPPβ-Br_4)Cl	0.013	2.0	155	-
Fe(TPPCl_8)Cl	0.019	5.0	260	89
Fe(TPPCl_8,β-Br_4)Cl	0.020	17.3	865	83
Fe(TPPF_{20})Cl	0.016	32.6	2040	90
Fe(TPPF_{20},β-Br_8)Cl	0.013	40.2	3090	89
Fe(TPP)N_3	0.013	1.7	130	92
Fe(TPPβ-Br_4)N_3	0.013	2.3	180	-
Fe(TPPCl_8)N_3	0.023	15.0	650	80
Fe(TPPCl_8,β-Br_4)N_3	0.023	21.5	930	82
Fe(TPPF_{20})N_3	0.016	33.0	2060	89
[Fe(TPP)]$_2$O	0.019	0	0	-
[Fe(TPPβ-Br_4)]$_2$O	0.013	0	0	-
Fe(TPPCl_8)OH	0.013	9.2	710	83
[Fe(TPPF_{20})]$_2$O	0.013	24.0	1,850	84
Fe(TPPF_{20})OH	0.013	29.2	2,245	82

[a] A solution of the catalyst in 25 ml benzene containing 6 grams of isobutane was stirred at 80°C under 100 psig of O_2 for 6 hours.
[b] moles O_2 consumed/mole catalyst used.
[c] (moles t-butyl alcohol produced/total moles liquid product) x 100

TABLE 2

Iron Haloporphyrin-Catalyzed Isobutane Oxidations[a]

Catalyst	T, C	Charge to Reactor			Reaction Products, mmoles				Conversion	Select.	
		t, Hrs.	$i\text{-}C_4H_{10}$	O_2	TBA	Acetone	CO_2	CO	$i\text{-}C_4H_{10}$,%	TBA,%[b]	TON[c]
$Fe(TPPF_{20}\text{-}\beta\text{-}Br_8)Cl$	80	3	1870	53	277	43	23	8.0	17	87	10,660
	80	3	1862	100	429	86	26	4.9	28	83	17,150
	80	3	1862	148	414	81	28	6.0	27	84	16,500
	80	3	1869	205	290	45	37	10.5	18	87	11,180
	60	3	1865	47	230	23	20	tr	14	91	8,420
	60	3	1874	139	184	18	17	tr	11	91	6,730
	25	71.5	1862	53	372	35	27	tr	22	92	13,560
$Fe(TPPF_{20})OH$	24	143	1871	53	332	17	18	0	18	95	12,150

a Isobutane was oxidized by an oxygen-containing gas mixture (75 atm, diluent = N_2) in the liquid phase (180 ml) for 3 hours. Oxygen added as consumed.

b (moles TBA/moles liquid product) X 100.

c moles (TBA + Acetone) produced/mole catalyst used.

When all twenty-eight of the hydrogens present in the TPP ring have been replaced by halogens, the perhaloporphyrin complex exhibits unprecedented oxidation activity. Table 2 shows that the perhaloporphyrin complex is a soluble catalyst for oxidation in neat isobutane. The alcohol is produced in over 90% selectivity at over 20% conversion at room temperature. The ligand system was virtually unchanged in over seventy hours of operation during which well over 13,000 turnovers (moles TBA produced/g.-atom iron used) had occurred.

As a result of extensive ring halogenation there is considerable electron withdrawal from the porphyrin ring. This has a number of effects which would be expected to enhance a reaction pathway such as that shown in Figure 3. As electron density is removed from the iron center, the position of equilibrium (a) would be shifted away from Fe(III)O_2 towards Fe(II) and both formation (b) and cleavage of Fe(III)OFe(III) should be enhanced. Electron-withdrawing halogen substituents should activate the ferryl intermediate formed by cleavage of the ferric peroxide (c) toward hydroxylating the alkane (d,e). The hypothetical pathway shown in Figure 3 is dependent upon a μ-oxo diiron(III) species which disproportionates in equilibrium with Fe(II) and Fe(IV)O, (f). Halogenation could shift the position of equilibrium from the μ-oxo diiron(III) species to some extent toward a low oxidation state iron(II) complex and a high oxidation state iron(IV) ferryl. Both steric and electronic factors could destabilize the diiron μ-oxo complex toward disproportionation. Electron withdrawal from the porphyrinato ligand should make it more difficult for oxidation of the ligand by electron transfer to the iron center. Thus, perhaps an iron(IV) ferryl species generated from symmetrical cleavage of the μ-peroxodimer (26) of a halogenated porphyrin could survive and be effective in alkane hydroxylation. Finally, by removing electron density from the ring, the halogens should make the porphyrin less susceptible to electrophilic attack by the ferryl species itself which could lead to destruction of the catalyst (27).

<u>In-situ</u> absorption spectroscopy has enabled us to examine metalloporphyrin catalyzed hydroxylation under reaction conditions. Figure 5 shows the spectra from a Fe($TPPF_{20}$)Cl-catalyzed reaction of isobutane with air producing <u>tert</u>-butyl alcohol. When 500 psig of air is pressed on 20 g. of isobutane in 80 g. of benzene containing 0.01 mmole of Fe($TPPF_{20}$)Cl at room temperature, the 415 nm absorption normal for the hemin diminishes and new peaks at 396 and 443 nm arise. The 443 nm peak, which persists throughout the entire 3 hours of reaction is similar to the Fe(II) spectrum reported by Suslick, et. al. (28) during the photoreduction of Fe(III)(TPP)Cl. The 396 and 559 nm bands are those that are expected for the μ-oxo di-iron species. After the experiment intact μ-oxo iron complex, Fe$(TPPF_{20})_2$O, can be recovered from solution.

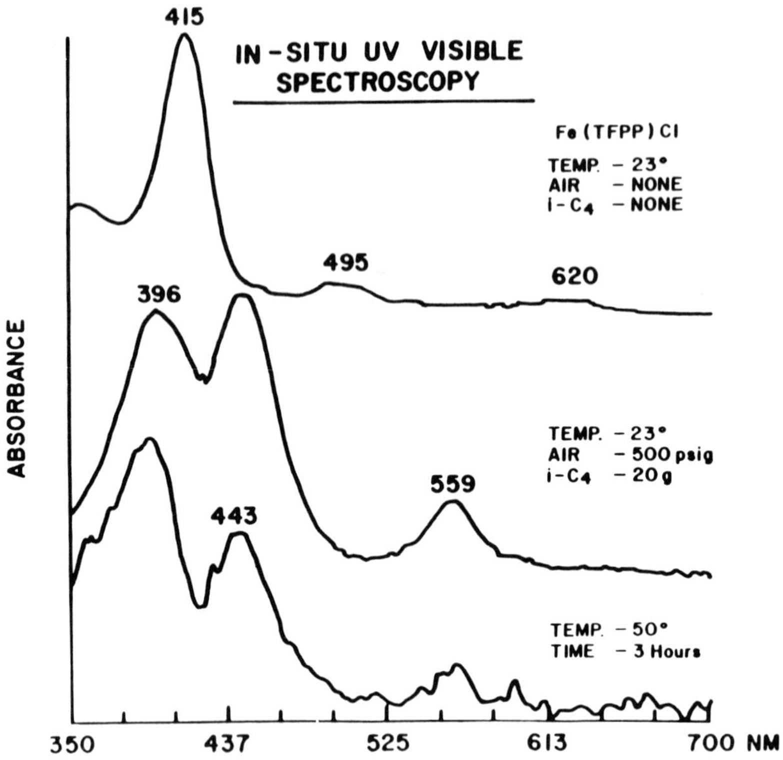

Figure 5. In-situ Absorption Spectra Under Oxidation Conditions

The activity enhancement found for iron complexes of halogenated porphyrins made it possible to oxidize propane under mild conditions in the liquid phase, Table 3. The same trend in increased activity with increased extent of halogenation of the iron porphyrin was found for propane as was found for isobutane. Because of the greater degree of difficulty of cleaving the secondary C-H bond in propane than the tertiary C-H bond in isobutane, rates in excess of a hundred turnovers/hour could only be obtained at reaction temperatures of 125°C or more. At these elevated temperatures even the halogenated porphyrins demonstrated catalyst life problems. The catalysts shown in Tables 1-3 were not effective for the rapid oxidation of methane and ethane which have very strong primary C-H bonds.

TABLE 3

Effect of Ring Halogenation on the Propane Oxidation Activity of Porphyrinato Iron (III) Complexes[a]

Catalyst	mmoles	T, Hrs.	TON[b]	IPA/Acetone
Fe(TPP)Cl	0.023	3	2	na
Fe(TPPβ-Br_4)Cl	0.023	3	0	-
Fe(TPPCl_8)Cl	0.023	6	0	-
Fe(TPPCl_8,β-Br_4)Cl	0.023	3	125	1.0
Fe(TPPF_{20})Cl	0.023	3	230	0.8
Fe(TPPF_{20},β-Br_8)Cl	0.023	4.5	470	1.0
Fe(TPP)N_3	0.023	3	0	-
Fe(TPPβ-Br_4)N_3	0.023	3	0	-
Fe(TPPCl_8)N_3	0.023	4.5	0	-
Fe(TPPCl_8,β-Br_4)N_3	0.023	4.5	250	0.8
Fe(TPPF_{20})N_3	0.023	3	330	0.8
Fe(TPPF_{20},β-Br_8)N_3	0.013	4.5	540	0.9
$[Fe(TPP)]_2O$	0.023	3	0	-
[Fe(TPPβ-Br_4)$]_2$O	0.023	4.5	0	-
Fe(TPPCl_8)OH	0.023	4.5	0	-
[Fe(TPPF_{20})$]_2$O	0.013	3	440	0.8
Fe(TPPF_{20})OH	0.013	3	270	0.6

[a] Stirred a solution of the catalyst in 60 grams of propane in 48 ml benzene at 125°C under 1000 psig air.
[b] moles (isopropyl alcohol + acetone) formed/mole catalyst used.

Alkane Oxidations Catalyzed by Polyoxometallates

In the search for a more rugged molecular environment for iron centers, we chose a series of polyoxometallates having the Keggin structure sometimes called "inorganic porphyrins" (29) because of the way in which oxidation-active metals which are placed in their framework respond as hydrocarbon oxidation catalysts. Although not as active as the perhalotetraphenylporphyrin complexes at low temperature, these catalysts had greater oxidative stablility and could be used at higher temperatures.

Another advantage of the Keggin structures is that we can incorporate more than one oxidation-active metal center in such a way that these centers can be proximate - separated only by μ-oxo bridges (30,31). This is of interest because of the apparent cooperativity of μ-hydroxo bridged diirons in methane monooxygenase, Figure 1.

We were able to synthesize a large number of Keggin structures incorporating three metal centers by reacting the trilacunary complex B-$Na_9[PW_9O_{34}]$ with a series of μ-oxo bridged tri-metal clusters as shown in Figure 6. Although structural analysis is in progress, we have not yet ascertained the extent to which the tri-metal cluster geometry is retained in the Keggin structure. Elemental analysis indicates that the three metal centers are incorporated into the compound. In addition to having good isobutane oxidation activity, these complexes were quite active for oxidizing propane at 150°C, Table 4. It is interesting to note that the diiron heteropolyacid containing Ni(II) - a metal center which would not be expected to oxidize - was the most active, while diiron complexes containing Co(II) and Mn(II) which would oxidize to Co(III) and Mn(III) under reaction conditions were the least active of the series. Triiron and diironnickel complexes even showed some activity for methane and ethane oxidation at elevated temperatures, although rate and selectivity were poor (conversion < 5%, selectivity < 33%), Table 4. The potassium salt of the diiron nickel complex, $K_7PW_9Fe_2NiO_{37}$, was shown to be active when used as a heterogeneous catalyst under the conditions given in Table 5 (800 psig, ghsv = 530, 410°C) and produced methanol in 42% selectivity at 4.1% methane conversion.

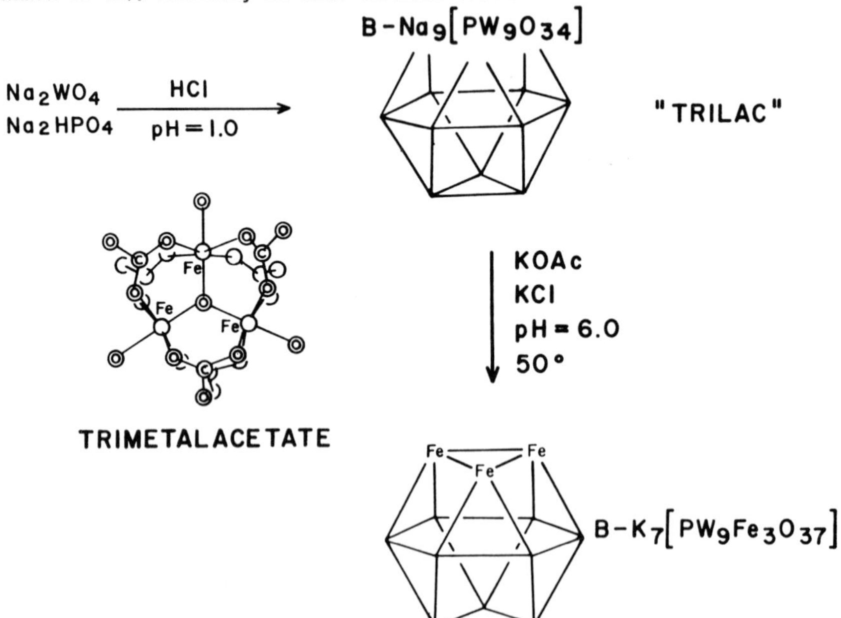

Figure 6. Synthesis of Tri-metal Substituted Keggin Ions

TABLE 4

Liquid Phase Oxidation of Alkanes Using Heteropolyacid Catalysts

Substrate	Catalyst	T,°C	T.O'S[a]	IPA/A[c]
Propane[b]	None	150	0	-
	$H_3PW_{12}O_{40}$	150	750	0.85
	$H_6PW_9Fe_3O_{37}$	150	8110	0.61
	$H_4PW_{11}FeO_{39}$	150	2034	0.52
	$Fe_3O(OAc)_6(H_2O)_3Cl$	150	190	0.53
	$H_7PW_9Fe_2NiO_{37}$	150	9730	0.71
	$H_7PW_9Fe_2ZnO_{37}$	150	5640	0.65
	$H_7PW_9Fe_2MnO_{37}$	150	5570	0.65
	$H_7PW_9Fe_2CoO_{37}$	150	3290	0.70
	$H_7PW_9Fe_2MnO_{37}$	125	158	0.30
	$H_7PW_9Fe_2CoO_{37}$	125	125	0.40
Ethane	$H_7PW_9Fe_2NiO_{37}$	200	240	
Methane	$H_6PW_9Fe_3O_{37}$	280	50	

a Moles propane/mole catalyst used.
b Propane, 1.36 moles was added to a solution of 5μ moles catalyst in 38 ml acetonitrile and heated under 1000 psig air for 3 hours. Reactions promoted by 5 mg NaN_3.
c Molar ratio of isopropyl alcohol/acetone produced.

Oxidations Catalyzed by Solid Silicometallates

Using metals in perhalogenated macrocycles or in polyoxo metallates, we have activated all of the light (C_1-C_4) alkanes with varying degrees of success in the liquid phase. Since it may be desirable to operate in the vapor phase, especially with methane or natural gas, we have implanted oxidation-active iron centers into the frameworks of siliceous molecular sieve matrices and examined the catalytic activity of these systems for the vapor phase oxidation of light alkanes with an emphasis on methane oxidation to methanol. A particularly effective catalyst was an iron sodalite, [Fe]SOD, (>10% by weight Fe), having iron in both framework and exchange positions.

Table 5 compares the results of the vapor phase oxidation of methane over two differently prepared and activated iron sodalites, a framework iron containing pentasil zeolite, hydroxysodalite, and a silica-supported iron oxide. Once activated, the iron sodalite shown in Table 5 retained its activity for over three months of daily operation before it was removed from the reactor. Our observations suggest that a relatively high iron content is required for high catalytic activity to develop in this series, but that a silica-supported iron oxide is much less selective towards methanol formation than is a crystalline iron sodalite despite their similar iron contents. Little initial activity is generally observed for non-calcined crystalline iron sodalite samples. High temperature calcinations or prolonged high temperature oxidation results in partial structural collapse and framework deferration occurs, and it is likely that during this period iron is driven into ion exchange sites associated with the residual framework iron. This conclusion is consistent both with PXRD measurements which show growth of an amorphous halo at the expense of a crystalline phase and ESR measurements which show a decrease in the peak intensity of the $g=4.3$ peak, ascribed to framework iron(III), and a corresponding increase in a peak at $g=2.0$, extra framework iron. Other experiments indicated that Fe(II) exchange of [Fe]SOD resulted in improved methanol selectivity when tested in a stainless steel reactor.

These observations can be accounted for by postulating that active proximate iron centers, might develop between an ion exchange iron position and a framework iron position in an activated iron sodalite. Extended oxo-bridged iron(III) arrays, as is likely a feature of the structure of the supported iron oxides used, are ineffective, however. Figure 6 suggests a possible pathway in which both framework and exchange site iron centers might act cooperatively to catalyze methane oxidation to a methanol-rich product mixture via surface ferryl intermediates. This hypothetical pathway uses two proximate iron centers to generate surface ferryl intermediates which oxidize methane by a route that avoids the formation of μ-oxo Fe(III)OFe(III) species on the surface.

TABLE 5

Vapor Phase Air Oxidations of Methane to Methanol[a]

Catalyst	GHSV h^{-1}	Bed T °C	CH_3OH Sel., %	CH_4 Conv., %	O_2 Conv.,%
[Fe]SOD[b]	530	407	64	4.6	76
		416	70	5.7	90
		418	68	5.4	91
		422	65	5.4	92
		432	63	5.5	90
		442	64	6.1	90
[Fe]SOD[c]	530	400	-	0.0	0
	1330	420	27	0.2	na
Hydroxysodalite	530	404	-	0.0	0
		430	9	0.3	4
		445	24	5.1	84
Fe_xO_y/SiO_2 (12.5%)	530	431	26	5.1	85
	700	398	10	0.5	11
		409	18	1.2	15
		420	20	2.7	74
		428	23	3.9	83
[Fe]ZEOL[d]	700	400	33	2.4	52
		430	34	3.8	74

[a] A 3/1 methane/air mixture (800psig) was passed over the catalyst in the reactor described previously (21).

[b] [Fe]SOD, binder added and calcined at 550°C, 1 hr. after synthesis (grey solid, 18-35 mesh, 10.1% Fe by weight).

[c] [Fe]SOD, binder added and calcined at 300°C, 1 hr. after synthesis (white solid, 18-35 mesh, 9.9% Fe by weight).

[d] A crystalline silicoferrate having low iron concentration.

Although it is believed that high oxidation state iron oxo species are capable of homolytically cleaving carbon-hydrogen bonds, low temperature reactions in gas or liquid phase may produce alcohols directly by collapse of metal-bound radicals (or metal alkyls) in a manner similar to that shown in Figure 3, step e. As we indicate in Figure 7, it would be expected that a significant fraction of the bound radicals formed at elevated temperatures would be expelled into the gas phase where they would react by homogeneous vapor phase pathways. It is not unlikely that radical pathways contribute to some extent in liquid phase oxidations as well.

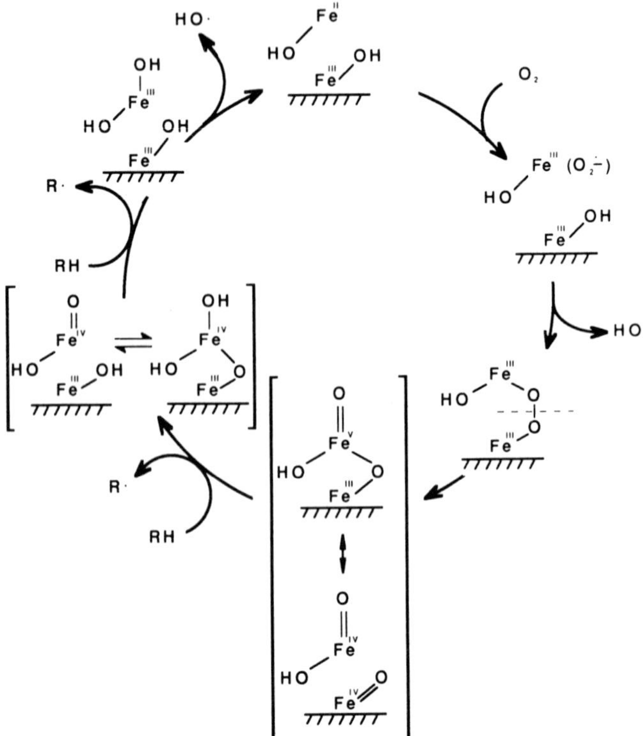

Figure 7. Hypothetical Catalytic Cycle for Silicoferrate Activity

CONCLUSIONS

Much effort has gone into attempts to avoid the formation of μ-oxo dimers by introducing steric hindrance (25) or surface isolation techniques (32) to prevent the two iron centers from being proximate. Although this strategy prevents what may be an inactive species from forming, and may be necessary to truly mimic biological processes, it often does not allow two iron centers to cooperate in activating a single dioxygen molecule. We have conceived of several molecular environments which allow two irons to be proximate enough so that they could cooperate in activating the dioxygen molecule but which have electronic properties that might favor ferryl intermediates over μ-oxo dimers. Although we cannot say conclusively at this time that pathways involving ferryl intermediates are operative, catalysts prepared with these design principles in mind are active for selective partial oxidation of alkanes to alcohol-rich product mixtures.

ACKNOWLEDGEMENT

Acknowledgement is made to the Department of Energy, the Gas Research Institute and the Sun Company for continued support of this work.

REFERENCES

1 K. Weissermel and H. Arpe, Industrial Organic Chemistry, Verlag Chemie, New York, 1978, p. 154.
2 D. Winkler and G. Hearne, Ind. and Eng. Chem., 53 (1961) 655.
3 (a) M. Ai, J. Chem. Soc. Chem. Commun., (1986) 786, J. Catalysis. 10 (1986) 389-315, (b) T. Komatsu, Y. Uragami, and K. Otsuka, Chem. Lett., (1988) 1903.
4 B.K. Hodnett, Catal. Today, 1987 1 (and references cited therein).
5 G.N. Kastanes, G.A. Tsigdinos and J. Schwank, Proc. 1988 A.I.Ch.E., Mtg, Paper 172 (1988) C4-C9.
6 D.A. Dowden and G.T. Walker, British Patent 1,244,001 (1968).
7 P.S. Yarlagadda, L.A. Morton, N.R. Hunter, and H.D. Gesser, Ind. Eng. Chem. Res., 27 (1988) 252.
8 T.J. McMurry and J.T. Groves in: P.R. Ortiz de Montellano (Ed), Cytochrome P-450: Structure, Mechanism and Biochemistry, Plenum Press, New York, 1986 pp. 1-28.
9 H. Dalton and J. Green, J. Biol. Chem., 264 (1989) 17698.
10 R.J. Cheng, L. Latos-Grazynski and A.L. Balch, Inorg. Chem., 21 (1982) 2412.
11 H.J. Callot, Bull Soc. Chim. Fr. (1974) 1492.
12 (a) E.B. Fleischer, J.M. Palmer, T.S. Srivastava and A. Chatterjee, J. Am. Chem. Soc., 92 (1971) 3162, (b) K. Jaydraj, A. Gold, G.E. Toney, J.H. Helms, W.E. Hatfield, Inorg. Chem., 25 (1988) 3516.
13 T.G. Traylor and S. Tsuchiya, Inorg. Chem., 26 (1987) 1338.
14 A.D. Adler, F.R. Longo, F. Kampos, and J. Kim, J. Inor. Nucl. Chem., 32 (1970) 2443.

15 (a) C. Tourne' and G. Tourne', Bull. Soc. Chim. Fr., (1969) 1124, (b) U.S. Patent 4,916,000 to Sun Co. (1990).

16 P.J. Domaille and G. Watunya, Inorg. Chem., 25 (1986) 1239.

17 (a) A. Earnshaw, B.N. Figgs and J. Lewis, J. Chem. Soc. A, (1966) 1556, (b) A.B. Blake, A. Yavari, W.E. Hatfield and C.N. Sethulekshmi, J. Chem. Soc. Dalton Trans. (1985) 2509.

18 R. Szostak and T.L. Thomas, JCS, Chem. Commun. (1986) 113.

19 G. Calis P. Frenken, E. DeBoer, A. Swolfs and H. Hefni, Zeolites, 7 (1987) 319.

20 V.A. Durante, D.W. Walker, S.M. Gussow, and J.E. Lyons, U.S. Patent 4,918,249 (1990).

21 V.A. Durante, D.W. Walker, W.A. Seitzer and J.E. Lyons, Preprints of 3B Symposium on Methane Activation, Conversion and Utilization, at the 1989 Internat. Chem. Congr. of Pacific Basin Societies, Dec. 17-20 (1989).

22 P.E. Ellis, Jr. and J.E. Lyons, Catalysis Lett., 3 (1989) 389.

23 J.E. Lyons and P.E. Ellis, Jr., Catalysis Lett., (1990) in press.

24 P.E. Ellis, Jr. and J.E. Lyons, Coordination Chem. Rev., 105 (1990) 181.

25 J.P. Collman, R.R. Gagne, C.A. Reed, T.R. Halbert, G. Lang and W.T. Robinson, J. Am. Chem. Soc., 97 (1975) 1427.

26 A.L. Balch, Y.W. Chan, R.J. Cheng, G.N. LaMar, L. Latos-Grazynski, and M.W. Renner, J. Am. Chem. Soc., 105 (1984) 7779.

27 M.J. Nappa and C.A. Tolman Inorg. Chem., 24 (1985) 4711.

28 D.N. Hendrickson, M.G. Kinnaird, and K.S. Suslick, J. Am. Chem. Soc., 109 (1987) 1243.

29 C.L. Hill in: Activation and Functionalization of Alkanes, Wiley, New York, (1989) pp 243-279.

30 F. Ortega, Doctoral Dissertation, Georgetown Univ., (1982) p. 149.

31 R.G. Finke, B. Rapko, R.J. Saxton, P.J. Domaille, J. Am. Chem. Soc., 108 (1980) 2947.

32 C.A. Tolman, J.D. Druliner, M.J. Nappa, and N. Herron, in: C.L. Hill (Ed.) Activation and Functionalization of Alkanes, Wiley, New York, (1989) pp 344-355.

R.K. Grasselli and A.W. Sleight (Editors), *Structure-Activity and Selectivity Relationships in Heterogeneous Catalysis*
© 1991 Elsevier Science Publishers B.V., Amsterdam

THE OXIDATIVE COUPLING OF METHANE OVER Sm_2O_3 AND La_2O_3

S.J. KORF, J.G. VAN OMMEN and J.R.H. ROSS

Faculty of Chemical Technology, University of Twente, P.O. Box 217, 7500 AE Enschede, The Netherlands

ABSTRACT

A comparison has been made of the behaviour in the oxidative coupling of methane of Sm_2O_3 in the cubic modification with that of Sm_2O_3 in the monoclinic form and with that of La_2O_3. Particular attention has been paid to the effect of residence time (W/F) on the conversion of methane and oxygen and on the rates of production of the various products. It is shown that the cubic and monoclinic forms of Sm_2O_3 have very different catalytic behaviour and that the change in surface area occurring during the phase change cannot account for this change. In contrast, experiments with La_2O_3 calcined at various temperatures showed that the rate of C_2 production was directly proportional to the surface area. The effect of residence time on the formation of products was studied for all three catalysts and the results are compared and contrasted with equivalent results for the Li/MgO and Ba/CaO systems. The primary C_2 product in all cases is ethane. The direct oxidation of methane (or of an intermediate methyl or CH_xO species) to give CO and CO_2 (i.e. a parallel route) is significant in the case of monoclinic Sm_2O_3 but is less important in the case of cubic Sm_2O_3; the activation energies for the production of ethane and CO_x are significantly different and the values also depend on the crystal modification. La_2O_3 and Ba/CaO also give substantial production of CO and CO_2 by a parallel route whereas Li/MgO has a high initial selectivity to ethane. A scheme for the reactions occurring on the various catalysts is proposed.

INTRODUCTION

The selective oxidation of methane to C_2 hydrocarbons has been studied extensively since Keller and Bhasin [1] showed that a range of oxides are selective for the formation of C_2 hydrocarbons. It has been found that the best selectivities are obtained with alkali-metal oxides associated with alkaline-earth, lead, manganese or rare-earth oxides [2-8]. Otsuka and his coworkers [4] have investigated the methane coupling reaction over various of the oxides of the rare earths. The best results were obtained with Sm_2O_3 and Dy_2O_3.

Sm_2O_3 can adopt two crystalline structures, cubic and monoclinic [9]. The monoclinic structure is the stable modification while the cubic structure is meta-stable, being that form present at temperatures up to about 900°C; above this temperature, an irreversible transition into the monoclinic structure takes place. We have shown that the cubic form of Sm_2O_3 is active for the oxidative coupling of methane but that transition to the high-temperature monoclinic form brings about a reduction of C_2 production [5]. Work in our laboratory on the doping of Sm_2O_3 has shown that the addition of Na, Li or Ca species has a major effect on the activity and selectivity of the resultant materials for the oxidative coupling of methane [5].

It was shown that compounds of Li and Na accelerate the phase transition of Sm_2O_3 from cubic to monoclinic while CaO stabilises the cubic structure [5,10].

The present paper presents an extension of this work with the aim of gaining a fuller understanding of the effect of the phase change in Sm_2O_3, the question being whether or not the change in surface area associated with the phase change is entirely responsible for the differences in catalytic properties. A comparison is made with results for La_2O_3 calcined at different temperatures as this is a material which does not exhibit a phase transition. In order to gain a better understanding of the relative importance of the parallel and consecutive reactions in the formation of CO_x (i.e. CO and CO_2) over these catalysts (see Figure 1) and to explain the

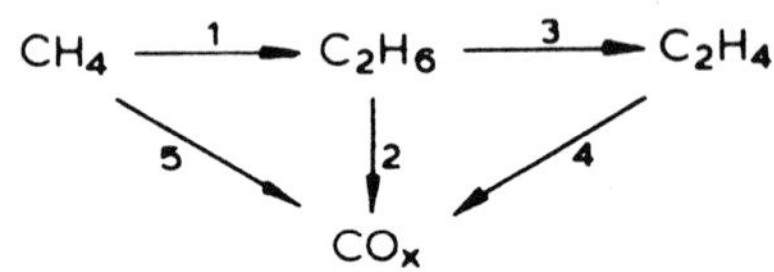

Figure 1: General reaction scheme for the oxidative coupling of CH_4 to C_2H_6 and C_2H_4.

different product distributions obtained, the effect of residence time on the rates of formation of product and hence the initial product selectivities have been studied for both the cubic and monoclinic forms of Sm_2O_3 and for La_2O_3; the results are compared with results previously obtained for Li/MgO and Ba/CaO materials [11] and a reaction scheme is proposed which is consistent with the results presented.

EXPERIMENTAL

Catalysts

The rare earth oxides were standard commercial samples of the oxides (Serva) and were calcined prior to use in air at 850°C and 1000°C for Sm_2O_3 and at temperatures ranging from 700°C to 1300°C for La_2O_3. The surface areas were determined using Ar adsorption at 78 K.

Catalytic Experiments

The catalytic experiments were carried out in a quartz reactor with an internal diameter of 4 mm. The gas feed consisted of methane (0.67 bar), O_2 (0.07 bar) and He (0.26 bar). The gases were analyzed by gas chromatography [12]. The effect of W/F was studied with a constant value of flow rate, F, of 1.67 $cm^3(STP)s^{-1}$ by variation of catalyst weight, W. For the measurements to show the effect of the temperature of pretreatment of Sm_2O_3 and La_2O_3, the flow rate of 1.67 $cm^3(STP)s^{-1}$ was also used. For the ageing experiment using La_2O_3 (T_c = 850°C), the catalyst weight W was 0.040 g and the flow rate was also 1.67 $cm^3(STP)s^{-1}$.

RESULTS AND DISCUSSION

The Effect of the Residence Time on Product Concentrations and Selectivities

Figures 2, 3 and 4 show the product concentrations (mol%) as a function of W/F for Sm_2O_3 (cubic), Sm_2O_3 (monoclinic) and La_2O_3, respectively, for a reaction temperature, T_R, of 670°C. Figures 5, 6 and 7 show the corresponding O_2 conversions, C_2 selectivities and yields and the (integral) selectivities to the various products. The temperature of 670°C was chosen partly because no reaction was detectable in the empty reactor at this temperature and partly because the conversions of oxygen for all the samples were relatively low. It can be seen from the figures that C_2H_6, CO and

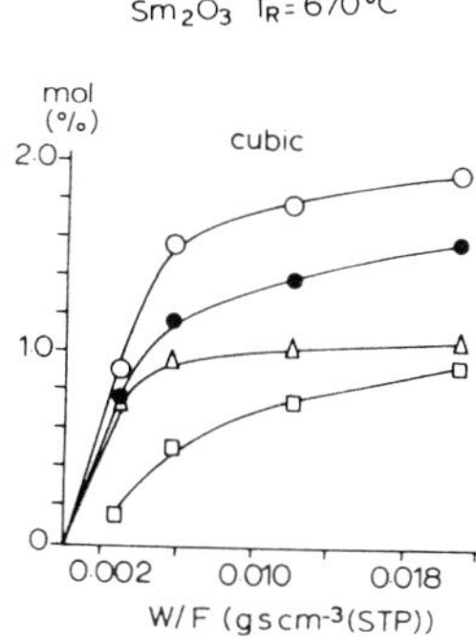

Figure 2:
The product concentrations (mol%) as a function of W/F for cubic Sm_2O_3 at 670°C.

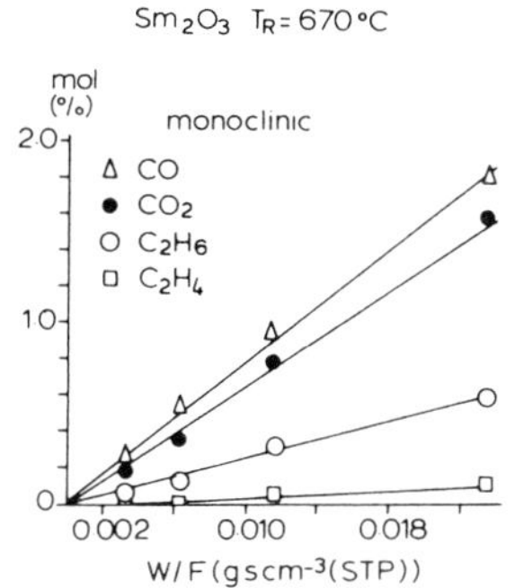

Figure 3:
The product concentrations (mol%) as a function of W/F for monoclinic Sm_2O_3 at 670°C.

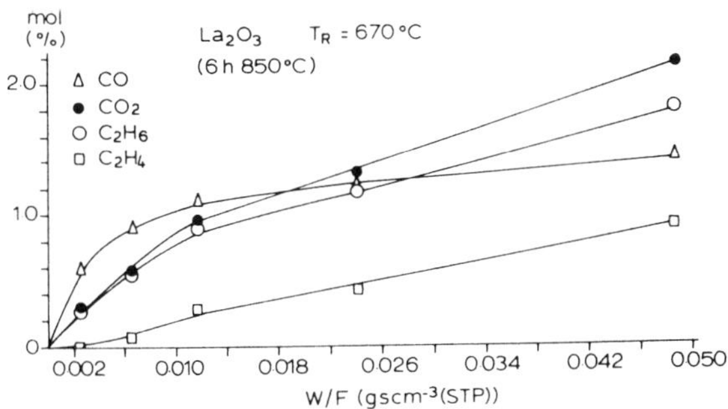

Figure 4: The product concentrations (mol%) as a function of W/F for La_2O_3 (T_c = 850°C) at 670°C.

CO_2 are all initial products for each of the catalysts but that C_2H_4 appears to be a secondary product. The product distributions for the different catalysts are different, the selectivity to C_2H_6 for the cubic modification of Sm_2O_3 being the highest. The initial rates of formation of each of the products (mol s^{-1} g^{-1}) have been calculated from the data of Figures 2-4 and are shown in Table 1, where they are compared with equivalent results for the Li/MgO and Ba/CaO catalysts calculated from data presented previously [11]. Using the surface areas measured for each of the

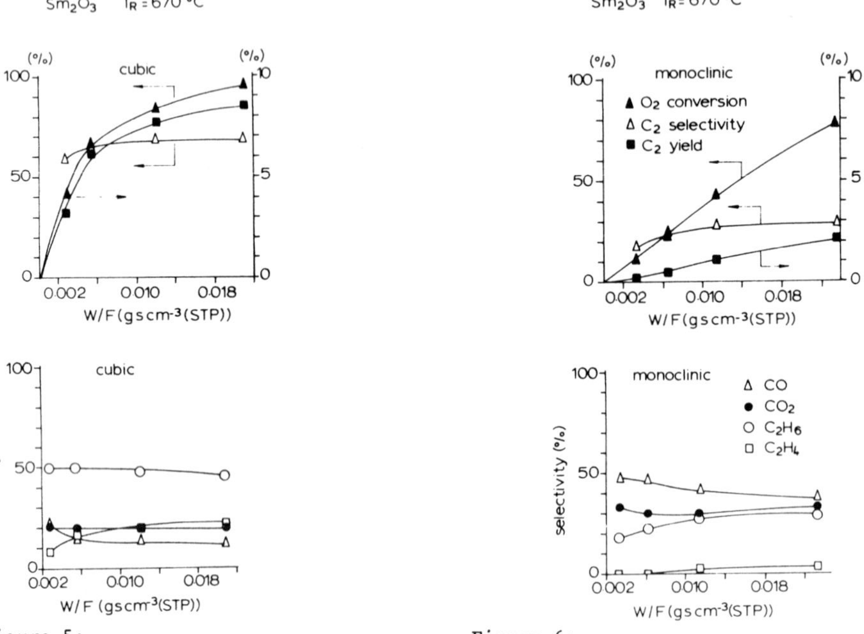

Figure 5:
Product selectivity, oxygen conversion and C_2 yield as a function of W/F at 670°C for cubic Sm_2O_3.

Figure 6:
Product selectivity, oxygen conversion and C_2 yield as a function of W/F at 670°C for monoclinic Sm_2O_3.

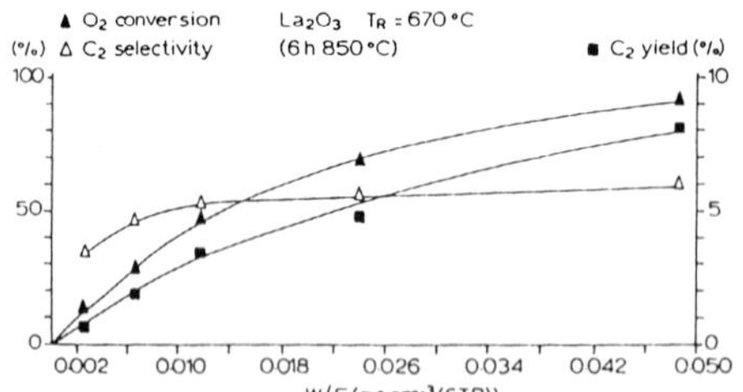

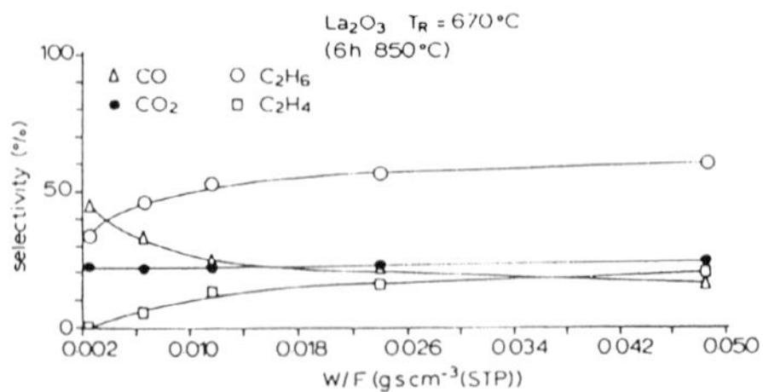

Figure 7: Product selectivity, oxygen conversion and C_2 yield as a function of W/F at 670°C for La_2O_3 (T_c = 850°C).

samples, which are also given in Table 1, the data have also been recalculated in units of mol s^{-1} m^{-2} and these values are also presented. The last column in Table 1 gives the initial selectivities to the various products, these having been calculated from the initial rates of formation of the various products.

Table 1 The initial reaction rates and initial selectivities for Sm_2O_3 (cubic), Sm_2O_3 (monoclinic) and La_2O_3 at 670°C and of Ba/CaO and Li/MgO at 710°C.

Catalyst	Area m^2g^{-1}	Product	Initial Rate 10^{-5} $mol.s^{-1}.g^{-1}$	Specific Rate 10^{-5} $mol.s^{-1}.m^{-2}$	Initial Selectivity %
Sm_2O_3 (Cubic)	6.6	C_2H_6	26.3	4.0	55.0
		C_2H_4	0	0	0
		CO	10.8	1.6	22.5
		CO_2	10.8	1.6	22.5
Sm_2O_3 (Monoclinic)	2.7	C_2H_6	1.3	0.5	18.8
		C_2H_4	0	0	0
		CO	3.3	1.2	47.8
		CO_2	2.3	0.9	33.3
La_2O_3	2.4	C_2H_6	9.1	3.8	37.4
		C_2H_4	0	0	0
		CO	10.1	4.2	41.6
		CO_2	5.1	2.1	21.0
Ba/CaO	2.0	C_2H_6	0.94	0.47	11.7
		C_2H_4	0	0	0
		CO	7.89	3.95	74.3
		CO_2	1.09	0.55	13.8
Li/MgO	0.9	C_2H_6	0.87	0.97	75.5
		C_2H_4	0	0	0
		CO	0	0	0
		CO_2	0.29	0.32	24.5

It can be seen from the results for Sm_2O_3 in Table 1 that the initial rates of formation of each of the products per gram of catalyst is greater for the cubic modification than for the monoclinic modification and that the selectivity to C_2 products of the former structure is also superior to that of the latter. That the difference between the samples cannot be attributed to differences in surface areas can be seen clearly from the data of the third column which shows that the specific initial rate of formation of the C_2 products is very much higher over the cubic structure than over the monoclinic structure; the initial rates of formation per unit area of both CO and CO_2 do not differ so much for the two modifications. Preliminary experiments were carried out to examine the effect of temperature on the initial rates of production of the various initial products and the activation energies calculated from the results are given in Table 2; this shows that there is a significant difference between the values of the activation energies for the

Table 2 Activation energies for the formation of products over cubic and monoclinic Sm_2O_3

Catalyst	Activation Energy for Formation of Product / kJ mol^{-1}		
	C_2H_6	CO	CO2
Cubic Sm_2O_3	120	32	26
Monoclinic Sm_2O_3	275	146	112

formation of ethane and of CO and CO_2 and that the change in crystal modification also has a significant effect on the values found.

We are thus tempted to speculate that reaction to form ethane is in some way favoured over the reaction to form the carbon oxides by the surface structure of the cubic modification and that the two reactions are thus independent of one another. A possible and rather speculative reaction scheme is shown in Fig. 8. In this, we show the formation of methyl groups on the surface by reaction with molecular oxygen to be the rate determining step in the reaction; this is in agreement with the kinetic measurements reported by Otsuka [13]. The methyl species are then shown to combine, in a rapid process, to form gaseous ethane, the primary product in the reaction; this ethane must readsorb on the surface, in a step not shown in Fig. 8, to form the secondary product, ethylene. The CO_x is shown as being formed by interaction of CH_xO species (formed from adsorbed methane and an adsorbed oxygen species) with surface oxygen rather than directly from methane or from surface CH_3 species; the latter reactions, which we cannot completely exclude (see below), are shown with dotted arrows in Fig.8. In the same way, the formation of ethane via gas-phase CH_3 radicals is shown via dotted arrows. Although we cannot definitely exclude a mechanism involving gaseous radicals, we feel that it is less likely than the surface combination reaction: a three-body collision would be necessary to remove the energy of the gaseous combination process, energy which is easily accomodated by the surface in a surface desorption process. The structure of the cubic modification of Sm_2O_3 is less closely packed than that of the monoclinic form and may therefore present a more favourable geometry for the approach of two adsorbed methyl groups to form ethane than that found on the monoclinic modification. The change in surface geometry may also give a change in the nature of the oxygen surface species responsible for the formation of the CH_xO intermediates and hence a change in the kinetics of this step. The scheme which we propose is also consistent with the observation that the activation energies for the formation of C_2 products and CO_x are different: these molecules are shown as being formed by completely independent routes. The alternative route to CO_x via CH_3 species is only possible if the reaction of CH_3 to give CO_x were also slow and the allover reaction to give CO_x had an apparent activation energy which had contributions from both steps in the reaction; if the formation of adsorbed CH_3 species were the rate-determining step, the activation energies for the production of ethane and CO_x would be the same, and it is difficult to conceive of a situation in which the formation of CO_x from CH_3 species could be rate determining without a contribution from the rate of formation of the CH_3 species. The route via CH_xO species is also favoured by our observation, using FTIR techniques [14], of CH_xO species on the surface of Sm_2O_3 catalysts under reaction conditions.

The initial specific rate of formation (i.e. per m^2) of C_2 products on La_2O_3 shown in Table 1 is of the same order of magnitude as that on the cubic form of Sm_2O_3 but the rates of formation of CO and CO_2 are higher, with the consequence that the initial selectivity is lower. Ethane is again the primary C_2 product. All the product concentrations rise almost linearly with time and this has the consequence that measurement at relatively high residence times can still be regarded as being approximately differential (see below). The initial rates of formation of all the products over Li/MgO and Ba/CaO at the reaction temperature of 710°C, which are also shown in Table 1 for comparison purposes, are much lower than with the rare earth oxides at the slightly lower reaction temperature of 670°C. However, the selectivity to C_2 products is much higher for the Li/MgO material than for the Ba/CaO material; the parallel route to CO_x over this catalyst would thus seem to occur to a minor extent. If the scheme of Fig.8 also applies to the Li/MgO catalyst, we can argue that

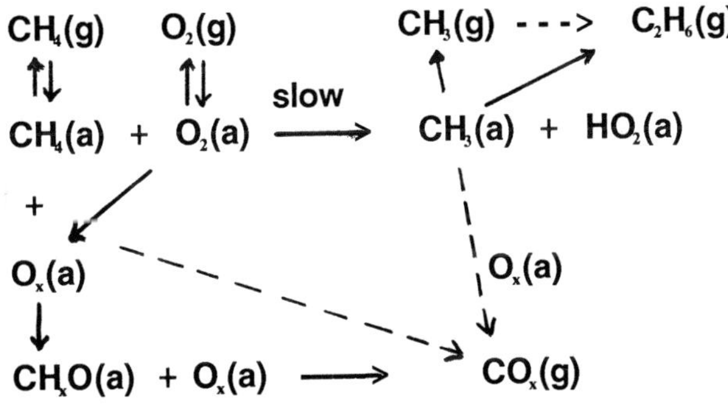

Figure 8: Possible reaction scheme for the formation of ethane and CO_x in the oxidative coupling of methane.

the surface geometry of this material favours the surface combination of CH_3 radicals as compared to the sequence of reactions to give CO_x species. We have shown elesewhere

[15] that, as with Sm_2O_3, the reaction of CH_4 with O_2 is also rate determining over this catalyst.

The results of Figures 5-7 show clearly that ethylene is formed as a product only after longer residence times. We therefore conclude that ethylene is a secondary product of the reaction over the rare earth oxides studied; this is in agreement with our earlier work using both plug flow reactors and a stirred tank reactor [11,16,17]. The selectivities to the C_2 products also increase with increasing residence time. The consecutive oxidation of the ethane and ethylene formed might be expected to give rise to extra CO_x and hence to a lower selectivity with increase in W/F; however, the CO_x formation reactions will become less and less important as the partial pressure of oxygen decreases at higher values of W/F and the coupling reaction will also have a more significant role. These changes therefore outweigh the contribution from the consecutive reaction.

Effect of T_c on the Surface Area and Catalytic Properties of La_2O_3

Table 3 shows the effect on the the surface area and catalytic properties

Table 3 The effect of calcination temperature of La_2O_3. T_R = 670°C.

T_c /°C	area /m^2g^{-1}	W /g	a m^2	Conversions /%		Selectivities /%		Y_2/%	R_{c2} 10^{-6} $mols^{-1}m^{-2}$
				CH_4	O_2	C_2H_x	CO_x		
700	3.4	0.040	0.136	11.8	84	60	40	7.0	23
850	2.4	0.040	0.096	8.9	71	55	45	4.9	24
1000	1.6	0.040	0.064	6.6	57	49	51	3.2	24
1150	0.6	0.040	0.024	4.6	38	34	66	1.6	31
1300	0.3	0.040	0.012	2.4	22	20	80	0.5	18
1000	1.6	0.085	0.136	12.3	92	60	40	7.3	24
1150	0.6	0.160	0.096	9.4	77	53	47	5.0	25

of La_2O_3 of increasing the calcination temperature, T_c, from 700 to 1300 °C. Increasing T_c gave a substantial decrease in the surface area of the sample, from 3.4 to 0.3 m^2 g^{-1}; X-ray diffraction showed that there was no change in the phase structure over this whole temperature range. It was therefore of interest to see whether or not the high-temperature treatment gave rise to any change in the catalytic behaviour other than that which could be ascribed to the decrease in area. The results of Table 3 show that although the methane conversion was decreased substantially by increase in calcination temperature, there was no appreciable change in the specific rate of formation of C_2 products. (The specific rates were calculated assuming that the results can be treated as being differential; see above). Compared with the results for the samples calcined at high values of T_c, samples calcined at low values of T_c resulted in relatively high C_2 selectivities and C_2 yields. However, it can be seen that the specific rate of production of C_2 products (i.e. based on R_{C2}) is the same for each sample of this catalyst. The higher C_2 selectivities resulting from lower calcination temperatures (with correspondingly higher surface areas) shown for the La_2O_3 catalysts (Table 3) can be explained in terms of changes of oxygen partial pressures as was done above to explain the changes of selectivity with W/F.

To confirm that the only differences were due to surface area and that the temperature of calcination had no secondary effect, two experiments were carried out with different weights of catalyst chosen so that the total area (a) of catalyst was the same as those of samples calcined at different temperatures: compare the last two rows of data of Table 3 with the first two rows of data. The conversions of methane and oxygen and the selectivities towards C_2 and CO_x production were the same, within the experimental error, for the different cases. We therefore conclude that the surface area is the only important factor determining the rate of C_2 production of the catalyst in the case of La_2O_3.

A further experiment was carried out with the La_2O_3 sample calcined at 850°C; the sample was exposed to the standard reaction mixture for 160 h at a reaction temperature of 770°C. The conversion of the methane during this period fell by some 20% and the selectivity towards C_2 products fell by some 6%. These changes are entirely consistent with a decrease of area, due to sintering, of approximately 20%. The fact that other authors [18] have claimed that La_2O_3 is totally stable under these conditions could be attributed to the fact that they worked with total oxygen conversion; this would mean that not all the surface available in the reactor was in use for the coupling reaction, with the consequence that the effect of sintering would not be visible.

We have shown above that ethylene is not a primary product of the coupling reaction with both Sm_2O_3 and La_2O_3. This result is in contrast with results reported by Hutchings et al. [19] who argued that both ethylene and ethane are primary products for Sm_2O_3 and La_2O_3. In their experiments, W/F was increased by decreasing the gas flow. As we have shown elsewhere [20], a decrease in rate of flow can result in a change of the flow pattern throughout the reactor, from plug flow at high values of F to ideal mixing behaviour at low values of F; back-mixing leads to a greater chance of further oxidation of the products formed. At a given residence time, a higher selectivity of C_2H_4 and CO_x will thus be found in a situation where back-mixing can occur, these products being formed from the primary product, ethane. The measurements of Hutchings et al. are almost certainly influenced by this effect. Furthermore, Hutchings et al. used relatively high residence times and these also favour the formation of ethylene (see Figs. 2-7).

CONCLUSIONS

1. The specific rate for the formation of C_2 products of Sm_2O_3 is much higher for the cubic phase than for the monoclinic phase.
2. Sintering causes deactivation of La_2O_3 catalysts. The surface area is the only factor which determines the rate of C_2 production of the catalyst.
3. The direct oxidation of methane or some intermediate species to give CO and CO_2 plays a significant role over monoclinic Sm_2O_3 but is less important for cubic Sm_2O_3. The production of CO_x probably occurs via different surface species than does that of the C_2 products.

ACKNOWLEDGEMENTS

S.J.K. thanks the Dutch Foundation for Scientific Research for financial support. We also thank the Non-Nuclear Energy programme of the European Community for partial support of the work (Contract No. EN3C-039-NL (GDF)).

REFERENCES

1 G.E. Keller and M.M. Bhasin, J. Catal., 73 (1982) 9.
2 T. Ito, J-X. Wang, C-H. Lin and J.H. Lunsford, J. Amer. Chem. Soc., 107 (1985) 5062.
3 S.J. Korf, J.A. Roos, N.A. de Bruijn, J.G. van Ommen and J.R.H. Ross, Catal. Today 2 (1988) 535.
4 K. Otsuka, K. Jinno and A. Morikawa, J. Catal., 100 (1986) 353.
5 S.J. Korf, J.A. Roos, J.M. Diphoorn, R.H.J. Veehof, J.G. van Ommen and J.R.H. Ross, Catal. Today, 4 (1989) 279.
6 C.A. Jones, J.J. Leonard and J.A. Sofranko, J. Catal., 103 (1987) 311.
7 W. Bytyn and M. Baerns, Appl. Catal., 28 (1986) 199.
8 C. Miradatos, G.A. Martin, J.C. Bertolini and J. Saint-Just, Catal. Today, 4 (1989) 301.
9 M.P. Rosynek, Catal. Rev. Sci. Eng. 16 (1977) 111.
10 S.J. Korf, J.A. Roos and J.M. Diphoorn, Preprints 196th ACS National Meeting, 33(3) (1988) 437.
11 S.J. Korf, J.A. Roos, J.W.H.C. Derksen, J.A. Vreeman, J.G. van Ommen and J.R.H. Ross, Appl. Catal., 59 (1990) 291.
12 J.A. Roos, A.G. Bakker, H. Bosch, J.G. van Ommen and J.R.H. Ross, Catal. Today, 1 (1987) 133.
13 K. Otsuka and K. Jinno, Inorg. Chim. Act., 121 (1986) 237.
14 J.G. van Ommen, G.J.M. Weierink, S.J. Korf, H.S. Swaan and J.R.H. Ross, to be published.
15 J.A. Roos, S.J. Korf, R.H.J. Veehof, J.G. van Ommen and J.R.H. Ross, Appl. Catal., 52 (1989) 131.
16 J.A. Roos, S.J. Korf, R.H.J. Veehof, J.G. van Ommen and J.R.H. Ross, Appl. Catal., 52 (1989) 147.
17 S.J. Korf, J.A. Roos, J.A. Vreeman, J.W.H.C. Derksen, J.G. van Ommen and J.R.H. Ross, Catal. Today, 6 (1990) 417.
18 A. Kooh, J.-L. Dubois, H. Mimoun and C.J. Cameron, Catal. Today, 6 (1990) 453.
19 G.J. Hutchings, M.S. Scurrell, J.R. Woodhouse, Chem. Soc. Rev., 18 (1989) 251.
20 J.A. Roos, S.J. Korf, A.G. Bakker, N.A. de Bruijn, J.G. van Ommen and J.R.H. Ross "Methane Conversion", ed. D.M. Bibby, C.D. Chang, R.F. Howe and S. Yurchak, Stud. Surf. Sci. Catal., 36 (1987) 535.

R.K. Grasselli and A.W. Sleight (Editors), *Structure-Activity and Selectivity Relationships in Heterogeneous Catalysis*
© 1991 Elsevier Science Publishers B.V., Amsterdam

EFFECTS OF METAL PARTICLE SIZE AND CARBON FOULING ON THE RATE OF HEPTANE OXIDATION OVER PLATINUM

R.F. HICKS, R.G. LEE, W.J. HAN AND A.B. KOOH

Chemical Engineering Department, University of California, Los Angeles, CA, 90024-1592

ABSTRACT

A series of platinum catalysts, with metal dispersions ranging from 10 to 81%, have been tested for heptane oxidation in 5% excess oxygen and at temperatures between 90 and 140°C. The rate of reaction declines exponentially with time because carbon fouls the metal surface. The deactivation rate is first order in the concentration of active sites. The rate of coke formation depends on the metal particle size, the support composition, and the density of metal particles on the support. Small particles are inactive for carbon deposition. On large particles, the amount of carbon deposited per surface platinum atom is higher on alumina than on zirconia, and increases as the number of particles decreases. Most of this carbon resides on the support. The rate of catalyst deactivation follows a trend opposite to that of coke formation. Small particles deactivate very slowly. On large particles, the deactivation rate is higher on zirconia than on alumina, and increases as the number of particles increases. Evidently, the faster the metal transfers carbon to the support, the slower it deactivates. The turnover frequency for heptane oxidation on sites not fouled by carbon (A sites) depends on metal particle size. On average, large crystallites are 23 times more active than small ones. At long reaction times, the turnover frequency for heptane oxidation is obscured by carbon fouling, and the rate appears insensitive to particle size.

INTRODUCTION

Catalytic oxidation is used extensively to reduce hydrocarbon emissions from automobiles (1,2) and industrial processes (3). However, there have been few published studies of the catalytic oxidation of hydrocarbons other than methane. Prior research on C_2-C_7 alkane oxidation has shown that the rate of reaction depends on catalyst structure (4-8). Platinum is more active than palladium, and large metal crystallites are more active than small ones. Nevertheless, it is difficult to tell the magnitude of these effects, because the rates were often measured at high conversion, where transport resistances may have influenced the results. Also, the effect of reaction time on the rate was not considered.

In this paper, we report the results of our study of heptane oxidation over supported platinum. Samples were prepared using alumina and zirconia as supports and with metal dispersions ranging from 10 to 81%. The intrinsic rate of reaction was measured as a function of time at low temperature in 5% excess oxygen and at low conversion. We found that the rates fell rapidly with time because carbon fouls the catalyst. The effect of the metal particle size on the rate of heptane oxidation and the rate of deactivation is described below. A more complete analysis of our results is presented elsewhere (9).

EXPERIMENTAL

The samples used in this study and their method of preparation are shown in Table 1. The supports were Degussa, flame-synthesized aluminum oxide "C" and zirconia. The alumina was calcined at 1000°C for 24 h and had a surface area of 83 m^2/g. The zirconia was calcined at 600°C for 24 h and had a surface area of 40 m^2/g. The metal was deposited by ion exchange on sample *a* and by incipient-wetness impregnation on samples *b* through *f*. The samples were dried at 125°C for 2 h, then calcined as shown in the Table. Metal loadings were determined by inductively coupled plasma emission spectroscopy. The platinum dispersion was measured by hydrogen titration of preadsorbed oxygen at 25°C, assuming an adsorption stoichiometry of 1.5 H_2 per PtO_s (10).

TABLE 1

Physical characteristics of the platinum catalysts.

Sample designation	Support	Metal salt	Calcined 2 h in air at (°C)	Metal loading (%)	Dispersion (%)
a	ZrO_2	$Pt(NH_3)_4Cl_2$	500	0.4	81
b	ZrO_2	$Pt(NH_3)_4Cl_2$	500	0.3	56
c	ZrO_2	$Pt(NH_3)_4Cl_2$	500	5.0	19
d	Al_2O_3	$Pt(NH_3)_4Cl_2$	500	0.3	58
e	Al_2O_3	H_2PtCl_6	600	0.8	13
f	Al_2O_3	H_2PtCl_6	600	5.0	10

The rate of heptane oxidation was determined in a fixed-bed microreactor, equipped with on-line gas chromatography. Carbon dioxide was the only reaction product. The method of testing the catalysts was as follows. Between 0.1 to 0.7 g of sample (32-60 mesh pellets) was loaded into a 4 mm-I.D. glass tube, oxidized in 50 cm^3/min oxygen at 500°C for 20 min, and reduced in 50 cm^3/min hydrogen at 300°C for 20 min. Then the sample was cooled to the reaction temperature and the feed switched to 25 Torr heptane, 282 Torr oxygen and 918 Torr helium at 235 cm^3/min (1 Torr = 133 N/m^2). The temperature and the amount of catalyst were adjusted to keep the conversion below 1%. The reaction was continued for 4 to 16 h. At the end of this period, the feed was switched to 30 cm^3/min helium, and the reactor was cooled to 60°C. The number of metal atoms exposed was determined by hydrogen-oxygen titration and carbon monoxide adsorption. On some samples, the amount of carbon deposited during reaction was determined by recording the carbon dioxide evolution upon heating in oxygen. In this case, the sample was cooled to 50°C in 50 cm^3/min helium. Then 22 cm^3/min oxygen and 120 cm^3/min helium was introduced, and the sample was heated at 5°C/min to 500°C. The amount of carbon dioxide evolved was measured every 3 min with the gas chromatograph.

RESULTS

Shown in Fig. 1 is the effect of time on the turnover frequencies for heptane oxidation over five platinum catalysts. The turnover frequencies decrease exponentially with time.

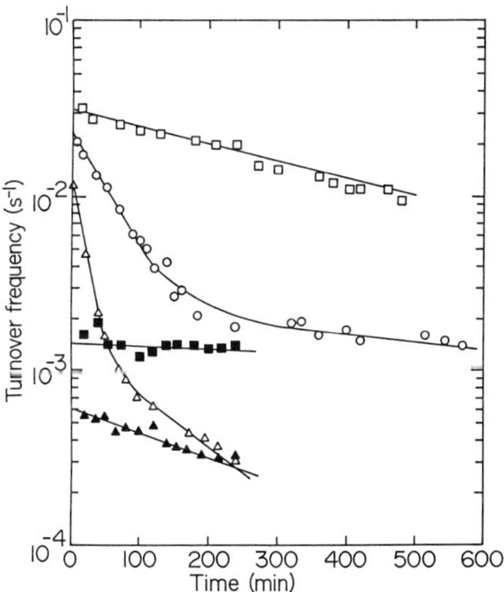

Fig. 1. The dependence of the turnover frequency on time at 90°C for samples *b* (■), *c* (△), *d* (▲), *e* (□), and *f* (○).

Over samples *b*, *d* and *e*, the slope of the line of log (turnover frequency) versus time remains constant over the whole run. However, over samples *c* and *f*, the slope of the line drops after a certain period of time. The decay of the rate with time can be attributed to poisoning of the active sites.

These sites are lost at a rate proportional to the number present, ie., first-order deactivation (11,12):

$$\frac{d\theta}{dt} = -k\theta, \quad [1]$$

which upon integration yields,

$$\theta = \exp(-kt), \quad [2]$$

where θ is the fraction of active sites not poisoned, t is time (s), and k is the apparent rate constant for deactivation (s^{-1}). The change in the slope of the line of log (turnover frequency) versus time, observed for some samples, suggests that there are two sites of high and low intrinsic activity:

$$TOF_{obs} = TOF_A\theta_A + TOF_B\theta_B, \quad [3]$$

where TOF is the turnover frequency (s^{-1}), and the subscripts obs, A and B refer to observed, A site and B site, respectively. The A site may be thought of as a site on the "clean" platinum surface, while the B site may be thought of as a site on the poisoned platinum surface. Substitution of Eq. 2 into Eq. 3 yields:

$$TOF_{obs} = TOF_A\exp(-k_At) + TOF_B\exp(-k_Bt). \quad [4]$$

The curves drawn through the data points in Fig. 1 are the best fit of Eq. 4 to the results.

Shown in Table 2 are the values of the turnover frequencies and rate constants for deactivation of the A and B sites, which are obtained from fitting Eq. 4 to the rate data. The turnover frequency of the A site varies over a wide range, from 0.06 $\times 10^{-2}$ s^{-1} for sample *d* to 3.2 $\times 10^{-2}$ s^{-1} for sample *e*. The change in the specific activity of the A site correlates with the platinum particle size. Samples containing small metal particles, ie., dispersions greater than 50%, exhibit low activity, while samples containing large metal particles, ie., dispersions less than 20%, exhibit high activity.

The influence of temperature on the reaction rate was determined on samples *b* and *f*. The apparent activations energies for heptane oxidation on samples *b* and *f* are 20 ± 4 and 19 ± 2.5 kcal/mole (1 kcal = 4.186 kJ) (9). These activation energies are the same within experimental error. This indicates that the large difference in specific activity of the samples is not due to a difference in the reaction mechanism. Instead, the specific rate of heptane oxidation on the A site increases with platinum particle size, because the large particles contain a greater fraction of active sites than the small ones.

TABLE 2

Rates of heptane oxidation and deactivation over the platinum catalysts.

Sample designation	Composition	Initial dispersion (%)	Turnover frequency ($x10^{-2}$ s^{-1}) at 90°C		Deactivation rate ($x10^{-4}$ s^{-1}) at 90°C	
			TOF_A	TOF_B	k_A	k_B
a	0.4% Pt/ZrO_2	81	0.07[a]	0.05[a]	0.41[a]	-0.02[a]
b	0.3% Pt/ZrO_2	56	0.14	0.00	0.07	
c	5.0% Pt/ZrO_2	19	1.10	0.14	11.00	1.10
d	0.3% Pt/Al_2O_3	58	0.06	0.00	0.53	
e	0.8% Pt/Al_2O_3	13	3.20	0.00	0.38	
f	5.0% Pt/Al_2O_3	10	2.10	0.24	3.20	0.16

[a]Corrected from a reaction temperature of 140°C to 90°C using an activation energy of 19 kcal/mole for the turnover frequency and 11 kcal/mole for the deactivation rate (9).

The turnover frequency of the B site does not change nearly as much as the turnover frequency of the A site. There is a fivefold increase in the B-site activity as the platinum dispersion decreases from 81 to 10%. Also, the turnover frequency of the B site is approximately equal to the turnover frequency of the A site on the small platinum particles. At long reaction times, the observed rate on small particles is close to the turnover frequency of the A site, while the observed rate on large particles is close to the turnover frequency of the B site. Since these turnover frequencies are nearly equal, the reaction rate at long times appears to be insensitive to the platinum particle size.

The rate constants for deactivation of the A and B sites vary widely over samples *a* through *f*. On the samples with platinum dispersion greater than 50%, the rate constant for deactivation of the A site is low. Whereas, on the samples with platinum dispersions less than 20%, the rate constant for deactivation of the A site depends on the metal loading and support composition. The deactivation rate constant increases 8.4 times as the amount of metal on the alumina increases from 0.8 to 5.0% at constant dispersion (samples *e* and *f*). For a metal loading of 5%, switching from an alumina to a zirconia support increases the deactivation rate constant by 3 times (samples *f* and *c*). These data indicate that the deactivation rate is sensitive to the platinum particle size, the nature of the support and the amount of platinum in contact with it.

Carbon deposits on the platinum samples during heptane oxidation. Shown in Table 3 is the amount of carbon deposited as determined by temperature-programmed oxidation before and after reaction. Before reaction, the samples were oxidized at 500°C and reduced at 300°C three times before recording the TPO spectrum. The reaction temperature was 90°C, and the reaction time was 4 h for the platinum catalysts and 2 h for the supports. The alumina and zirconia supports accumulate a significant quantity of carbon, but do not convert any heptane to carbon dioxide during exposure to heptane oxidation. The amount of carbon oxidized off the supports after reaction equals 7.5 $\times 10^{-5}$ mole/g. The extent of carbon deposition on the platinum catalysts depends on the metal particle size. On samples *b* and *d*, with dispersions of 56 and 58%, the amount of carbon deposited during reaction is not much greater than that deposited on the supports. By contrast, on samples *c*, *e* and *f*, with dispersions between 10 and 19%, the carbon deposited during reaction is ten times greater than that deposited on the supports. Evidently, small platinum crystallites are inactive for the conversion of heptane into coke. As shown above, small crystallites are also inactive for the conversion of heptane into carbon dioxide.

TABLE 3

The amount of carbon deposited on the platinum catalysts.

Sample designation	Composition	Initial dispersion (%)	Metal exposed ($\times 10^{-5}$ moles/g)	Carbon oxidized ($\times 10^{-4}$ mole/g)	
				Before reaction	After reaction
b	0.3% Pt/ZrO_2	56	0.9	0.3	2.4
c	5.0% Pt/ZrO_2	19	4.9	1.1	6.4
d	0.3% Pt/Al_2O_3	58	0.9	0.2	1.0
e	0.8% Pt/Al_2O_3	13	0.5	0.0	8.9
f	5.0% Pt/Al_2O_3	10	2.6	1.1	11.6
	ZrO_2			0.3	0.8
	Al_2O_3			0.3	0.7

In Table 3, the moles of metal exposed on each sample is compared to the amount of carbon deposited. In every case, the carbon accumulated during reaction greatly exceeds the amount of metal exposed. Most of this carbon is probably on the support. On the three active platinum catalysts, samples *c*, *e* and *f*, the mole ratio of carbon to surface platinum equals 13, 178 and 45, respectively. The mole ratio of carbon to surface platinum is higher on Pt/Al_2O_3 than on Pt/ZrO_2, and it increases as the platinum loading goes down. These ratios may be compared to the rate constants for deactivation of the A sites, which are 11.00, 0.38 and 3.20 $\times 10^{-4}$ s^{-1}, respectively. These rate constants follow an inverse correlation with the mole ratio of carbon to surface platinum.

This correlation can be rationalized as follows. Carbon produced on the metal migrates to the support. The surface diffusion of carbon is consistent with an observed square-root dependence of the amount of carbon deposited on reaction time (9). If the rate of carbon migration to the support is high, as evidenced by a high mole ratio of carbon to platinum exposed, then deactivation is slow. Conversely, if the rate of carbon migration is slow, as evidenced by a low mole ratio of carbon to platinum exposed, then deactivation is fast. The rate of carbon migration appears to be faster on alumina than on zirconia, and increases as the number of platinum particles on the support decreases.

Titration of the catalysts after reaction indicates that the metal surfaces are covered by carbon and a small amount of oxygen (9). Between 30 and 100% of the carbon on the metal is displaced by hydrogen and carbon monoxide. The fraction displaced increases with decreasing particle size. After oxidation at 500°C and reduction at 300°C, the adsorption capacities of all the catalyst samples recover to their initial values. The number of sites which can be titrated with hydrogen, oxygen and carbon monoxide does not change with reaction time. This is true in spite of the fact that carbon continuously builds up throughout the run. The ability of hydrogen and carbon monoxide to displace the carbon from the platinum suggests that only one layer of carbon sits on the metal surface. This carbon is a small fraction of the total amount deposited on the catalyst. Most of the carbon migrates to the support. The distribution of carbon and oxygen on the platinum surface may depend on the particle size and the reaction time. However, this distribution could not be determined by the titration experiments, because uncontrolled amounts of heptane and oxygen were exposed to the catalyst upon switching from reaction to gas-pulsing (9).

DISCUSSION

The deactivation of the catalysts is most likely due to carbon fouling of the metal particles. First-order deactivation, Eq. 1, is consistent with the surface sites becoming progressively covered by some species during reaction (12). Titration of the samples after reaction reveals that the surface of the platinum crystallites is covered with carbon. Once the carbon has been removed by oxidation and reduction, the adsorption capacity of the platinum recovers to its initial value, and so does the rate of heptane oxidation.

The rate of carbon deposition depends on the metal particle size. Large platinum crystallites are active for converting heptane into coke, whereas small platinum crystallites are not. The rate of coke deposition during hydrocarbon reforming over platinum also increases with increasing particle size (13). However, both large and small particles deposit significant amounts of carbon during reforming. On the large crystallites, the rate of carbon accumulation per exposed metal atom depends on the support composition and the number of metal particles in contact with the support.

This is because coke formation is a self-poisoning reaction. Carbon remaining on the metal surface blocks the sites for converting heptane into coke. However, if the support contains sites for carbon adsorption, then the carbon can migrate to these sites, and free up the metal surface for further reaction. Decreasing the number of metal particles, increases the number of support adsorption sites relative to the metal atoms exposed, and increases the rate of carbon migration from the metal to the support.

The rate of deactivation of the catalysts also depends on metal particle size, support composition, and the number of metal particles in contact with the support. Small platinum crystallites are inactive for carbon deposition and consequently, deactivate slowly. Large platinum crystallites are active for carbon deposition, and in this case, the rate of deactivation depends on how fast the carbon migrates to the support. On sample *e*, the platinum deactivates slowly because the low concentrations of metal on the alumina promotes a rapid rate of carbon diffusion to the support. Conversely, on sample *c*, the platinum deactivates quickly because the high concentration of metal on the zirconia promotes a slow rate of carbon diffusion to the support. These results have important implications for the design of hydrocarbon oxidation catalysts. To improve the resistance of the catalyst to carbon fouling, the surface area of the support should be increased as much as possible, and the support should contain a maximum number of carbon adsorption sites.

The turnover frequency for heptane oxidation on the A site (ie., the surface not fouled by carbon) is moderately affected by the platinum particle size. Catalysts with dispersions below 20% are on average 23 times more active than catalysts with dispersions above 50%. These results qualitatively agree with other studies of hydrocarbon oxidation over platinum (4,5).

The rate of heptane oxidation and the rate of coke deposition show similar dependencies on platinum crystallite size. This suggests that these products come from common intermediates on the metal surface. The slow step in both mechanisms may be the reaction of adsorbed oxygen with adsorbed carbon. Both carbon and oxygen are present on the metal surface during reaction. The rate law for hydrocarbon oxidation is zero order in hydrocarbon and oxygen partial pressures, and is consistent with a surface reaction between carbon and oxygen (5,9). Also, very little carbon accumulates on the platinum catalysts when they are exposed to heptane at 90°C and no oxygen. Thus, the effect of particle size on the rates of heptane conversion to carbon dioxide and coke may be to alter the distribution of carbon and oxygen on the metal surface.

The dependence of the rate of heptane oxidation on the platinum particle size is obscured by carbon fouling. At long reaction times, the observed turnover frequency is close to the turnover frequency of the B site. The B-site activity is not very sensitive to particle size, increasing by a factor of five as the dispersion decreases from 81 to 10%. These results show that the deactivation kinetics must be accounted for in order to correctly interpret the effect of catalyst structure on the hydrocarbon oxidation rate.

ACKNOWLEDGEMENT

This work was supported by the National Science Foundation Engineering Research Center for Hazardous Substance Control at UCLA.

REFERENCES

1 J. Wei, Advan. Catal., 24 (1975) 57.
2 G. Kim, Ind. Eng. Chem. Prod. Res. Dev., 21 (1982) 267.
3 J.J. Spivey, Ind. Eng. Chem. Res., 26 (1987) 2165.
4 M.L. Carballo and E.E. Wolf, J. Catal., 53 (1978) 366.
5 Y.F. Yu Yao, Ind. Eng. Chem. Prod. Res. Dev., 19 (1980) 293.
6 V.A. Drozdov, P.G. Tsyrulnikov, V.V. Popovskii, N.N. Bulgakov, E.M. Moroz and T.G. Galeev, React. Kinet. Catal. Lett., 27 (1985) 425.
7 J. Volter, G. Lietz, H. Spindler and H. Lieske, J. Catal., 104 (1987) 375.
8 S.K. Gangwal, M.E. Mullins, J.J. Spivey and P.R. Caffrey, Appl. Catal., 36 (1988) 231.
9 A.B. Kooh, W.J. Han, R.G. Lee and R.F. Hicks, (J. Catal.) submitted for publication.
10 J.E. Benson and M. Boudart, J. Catal., 4 (1965) 704.
11 O. Levenspiel, Chemical Reaction Engineering", Wiley, New York, 1972.
12 J.B. Butt and E.E. Petersen, Activation, Deactivation and Poisoning of Catalysts, Academic Press, San Diego, 1988.
13 J. Barbier, Appl. Catal., 23 (1986) 225.

R.K. Grasselli and A.W. Sleight (Editors), *Structure-Activity and Selectivity Relationships in Heterogeneous Catalysis*
1991 Elsevier Science Publishers B.V., Amsterdam

Structure/Function Relations in Transition Metal Sulfide Catalysts

R. R. Chainelli, M. Daage
Exxon Research and Engineering
Annandale, NJ 08801

Introduction

In a recent article we have presented a brief review of the status of our current fundamental understanding of the TMS (Transition Metal Sulfide) based catalysts which will play an increasingly important role in the petroleum, synthetic fuels and chemical industries (1). The fundamental origins of the catalytic properties of the TMS are completely contained in the unsupported active TMS phases with the support playing a secondary role in enhancing properties required for industrial application. Many advances have been make in understanding the properties of TMS based catalysts, but some basic problems prevent further progress. The major impediment to further progress arises from the nature of MoS_2 itself, in its highly anisotropic structure (Figure 1). Here we review some of our recent research in understanding how the anisotropic crystal structure of MoS_2 is fundamentally related to its catalytic properties. The structural anisotropy of $MoS_2(WS_2)$ is a consequence of the chemical bonding. Within one layer, the structure can be viewed as a two-dimensional macromolecule. Each metal is bound to three metal atoms. Because the sulfur is so tightly bound, its interaction with the next layer of sulfur above it is extremely weak. This crates the "van der Waals" gap which is the main feature of interest in regard to intercalation and lubricity properties (2). Thus, although the basal planes (002 have been the general focus of studies in the vast intercalation literature, the "edge" planes (100) of the layered TMS become the focus to catalytic studies.

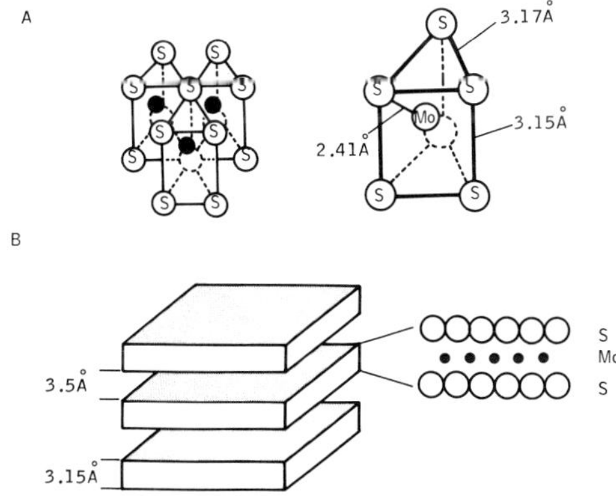

Figure 1. Schematic of the structure MoS_2.

The potential importance of MoS_2 edge planes in hydrotreating catalysts has long been recognized and some examples are cited above. Further evidence for the reactivity of the edge planes in MoS_2 can be found in the linear correlation between O_2 chemisorption and the HDS of dibenzothiopene (3). In general, HDS activity does not correlate to N_2 BET surface area measurement. This is because the basal plane area contributes to the total surface area but not to the catalytic activity. Therefore, MoS_2 catalysts make by a variety of preparative methods will have widely different edge to basal plane area ratios and only O_2 chemisorption will give a good correlation to activity. If the preparative method is constant, however, the basal plane are can be proportional to the edge area and a good correlation between total surface area and activity can be obtained (4).

A basic problem with O_2 chemisorption arises from the fact that O_2 chemisorbs corrosively, i.e., monolayer coverage at the edges are not achieved unless very mild conditions are used. If mild conditions are not used, oxidation penetrates deeper into the bulk and the amount of O_2 absorbed is the general only proportional to the number of edge sites (5). Furthermore, the presence of the promoter phase further complicates O_2 chemisorption studies and there is no general agreement as to its utility for supported catalysts. However, the technique has been widely applied, most recently to the supported WS_2 system using mile (low temperature) conditioners. (6) The most quantitatively detailed models of these catalysts systems come from a combination of activity data and chemisorption data. The recent geometric model of Kasztelan, et at., is a good example. (7,8) Using a geometrical model based on assumed shapes of MOS_2 and WS_2 sites. In their model, small slabs of MOS_2 consist of basal, edge or corner sites. By fitting activity curves with different shapes and numbers of theses sites, the authors concluded that hexagonal or rhombohederal single layered crystallites of about 10-20. A gave the best fit. Furthermore, they concluded that the edge sites were the active sites, that promotion occurred through enhancement of the quality of the sites and this promotion factor was calculated as being a factor of $4.4 \rightarrow 5.2$. Again, this procedure leads to a model which fits well with the edge-decoration model but does not give a detailed picture of the "promoted" sites. In order for this to be accomplished more detailed physical chemical and theoretical work is needed.

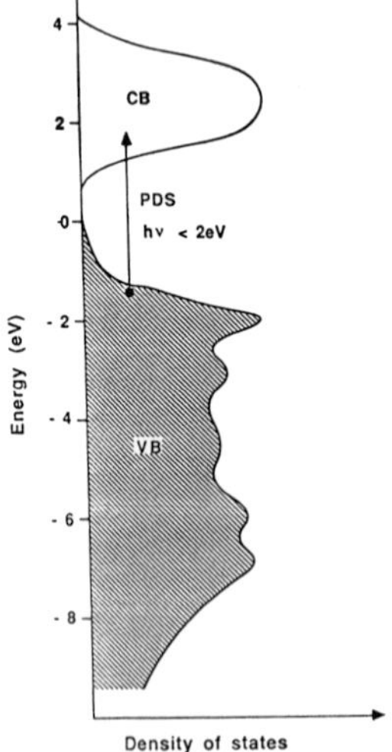

Figure 2. Schematic representation of density of states in MoS_2. Reproduced from reference 1 with permission of the authors.

The effect of O_2 on the d_2 tail states of MoS_2 was described in a recent publication (9). In this work, UPS studies showed the existence of surface states above the dz^2 band near the Fermi level (Figure 2). Furthermore, these tail state were reversibly quenched with 900L O_2 but irreversibly quenched at 1 at. of O_2. The irreversible quenching occurred with an accompanying appearance of bulk oxide states in the UPS spectrum. This result not only demonstrates the problems with O_2 chemisorption but also shows the relation between the bulk electronic states of MOS_2 and the active surface states. The dz^2 orbitals are the highest occupied molecular obitals of MoS_2 and the crucial catalytic electronic states lie just above them arising from the surface termination of the bulk states. Previously presented calculated bulk electronic trends and their correlation to activity (1) may now be understood in terms of the bulk electronic structure providing an "electronic support" for the catalytically important surface electrons.

In a recent paper, the optical properties of these "tail-state" were examined catalytically and optically (10). The optical properties of MoS_2 powders and platelets were measured in the near infrared using photothermal deflection spectroscopy (PDS). PDS is a technique well suited for the measurement of the official properties of black highly catalytic powders because it is insensitive to optical scattering (11). The absorption as measured by this techniques for crystalline samples is shown in Figure 3. This absorption A can be related to the absorption coefficient by α

$$A = 1 - e^{-\alpha V}$$

where ℓ is the average sample thickness. The absolute value of the absorption is known for all samples because they can be normalized to the strongly absorbing excitonic region where a − >>1. The spectra of small (1.7μm mean diameter) and large (36μm mean diameter platelets are compared against the spectrum of a single crystal of MoS_2 in Figure 3. The absorption for the single crystal begins strongly at 1.2 eV and increases toward higher energy due to the indirect bandgap (12). The flat lower energy absorption in the single crystal is from defects in the material and varies strongly from sample to sample.

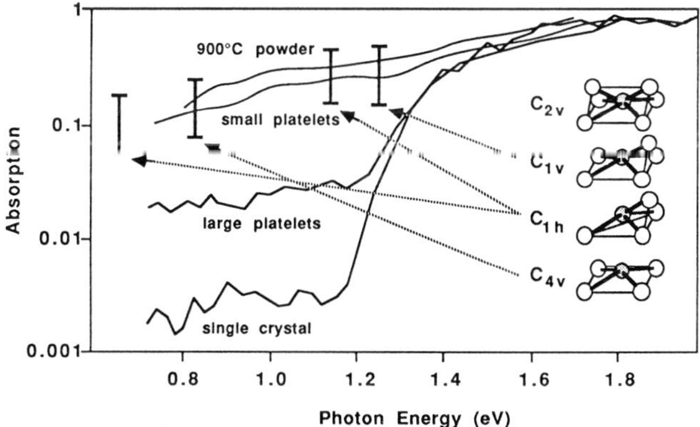

Figure 3. PDS measure spectra of MoS_2 single crystal and microcrystalite platelets. Also included are calculated positions for MoS_2 defects occurring on edge planes after reference 13.

The spectrum of the large platelets is seen to be very similar to that of the single crystal, except that the defect absorption below 1.2 eV is an order of magnitude higher. The striking similarity of the spectrum of the single crystal and large platelets between 1.3 and 1.6 eV shows that the large platelets are indeed single crystals with and average thickness of 5± 1μm because the magnitude of this absorption agrees with that of the 5μm thick single crystal. The absorption spectrum of the smaller platelets is also shown as the upper curve in Figure 3. In this case, the spectrum must be corrected for the difference in thickness by normalizing the spectrum at 1.5eV. The low energy absorption due to defects is an order of magnitude greater in the small platelets than it is in the large ones.

From these data it is evident that the optical absorption observed below 1.2 eV in the platelets due to the exposed edge planes. This is because SEM studies of the small and large platelets revealed that the small platelets have a greater edge plane area per gram that of the large platelets. In fact, a statistical study of micrographs of these samples showed that the "edge site" density of the small platelets was 6.1×10^{17} sites/gm and that of the large platelets was 7.2×10^{16} sites/gm (10). "Dangling bonds", vacancies, or other similar surface defects would be expected to have electronic states in midgap and thus increase the optical absorption in this region. From the known density of edge sites (N) the average optical absorption (a) of a single edge site can be calculated by

$$A = N\sigma$$

yielding 6.1×10^{-17} cm^2 for the small platelets and 8.4×10^{-17} cm^2 for the large platelets. The agreement between these two numbers is excellent and is consistent with the low-energy absorption is indeed proportional to the edge area.

The catalytic activity of the microplatelets could be determined directly (10). The HDS of dibenzothiophene (DBT) was measured. Biphenyl was the only product observed with no hydrogenation occurring. Conversion of DBT with time (350°C and 450 p.s.i. H_2) yielding a straight line below 15% conversion, and an HDS rate = 4.8×10^{16} molec/g-s was determined from the slope of this line. From this rate and the density of edge sites determined above, a turnover frequency of 7.9×10^{-2} molec/edge site -s was determined. This calculation assumes that each exposed Mo atom is catalytically active in which case the appropriate turnover number would be higher. Nevertheless, we believe that this is the only turnover number fro MoS_2 which has been determined without an ambiguity in the edge plane dispersion. Because of this, this number becomes the basis for further studies.

The above result has been extended to MoS_2 unsupported powders, where, because of disorder, knowledge of edge area has been limited to oxygen chemisorption studies. A series of powders was prepared by decomposing $(NH_4)_2MoS_4$ at different temperatures from 350°C to 900°C. The optical spectra of these samples showed a strong broad adsorption tail below the band-to-band adsorption which is dependant on the anneal temperature. This adsorption is very similar to that observed from edge plane defects in the platelets with a slight difference in shop due to disorder. The catalytic activity of these powders for the HDS of DBT was measured and a linear correlation between the activity and the absorbance was observed. Assuming that the absorption cross section is the same in all materials, the turnover frequency calculated from the slope of the Absorption/Activity plat was 3×10^{-2} mole/edge sites. This value is approximately two times lower than that obtained from the platelets, an agreement which is reasonable given uncertainties in the size and density of the disordered materials. It is also possible that disordered materials. It is also possible that disorder induces sites which, while counted by the PDS method, are not catalytically as effective of as accessible as those on well ordered materials.

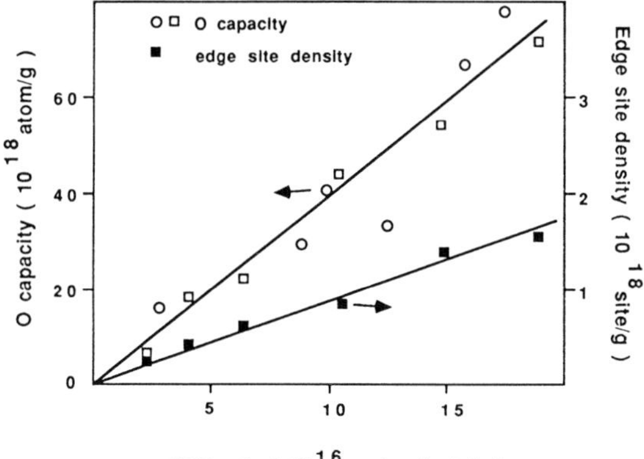

Figure 4. HDS of DBT vs 02 chemisorption and PDS edge site density. Reproduced from reference (1) with permission of the authors.

The similarity in turnover frequency between the disordered and micro crystalline materials indicates that the active sites for desulfurization in each are similar and are located on the edge surfaces. Such defects which are catalytically active, would generally be expected to have energy levels lying between the conduction and valence bands and thus absorb photons with below bandgap energies. This is indeed the behavior observed, and the $\sim 10^{-16}$ cm^2 cross section observed is typical of such defects. In fact, a recent set of $X\alpha$ calculations which modeled different types of sulfur vacancies which could occur at MOS_2 edges, showed that allowed optical transitions for these defects fall into the observed energy ranges below 1.2eV[13]. These results suggest that sulfur vacancies are responsible for the optical absorptions measured for the edge planes.

It is also noted that for a similar set of samples turnover frequency of 1.2×10^{-2} molec/site -s using O_2 chemisorption was obtained (Figure 4). Again this emphasizes that more molecules of O_2 are chemisorbed per active site due to the bulk oxidation. Furthermore, it was noted that the turnover number for the platelets was based on production of biphenyl only. In the powders as much as 50% cyclo-hexylbenzene was produced indication multiple sites. Therefore, at the writing, we feel that the highest turnover number on geometrically well determined material production a single product is the most reliable measurement. This work presumably can be extended to supported catalysts as well. However, the extension to promoted systems is not quite as straight forward because the presence of Co of Ni modifies the semi-conduction properties of MoS_2, confusing the interpretation of the measured optical spectra.

The above studies were performed on conventionally prepared microcrystalline materials. These materials are difficult to study because they have relatively low edge area due to a growth which occurs primarily in the direction parallel to the layers. A well-ordered edge surface is difficult to create by cutting or polishing because the layers fold and break irregularly. However, we recently reported a new way of preparing chemically reactive surfaces by using lithographic fabrication methods (14). Single crystals of MoS_2 prepared in this way have a surface that consist primarily of edge planes, which allows exceptional control of the surface morphology as indicated in Figure 5. In this figure, well ordered stacked planes of MoS_2 with 6.1 Å spacing can clearly be seen near the edge of the structure. These microstructures are ideal for the fundamental studies of edge planes surface properties described above. However, care must be taken to assure that the surfaces are prepared freshly in vacuum or exposed to an appropriate environment. For example, O_2 exposed samples, as seen in Figure 6, clearly show disorder and lattice expansion at the edge. Another example can be seen in Figure 7, in which freshly prepared posts are treated in H_2.15%H_2S at 350°C. in this case, the disorder induced by the H_2/15%H_2S treatment is quite severe. However, the properties of these "active" edges can still be readily studied and this state of the MoS_2 edge is probably most relevant to the HDS environment.

Figure 5. TEM micrograph of "lithographically" prepared MoS_2 edges.

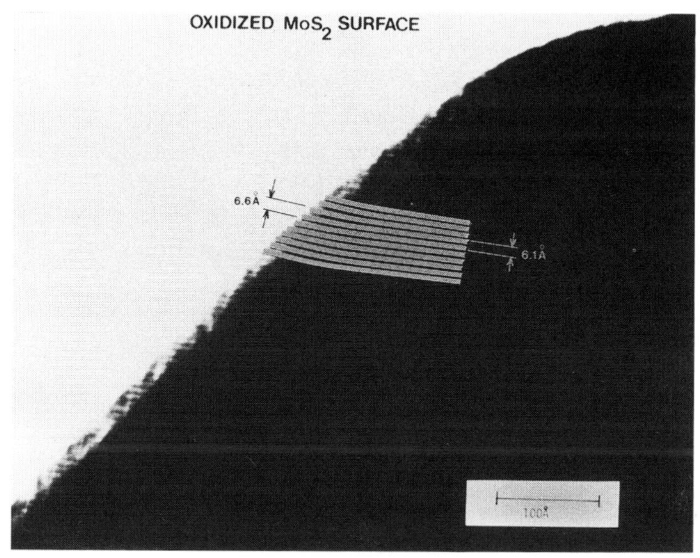

Figure 6. O_2 exposed "lithographically" prepared edges.

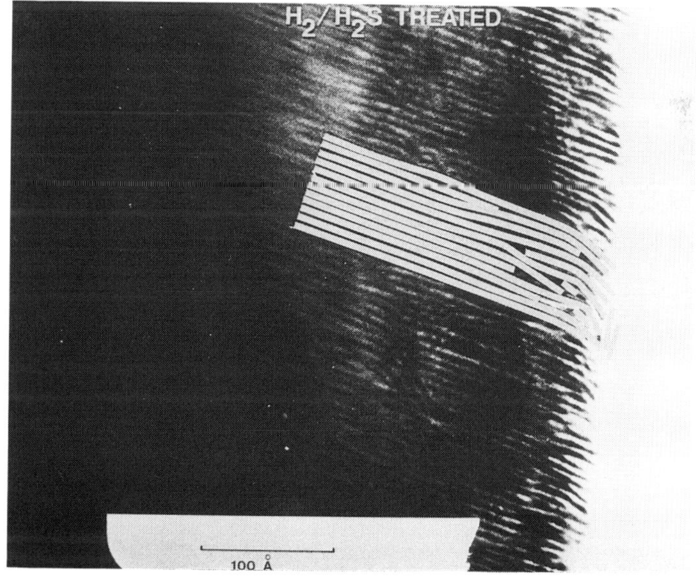

Figure 7. $H_2/15H_2S$ exposed "lithographically" prepared edges.

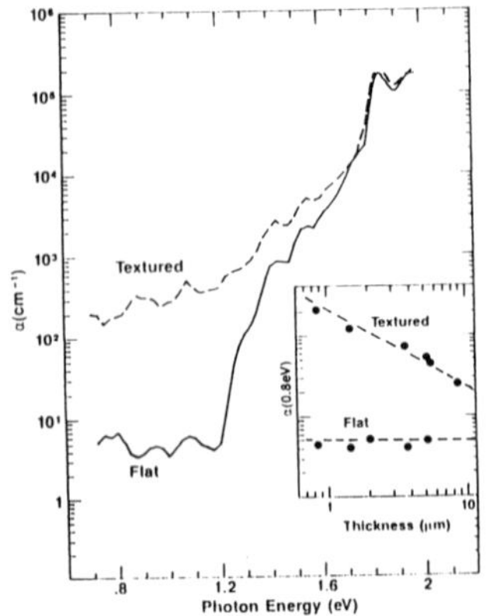

Figure 8. PDS measured spectra of textured MoS_2. Reproduced from reference 1 with permission of the authors.

PDS studies of samples of MoS_2 prepared in this manner are shown in Figure 8. In this figure, the single crystal spectrum is shown above it a textured sample from the same crystal. Again, we see the creating edge plane creates the same defect absorption below 1.2eV that is described above. The edge defects were also observed in x-ray photo emission spectroscopy. Figure 9 shows the Mo3d core levels of a textured and flat crystal. The textured crystal was treated in H_2/H_2S at 350°C to reduce and resulfide the surface. The edge surface spectrum is considerably broader than the spectrum of the basal surface spectrum of the basal surface and is also shifted to lower energy. The spectrum of the textured sample has an additional component that is shifted 0.8eV to lower binding energy. These two components gave a good fit to the entire spectrum and showed that the edge defects contain Mo that is reduced relative to the bulk. The shift of 0.8eV is about that expected from reduction of Mo^{+4} to Mo^{+3} in sulfide compounds. UPS measurements also showed that the Fermi level shifted 0.8eV closer to the valence band upon texturing. Because the shift is nearly as large as the band gap (1.2eV) the Fermi level of the edge surface must be within ~0.3eV of the balance band maximum. This implies that most of the edge surface defects within the band gap would be unoccupied, and that optical transitions would involve the excitation of electrons out of the valance band into the defect level. Such transitions would lead to the monotonic increase in absorption with photon energy and absorption cross section observed.

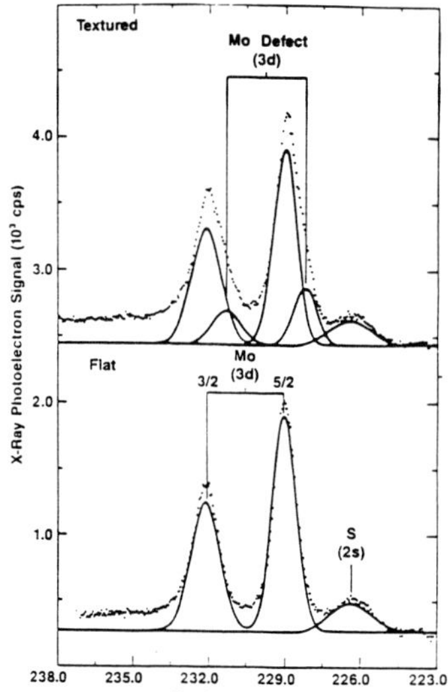

Figure 9. Mo 3d levels of a textured and a flat crystal. Reproduced from reference 1 with permission of the authors.

Summary

Complete understanding of the catalytic properties of MoS_2 (WS_2) requires more knowledge of the "edge planes" which terminate the anisotropic layers and are the location of the catalytically active "sites". Physical "edge plane based", turnover frequencies derived from chemisorption and activity measurements, exist which fit observed data well. But again, absence of absolute knowledge of MoS_2 dispersion tends to lead to models which greatly underestimate crystallite sizes of MoS_2. The bulk electronic structure of MoS_2 is related to the catalytically active surface states. The dz^2 orbitals of MoS_2 in the +4 state; the catalytic states, created by edge termination, lie just above them and are probably in the +3 state when operating in a catalytic environment. Optical and electron spectroscopic techniques directly measure these defect states and catalytic measurements on geometrically well determine catalysts yield an "HDS edge plane turnover frequency" which does not suffer from an ambiguity in dispersion. This turnover number now enables MoS_2 dispersion to be determined in all umpromoted MoS_2 catalysts. This result should lead to more precise models of promoted MoS_2 catalysts in the future.

Acknowledgements

The authors would like to thank C. B. Roxolo, H. W. Deckman, A. F. Ruppert and J. Gland for useful discussions regarding this paper.

Literature Cited

1.) Chianelli, R. R., and Daage, M., Fall AiChE Meeting, Washington, D. C., 1988.

2.) Whittingham, M. S., and Jacobson, A. J., eds, in: Intercatation Chemistry, Academic Press, New York, (1982).

3.) Tauster, S. J., Pecoraro, T. A., and Chianelli, R. R., J. Catal.,63, 515 (1980).

4.) Fretz, R., Breysse, M., Lacroix, M., and Vrinat, M., In: Second Workshop on Hydrotreating Catalysts Louvain, la Neuve, (Oct. 1984).

5.) Chianelli, R. R., Ruppert, A. F., Behal, S. K., Kear, B. H., Wold, A., and Kershaw, R., J. Catal., 92, 56 (1985).

6.) Nag, N. K., Sai Presada Rae, K., Chary, K. V. R., Rama Rao, B., and Subranhmanyam, V. S., Applied Cat., 41, 165-176 (1988).

7.) Kasztelan, S., Toulhoat, H., Grimbolt, J., and Bonnelle, J. P., Bull. Soc. Chem. Belg., 89, 807 (1984).

8.) Kasztelan, S., Toulhoat, H., Grimbolt, J., and Bonnelle, J. P., Applied Catal., 13, 127 (1984).

9.) Liang, K. S., Hughes, G. J., and Chianelli, R. R., J. Vac. Sci. Tech A2(2), 991-994 (1984).

10.) Roxlo, C. B., Daage, M., Ruppert, A. F., and Chianelli, R. R., J. Catal. 100, 176-184 (1986).

11.) Jackson, W. B., Amer, N. M., Baccara, A. C., and Fournier, D., Appl. Opt., 200W, 1333 (1981).

12.) Goldberg, A. M., Beal, A. R., Levy, F. A., and Davis, E. A., Phil. Mag., 32, 367 (1975).

13.) Horsley, J., Klier, K., personal communication.

14.) Roxlom C. B., Deckman, H. W., Gland, J., Cameron, S. D., and Chianelli, R. R., Science, 235, 1629-1631 (1987).

R.K. Grasselli and A.W. Sleight (Editors), *Structure-Activity and Selectivity Relationships in Heterogeneous Catalysis*

© 1991 Elsevier Science Publishers B.V., Amsterdam

Enantioselective Hydrogenation of Ethyl Pyruvate: Effect of Catalyst and Modifier Structure

H.U. Blaser*, H.P. Jalett, D.M. Monti, A. Baiker+ and J.T. Wehrli+
Central Research Laboratories, Ciba-Geigy AG,
CH-4002 Basel, Switzerland
+Department of Industrial and Engineering Chemistry
Swiss Federal Institute of Technology, ETH-Zentrum,
CH-8092 Zürich, Switzerland

ABSTRACT

The effect of catalyst and modifier structure on selectivity and activity for the enantioselective hydrogenation of ethyl pyruvate with cinchona modified Pt/Al_2O_3 catalysts has been investigated. It has been found that the platinum dispersion and the method of catalyst preparation have a decisive influence on the catalytic performance. In order to get high optical yields, the platinum dispersion should be lower than 0.2 - 0.3. The texture of the support is less important but must be optimized with respect to pore size in order to get high enantioselectivity. Modifiers derived from cinchonidine lead to an excess of (R)-ethyl lactate while cinchonine gives preferentially the S enantiomer. Changes in the O-C_9-C_8-N part of the cinchona alkaloid have a decisive influence both on sign and size of the optical induction while the nature of the substituent in the quinuclidine part, R_1, has only a minor effect. Partial hydrogenation of the quinoline rings results in a decreased enantioselectivity. Conclusions concerning the mode of action of the enantioselective catalyst are discussed.

INTRODUCTION

The enantioselective hydrogenation of carbonyl groups using modified heterogeneous catalysts is a topic of interest both from a preparative and mechanistic point of view. Results have been reported for the hydrogenation of various ketones using catalysts containing Ni, Pt, Pd, Ru, Co, Cu and some alloys, modified with several types of chiral compounds [1, 2]. Up to now, only two catalytic systems have been found which lead to products with a high enantiomeric excess (ee). The first to be described was tartrate-modified Raney Nickel which catalyzes the hydrogenation of β-dicarbonyl compounds with ee >80% [1a]. For the selective hydrogenation of α-ketoesters the best systems are Pt/Al_2O_3 catalysts modified with cinchona alkaloids [3, 4]. While enantioselectivities up to 90% have been observed, this can be achieved only when the catalytic system (substrate, catalyst, modifier, solvent, reaction conditions) is carefully optimized. A very important factor is the structure of the modified catalyst. Because the enantioselective catalyst is the result of a combination of a "normal" catalyst with a chiral modifier, it is necessary to discuss the influence on the catalytic performance of the *catalyst parameters* as well as of the *modifier structure*.

The present investigation deals with the influence of the structure of the catalyst and with the effect of structural variations of the cinchona modifier on the enantioselectivity and activity in the hydrogenation of ethyl pyruvate which serves as our model substrate.

$$CH_3COCOOC_2H_5 + H_2 \xrightarrow[\text{MODIFIER}]{\text{CATALYST}} \text{(R)-}CH_3CH(OH)COOC_2H_5 + \text{(S)-}CH_3CH(OH)COOC_2H_5$$

Ethyl Pyruvate → (R)-Ethyl Lactate + (S)-Ethyl Lactate

EXPERIMENTAL

Catalyst characterization

BET surface areas were measured at 77 K using nitrogen. The pore size distribution and the total pore volume were calculated from the desorption branch of the hysteresis curve using the method of Pierce [5]. Mean platinum particle sizes were determined using CO pulse chemisorption at 298 K on samples pre-treated in hydrogen at 673 K. The degree of dispersion and the mean particle size (spherical model) were estimated from the measured CO uptake assuming a cross-sectional area for a platinum atom of 8.93×10^{-20} m^2 and a stoichiometric factor of one [6]. The TEM investigation was performed on a JEOL JEM 200CX microscope; samples were prepared by suspending the catalyst powder in hexane and placing a droplet on an amorphous carbon grid.

Preparation of cinchona derivatives

The cinchona derivatives used were prepared by modification of the commercially available compounds cinchonidine and cinchonine. The structures given are in good agreement with their elemental analysis and their UV, IR, MS and NMR spectra. Details will be reported elsewhere.

Hydrogenation of ethyl pyruvate

The experimental procedure for the hydrogenation has been described in detail in a previous report [4]. The reaction was carried out in a 50 ml three-phase-slurry reactor with magnetic stirring (ca. 1000 rpm). All catalysts were freshly reduced for 2 hours in hydrogen at 400°C. For the test of the various modifiers the following conditions were used: 10.4 g (0.09 mole) ethyl pyruvate (freshly distilled); 100 mg 5% Pt/Al_2O_3 (E 4759); 10 mg modifier; 20 ml ethanol; temperature 25-30°C; H_2-pressure 70-100 bar. The reaction was run to completion and the conversion checked by gas chromatography (column: OV 101, 2m, 50°C). Initial turnover frequencies (TOF; [s^{-1}]) were calculated in terms of accessible platinum atoms obtained from CO adsorption data.

Optical yields were either determined by gas chromatography on a chiral capillary column (Chirasil-(L)-Val, 50 m, 150°C) after derivatization of the enantiomers with isopropyl-isocyanate or by measuring the optical rotation using $[\alpha]^{25}_{546} = 12.4°$ (neat). The optical yield is expressed as the enantiomeric excess (ee) of the hydroxyester:

$$ee\ [\%] = 100 \times |[R] - [S]| / ([R] + [S])$$

RESULTS AND DISCUSSION

Influence of catalyst structure

In order to determine the influence of catalyst parameters the following strategy was chosen: First we compared the catalytic performance of a large number of Pt/Al_2O_3 catalysts of known preparation method. The details of this investigation have already been reported [7] and the most important results are presented in Figure 1, which shows a correlation of the selectivity (ee) and the activity (TOF) of various catalysts with their platinum dispersion.

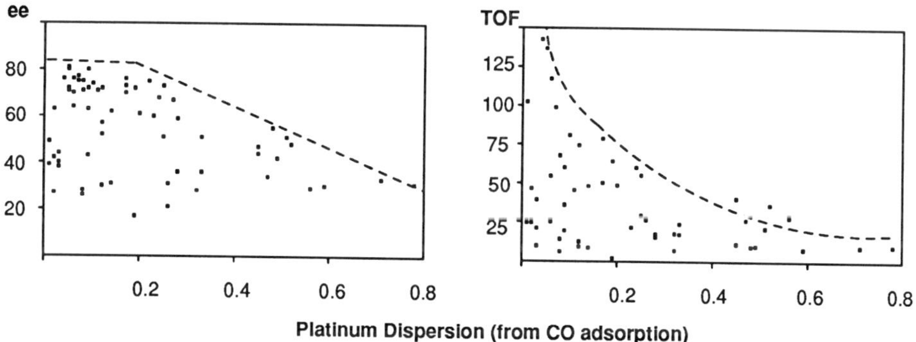

Figure 1: Dependence of optical yield (ee) and turnover frequency (TOF) on the platinum dispersion. Hydrogenation of ethyl pyruvate in ethanol in presence of cinchonidine at 20°C and 70 bar using different Pt/Al_2O_3 catalysts (data from [7]).

By varying the preparation procedures (support material, platinum precursor, platinum loading and reduction procedure) it was possible to obtain platinum dispersions between < 0.05 and 0.78. The enantioselectivity for the ethyl pyruvate hydrogenation increased with decreased platinum dispersions, reaching 80% at dispersions ≤0.2. The activities of the different catalysts showed a similar trend. In both cases, a strong scattering of the resulting values is observed, indicating that the platinum dispersion is by no means the only important catalyst parameter. In addition, an interesting interrelationship between optical yield and turnover frequency for the different catalysts was found. In general, good enantioselectivity is observed for catalysts with high turnover frequency.

Secondly we compared two well characterized commercial catalysts with similar platinum dispersions but a different texture. Their textural properties and their catalytic performance for the hydrogenation of two different α-ketoesters are presented in Figure 2 and Table 1 .

Catalyst	Pt dispersion	S_{BET} m^2/g	Pore volume ml/g	Particle diam. µm	Optical yield %	TOF 1/s
E 4759	0.24	168	0.27	50-120	84[a] 77[b] 80[c]	34[a] 17[b] 8[c]
5 R 94	0.22	131	0.93	10-30	91[a] 81[b] 85[c]	95[a] 37[b] 28[c]

a) ethyl pyruvate; toluene; 100 bar b) ethyl pyruvate; EtOH; 75 bar c) ethyl 2-oxo-4-phenylbutyrate; toluene; 70 bar. All reactions with 10,11-dihydrocinchonidine at 20°C.

Table 1: Textural and catalytic properties of two commercial 5% Pt/Al_2O_3 catalysts.

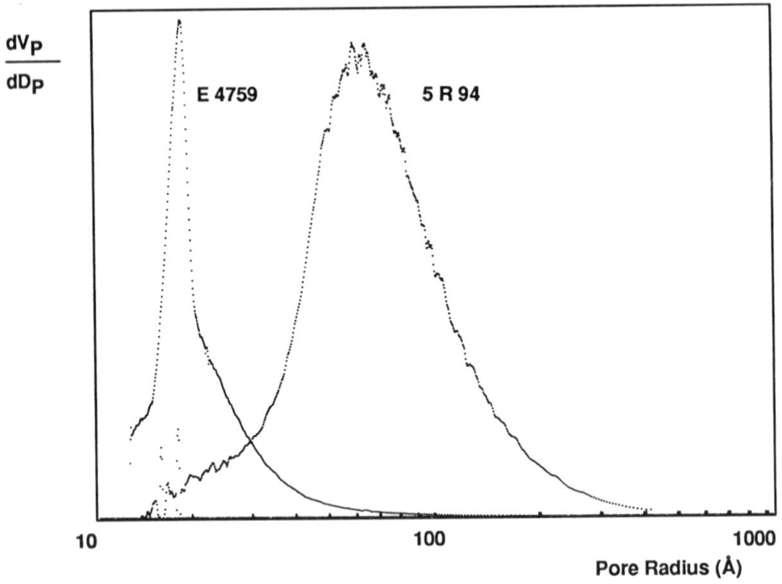

Figure 2: Pore size distribution of two commercial 5% Pt/Al_2O_3 catalysts.

It is apparent immediately that catalyst E 4759 has rather small pores and a low pore volume while 5 R 94 is a wide-pore catalyst with a large pore volume. In addition, HRTEM (Figure 3a; 3b) and XRD studies revealed that catalyst E 4759 consists of γ-alumina and has a well ordered, layered structure while 5 R 94 is a mixture of γ- and Θ-alumina where the alumina crystallites are of irregular shape with larger interstices. The catalytic performance was tested under different conditions for two α-ketoesters. In every case catalyst 5 R 94 showed a consistently higher enantioselectivity and a 2-3 fold higher turnover frequency.

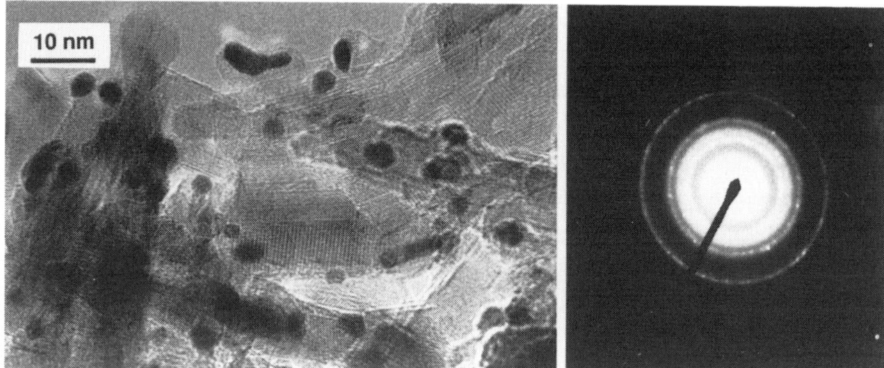

Figure 3a: High resolution transmission electron micrograph and electron diffraction pattern of catalyst 5 R 94.

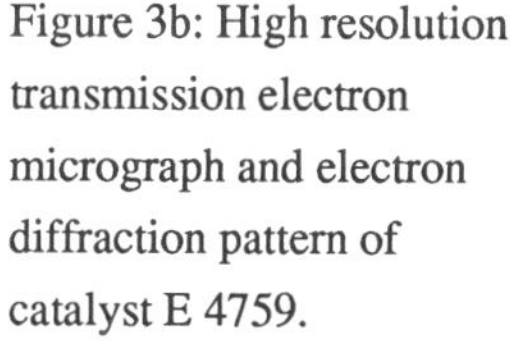
Figure 3b: High resolution transmission electron micrograph and electron diffraction pattern of catalyst E 4759.

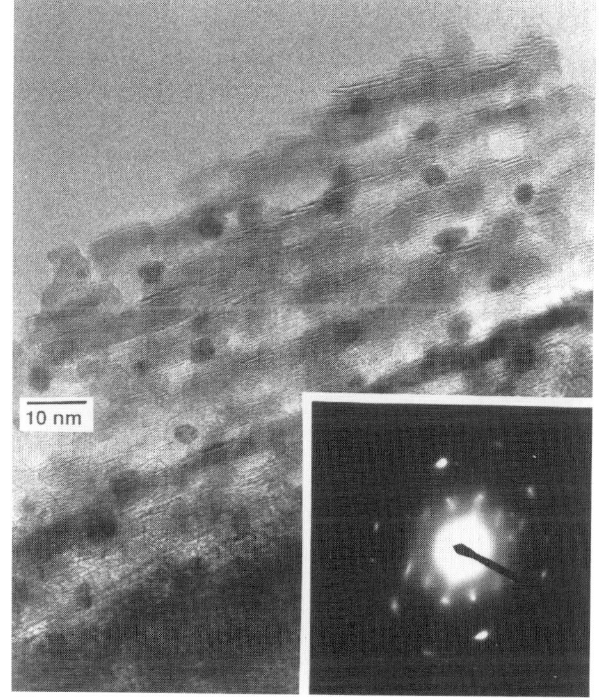

From the results described above, and in more detail in [7], the following conclusions on the influence of the catalyst structure can be drawn:

- *platinum dispersion (i.e. mean platinum particle size)* is the most important catalyst parameter; in order to obtain high optical yields dispersion should be ≤0.2 (see Figure 1).
- the *texture of the support* generally has a limited influence on the selectivity but in some cases the activity is changed by more than a factor of 2. Best results are observed using aluminas with relatively low S_{BET}, high pore volume and rather large pores (or without micropores). At the present time it is not clear whether intraparticle transport effects are responsible for this difference in activity or whether the degree of modification for the two catalysts is different, leading to both higher TOF and ee.
- in addition, the *method of catalyst preparation* has a large influence on both activity and enantioselectivity (platinum salt, reduction procedure, pre-treatments)[7, 8]. These observations suggest that factors such as morphology and size distribution of the platinum particles as well as contaminants and residues originating from the platinum and alumina precursors may affect the catalyst performance as well.

Similar results and conclusions have been reported for tartrate-modified ruthenium and nickel catalysts for the enantioselective hydrogenation of β-ketoesters by Klabunovskii et al. [9], Nitta et al. [10, 11, 12] and Sachtler et al. [13]. The fact that not only catalyst parameters but also the method of catalyst preparation has an influence on the catalytic properties seems to be a common feature of the different heterogeneous enantioselective hydrogenation catalysts.

Influence of modifier structure

In order to investigate the effect of modifier structure, two naturally occuring cinchona alkaloids were altered at various positions (see Figure 4) and then tested in the hydrogenation of ethyl pyruvate under standard conditions. Results are summarized in Table 2.

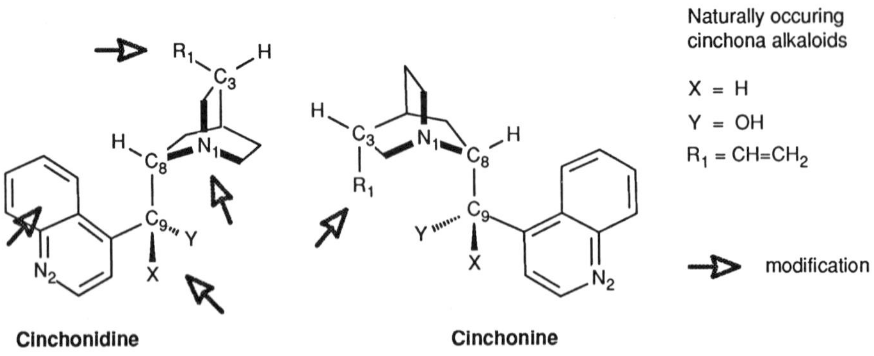

Naturally occuring cinchona alkaloids

X = H

Y = OH

R_1 = CH=CH$_2$

modification

Cinchonidine

Cinchonine

Figure 4: Relative and absolute configuration of the cinchona alkaloid derivatives prepared.

Modifier structure				Hydrogenation of ethyl pyruvate			Comments
X	Y	R_1	other changes	Solvent	ee (%)	major enantiomer	
Cinchonidine series							
H	OH	$CH{=}CH_2$		EtOH	76	R	$CH{=}CH_2$ hydrogenated
H	OH	CH_2CH_3		EtOH	79	R	
H	OH	"		AcOEt	81	R	
H	OH	"		Toluene	83	R	
H	OCH_3	"		EtOH	78	R	
H	OH	"		AcOEt	80	R	
H	OH	"		Toluene	81	R	
H	OAc	"		EtOH	20	R	
Cl	H	"		EtOH	44	R	
H	H	"		EtOH	44	R	
H	OH	CH_2OH		EtOH	ca.80	R	
H	OH	CH_2CH_3	a)	EtOH	73	R	N-O hydrogenated
H	OH	$CH{=}CH_2$	b)	EtOH	0		
H	OH	CH_2CH_3	c)	EtOH	30-50	R	
Cinchonine series							
H	OH	$CH{=}CH_2$		EtOH	56	S	$CH{=}CH_2$ hydrogenated
H	OH	"	d)	MeOH	58	S	$CH{=}CH_2$ hydrogenated
H	OH	CH_2CH_3	d)	MeOH	54	S	

a) N_1-oxide of dihydrocinchonidine b) N_1-benzyl-cinchonidinium chloride
c) quinoline hydrogenated (mixture of products, preliminary results) d) substrate methyl pyruvate

Table 2: Effect of modifier structure on the enantioselective hydrogenation of ethyl pyruvate under standard conditions.

In the original paper by Orito et al. [3] it was already reported that cinchona alkaloids with the same absolute configuration as cinchonidine induced preferentially the R-configuration of the α-hydroxyester, while those with cinchonine configuration produced an excess of the S-enantiomer. This is confirmed by our results. Because the two compounds differ only in the absolute configuration at C_8 and C_9, this strongly suggests that the interaction of the substrate with this part of the modifier determines the product distribution. We have shown that the reaction conditions, e.g. the concentration of the modifier, can influence the selectivity of the catalytic system [4], therefore the extent of the enantiomeric excesses reported in Table 2 could partially be due to non-optimal conditions for an individual modifier.

The following points are noteworthy:

- If N_1 is alkylated, optical induction is lost completely. The N-oxide is probably reduced very fast and then acts like dihydrocinchonidine.
- Changes at C_9 in most cases result in lower optical yields, but (R)-ethyl lactate is always formed in excess. In order to get very high optical yields the substituent Y at C_9 has to be either OH or OCH_3.

- Hydrogenation of the quinoline nucleus leads to lower enantioselectivities. We have found that some ring hydrogenation occurs under our reaction conditions as well, but it is much too slow to influence the optical yields.
- The nature of R_1 has little effect on the optical yield. Since the double bond is hydrogenated in the first minutes of reaction, it is difficult to determine its influence on the optical induction.

CONCLUSIONS

The results concerning the influence of catalyst and modifier structure make it possible to draw some conclusions concerning the mode of action of the modified catalyst. If we assume a classical Langmuir-Hinshelwood reaction mechanism [14], the observed enantioselection and acceleration can be explained assuming a very specific interaction between substrate, modifier and platinum surface. We have postulated that an enantioselective active site is formed dynamically by adsorption of one cinchona molecule on well defined platinum ensembles [15]. The effect of particle size on rate and enantioselectivity indicates that not all surface platinum atoms are suited for this coordination. The observed influence of the structure of the modifier molecule make it possible to assign a specific function to the different parts of the modifier: we suggest that the absolute configuration at C_8 determines which enantiomer is formed preferentially and that the decisive interaction of the substrate takes place with N_1, while the OH-group at C_9 does not play a essential role. This is surprising because it has been proposed that the formation of a hydrogen bond is important in other cinchona catalyzed reactions where ketones are involved [16]. At the present time there is little information on the exact nature of the substrate-modifier interaction, nor is it clear why replacing OH with OCH_3 leaves the optical induction unchanged while substitution by OAc or H results in a much lower enantioselectivity. The decrease in optical yield observed for the partially hydrogenated hetero-aromatic part of the cinchona molecule can be explained by a weaker adsorption on the platinum surface. We do not know where the hydrogen is activated, but we think that the α-ketoester interacts preferentially with the *modified sites* on the platinum surface where the enantioselective reaction takes place. Whether this interaction really occurs as a 1:1 complex between adsorbed modifier and substrate as proposed above and also for the nickel-tartrate catalyzed hydrogenation of β-ketoesters [1a] or whether it is rather of the nature proposed by Thomas ("cinchonidine forms an ordered, well spaced array of interstices in a sorbed layer") [17], still remains an open question and is the topic of further investigations.

ACKNOWLEDGMENTS

We would like to thank Dr. W. Lottenbach, Ciba-Geigy, for the preparation of the cinchona derivatives and Dr. A. Reller, University of Zürich, for the HRTEM investigation.

REFERENCES

[1a] Y. Izumi, Adv. Catal. 32 (1983) 215

[1b] M. Bartok, *Stereochemistry in Heterogeneous Metal Catalysis,* Wiley, New York 1985, p. 511

[2] E.I. Klabunovskii, J. Phys. Chem. (Russian), 47 (1973) 765.

[3] Y. Orito, S. Imai and S. Niwa, J. Chem. Soc. Jpn., (1980) 670.

[4] H.U. Blaser, H.P. Jalett, D.M. Monti, J.F. Reber and J.T. Wehrli, Stud. Surf. Sci. Catal., 41 (1988) 153.

[5] S.J. Gregg and K.S.W. Sing, Surface and Colloid Science, 9 (1976) 231.

[6] A. Renouprez, C. Houng-Van and P.A. Compagnon, J. Catal., 34 (1974) 411.

[7] J.T. Wehrli, Ph.D. Thesis No. 8833, Swiss Federal Institute of Technology (1989). J.T. Wehrli, A. Baiker, D.M. Monti and H.U. Blaser, J. Mol. Catal., in press.

[8] J.T. Wehrli, A. Baiker, D.M. Monti and H.U. Blaser, J. Mol. Catal., 49 (1989) 195.

[9] A.A. Vedenyapin, E.I. Klabunovskii, Y.M. Talanov and G.K. Areshidze, Izv. Akad. Nauk. SSSR, Ser. Khim., 11 (1976) 2628.

[10] Y. Nitta, O. Yamanishi, F. Sekine, T. Imanaka and S. Yerashi, J. Catal., 79 (1983) 475.

[11] Y. Nitta, M. Kawabe, H. Kahita and T. Imanaka, Chem. Express, 1 (1986) 631.

[12] Y. Nitta and T. Imanaka, Bull. Chem. Soc. Jap. 61 (1988) 295.

[13] L. Fu, H.H. Kung and W.M.H. Sachtler, J. Mol. Catal., 42 (1987) 29.

[14] ref. [1b], p. 335.

[15] J.T. Wehrli, A. Baiker, D.M. Monti, H.U. Blaser and H.P. Jalett, J. Mol. Catal., 57 (1989) 245.

[16] H. Wynberg, Topics in Stereochemistry, Vol.16, Wiley-Interscience, New York, 1986.

[17] J.M. Thomas, Angew. Chem. Adv. Mater., 101 (1989) 1105.

R.K. Grasselli and A.W. Sleight (Editors), *Structure-Activity and Selectivity Relationships in Heterogeneous Catalysis*
© 1991 Elsevier Science Publishers B.V., Amsterdam

Non-Stoichiometry, a Key to Modify the Activity and Selectivity of Spinel-type Catalysts for Hydrogenation Reactions

F. TRIFIRO' and A. VACCARI

Dept. Industrial Chemistry and Materials, Viale del Risorgimento 4, 40136 BOLOGNA (Italy)

ABSTRACT

Possible effects induced by non-stoichiometry in different mixed oxides were examined. Emphasis was placed on how such effects can be successfully controlled and applied to the preparation of catalysts for the hydrogenation of carbon monoxide and/or different organic molecules. Non-stoichiometric phases may be obtained by low-temperature methods, and form by different mechanisms as a function of the nature of the elements present. All these phases are metastable and evolve towards stoichiometric forms with increasing temperature. Binary systems (Zn/Cr, Cu/Cr, and Co/Cr) and ternary and quaternary systems (containing Co, Cu, Zn, and Cr) were examined to show how multicomponent catalysts may be obtained, whose properties can be regulated by proper selection of the component cations and appropriate adjustment of the composition. We report examples of synergic effects, related to the presence of the different cations in the same structure, which result in a considerable increase in the catalytic activity in the hydrogenation of both CO and organic molecules. The poisoning of low-temperature methanol catalysts by small amounts of cobalt also may be attributed to a specific interaction.

INTRODUCTION

Mixed oxides are widely employed by the chemical industry as both heterogeneous catalysts and materials with specific properties (1-4). The preparation of specific tailor-made mixed oxides able to perform complex functions is one of the most current topics in solid state chemistry (5). In recent years there has been increased interest in preparation methods at low temperature, which allow solids to be obtained with defect structures whose properties are very different from those of the same solids synthetised using ceramic methods.

One of the main examples of these unusual solids are the non-stoichiometric spinel-type compounds, which have applications as both solid state gas sensors (6) and catalysts for different hydrogenation reactions (of CO to methanol, methanol-higher alcohol mixtures, hydrocarbons and many organic molecules) (7-12). Their peculiar physicochemical, reactivity and catalytic properties depend considerably on the presence of a non-stoichiometry, which allows, for example, the presence in the same structure of an M^{2+}/M^{3+} ratio higher than 0.5 (M= metal), or the stabilization of ions with unusual coordination (13-15).

The aim of this work was to make a contribution to the understanding of the structure and reactivity of these non-stoichiometric phases by investigating several spinel-type catalysts as a function of the composition and nature of the different elements present in the structure.

EXPERIMENTAL

All the catalysts were prepared by coprecipitation at pH= 8.0 ± 0.1 from a solution of nitrates of the elements with a slight excess of $NaHCO_3$, washed until the sodium concentration was lower than 0.05% (as Na_2O) and dried at 363K. The precursors were heated for 24h at different temperatures and in different atmospheres to elucidate the mechanism of formation of non-stoichiometric phases. The XRD powder patterns were recorded using Ni-filtered Cu K_α radiation (λ= 0.15418nm) or Fe-filtered Co K_α radiation (λ= 0.17889nm) and a Philips goniometer automated by means of a General Automation 16/240 computer. The quantitative determination of the crystalline ZnO was carried out according to the method of Klug and Alexander (16). When the diffraction patterns showed broad and/or overlapped diffraction lines, both phase composition and crystal size were determined by means of an X-ray full profile-fitting method (10,14,15). A C.Erba Sorptomatic 1826 appparatus with N_2 adsorption was used to measure the surface area.

IR spectra were recorded using a Perkin-Elmer 1700 Fourier-transform spectrometer. Microcalorimetric analysis were performed using a Tian-Calvet calorimeter connected to a volumetric apparatus, which allowed simultaneous determination of the amounts of CO adsorbed and related heats (17,18). The XPS and TPD of methanol (Temperature Programmed Desorption) tests were performed using a Perkin-Elmer PHI 5400 ESCA system (17) and a laboratory apparatus (19,20), respectively. The amounts of chromates and "free" copper ions were determined spectrophotometrically after extraction with a NH_4OH/NH_4NO_3 solution (21).

The catalytic tests of CO hydrogenation were performed in different copper-lined tubolar reactors, operating in the 530-630K temperature range and up to 6.0MPa, using a H_2:CO:CO_2= 65:32:3 (v/v) gas mixture (7,10). Prior to the catalytic tests the catalysts were activated in-situ by hydrogen diluted in nitrogen; the hydrogen concentration and temperature were progressively increased during this pretreatment. Outlet gases were monitored on-line by gas-chromatography. The liquid products were analyzed on-line by gas chromatography or, alternatively, condensed in a cold trap during the time on stream (6h), then wheighed and analyzed off-line. The catalytic tests of hydrogenation of oxo-aldehyde mixtures were performed in an autoclave operating at 7.0MPa and 400-500K temperature range after previously activating the catalysts (12). The reaction was followed by periodically drawing of small samples, that were analyzed by gas-chromatography. After reaction, all catalysts were cooled at r.t. under a nitrogen flow.

RESULTS AND DISCUSSION

Binary systems

Zn/Cr mixed oxides in the range from 33:67 to 50:50 are unambiguously monophasic and constitute examples of excess zinc non-stoichiometric spinel-type phases. Above the 50:50 ratio, the systems are biphasic (spinel + ZnO). However, for all samples XRD quantitative determinations show that the amount of crystalline ZnO is smaller than that expected for a simple phase composition ZnO and $ZnCr_2O_4$. Taking into account the analogies observed between the XRD patterns of these phases and those reported in the literature for some non-stoichiometric mixed oxides (22,23) and assuming that the undetected ZnO was inside the spinel-type phase, the general formula $Zn_xCr_{2/3\,(1-x)}O$ was adopted. The difference between the value of x and 0.25 (the value of the stoichiometric spinel $ZnCr_2O_4$) can be taken as an index of the degree of non-stoichiometry.

By means of an X-Ray full profile-fitting method (14,15), it was found that the excess Zn^{2+} ions are located in B-type sites of the lattice (i.e. in octahedral coordination typical of trivalent cations in normal AB_2O_4 spinel) with the nearest tetrahedral sites left vacant. This implies a progressive structural change from the normal spinel lattice towards a rock-salt type structure, with a corresponding increase in the metal/oxygen ratio from 3/4 to 1. This evolution is characterized in the XRD powder patterns by a decrease in the intensities of the odd-indexed lines and an increase in those of the even ones, with a change in the intensity ratio of the lines corresponding to the [400] and [440] planes.

Zn/Cr non-stoichiometric spinel-type phases may form in two different ways: 1) In air, by a redox reaction via chromate intermediates. 2) In N_2 or reducing atmosphere by direct reaction between the oxides. However, the properties of the solids obtained are similar, with only small differences in the crystal size (24). These phases are non-equilibrium phases and evolve with increasing temperature towards stoichiometric $ZnCr_2O_4$ and parallel segregation of ZnO. It should be pointed out that ZnO segregation is less marked for the samples with a Zn/Cr ratio $\leq$ 50:50, i.e. for the monophasic non-stoichiometric spinel-type catalysts (17).

The presence of excess zinc corresponds to remarkable modifications in the solids, seen in both the bulk and surface properties. However, bulk data indicate a progressive variation of the properties away from those of the stoichiometric spinel. For example, the lattice parameter *a* plotted as a function of the zinc content (Fig. 1) shows a regular trend away from the value of $ZnCr_2O_4$ (ASTM 22-1107), indicating a regular expansion of the spinel-type cell. On the other hand, the surface properties, as for example the surface area, show a remarkable variation as soon as departure from stoichiometry occurs (17,18).

These catalysts have very different reactivities than ZnO and $Zn_2Cr_2O_4$: heterolytic dissociative adsorption of H_2 occurs only on non-stoichiometric spinel-type phases, involving octahedrally coordinated surface Zn^{2+} ions (25). On the other hand, the CO adsorption capacity shows a maximum for the sample at the beginning of the departure from stoichiometry, and decreases with increasing zinc content of the samples (18,20). Furthermore, data from TPD of methanol show that the deviation from the stoichiometry decreases the oxidizing power of the surface and creates new stable active sites, identified as surface zinc species (19,20).

The catalytic data confirm that non-stoichiometric spinel is the active phase for methanol synthesis. The maximum productivity was observed for monophasic non-stoichiometric catalysts with a Zn/Cr ratio near to one, whereas the samples richest in zinc, for which a side phase ZnO was also detected, showed a considerable decrease in activity (Fig. 2).

Non-stoichiometric spinels also were stable in our reaction conditions, in which zinc surface enrichment was not detected by XPS analysis (17). This is confirmed by the regular trend of the lattice parameter *a* of the spinel-type phases after reaction (Fig. 1), with the exception of the catalyst richest in zinc, for which a strong ZnO segregation took place (17).

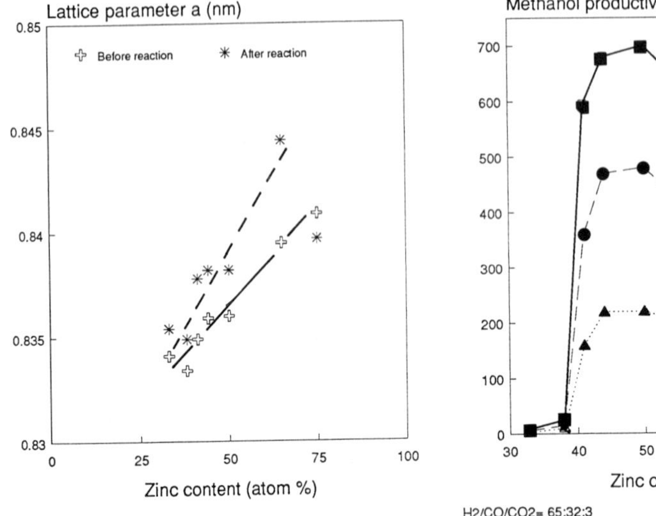

Fig. 1 Lattice parameter *a* of the spinel-type phase as a function of zinc content.

Fig. 2 Catalytic behaviour of the Zn/Cr catalysts. *P*= 6.0MPa; *GHSV*= 15000h^{-1}.

Also the Cu/Cr system may form cubic non-stoichiometric phases, characterized again by an excess of bivalent cations partially located in octahedral sites (26). As reported in the literature (27,28), the presence of Cr^{6+} ions is necessary for the stability of these phases. These ions, associated with cation vacancies, are localized in octahedral sites, while Cu^{2+} and Cr^{3+} ions are distributed in both octahedral and tetrahedral sites.

One difference between the Zn/Cr samples and the Cu/Cr samples is that non-stoichiometric Cu/Cr phases form at 653K only by calcination, whereas biphasic samples with a severe segregation of CuO are obtained by heating in the absence of oxygen. Further information may be obtained from the IR spectra, that show, for the Cu/Cr samples, the presence at low frequencies of a broad absorption band with a maximum at 554 cm^{-1}, attributable to the overlapping of the CuO and $CuCr_2O_4$ absorption bands (29,30).

However, the Cu/Cr non-stoichiometric phases also are metastable and evolve with increasing temperature towards monoclinic CuO (ASTM 5-0661) and tetragonal $CuCr_2O_4$ (ASTM 34-424), with a corresponding decrease in the amounts of both chromates and copper ions extracted by a NH_4OH/NH_4NO_3 solution (Fig. 3). Furthermore, when the samples are heated at 953K in air or at 753 in N_2, the tetragonal $CuCr_2O_4$ forms $CuCrO_2$ (ASTM 26-1113) (30), with the reduction of the Cu^{2+} ions to Cu^+ ones.

These catalysts after reduction have been described by Bonnelle et al. (28,31) as metallic copper supported on a residual defect spinel phase. The formation of metallic copper was attributed to the reduction of a part of the Cu^{2+} ions localized in tetrahedral sites, whereas the Cu^{2+} ions in octahedral sites were partially reduced to Cu^{+} in the same environment. However, it should be pointed out that we also detected the presence of quasi-amorphous Cr_2O_3 in samples prepared by heating in a reducing atmosphere or in the catalysts examined after the catalytic tests, in agreement with the data of Iimura et al. (32).

Cu/Cr catalysts showed high hydrogenating activity towards many organic molecules, attributed to their hydrogen reservoir capacity and related to the presence of cuprous ions in an octahedral environment (11,28). Figure 4 illustrates that preparation of the catalyst via non-stoichiometric phases increases the catalytic activity in the hydrogenation of oxo-aldehydes in comparison with that of a catalyst with the same composition, prepared by decomposition of a basic ammonium salt (33). On the other hand, it should be noted that Cu/Cr catalysts are also highly selective, but with low activities, in the methanol synthesis from syn-gas (34,35), with a maximum of activity for a Cu/Cr= 3 ratio (34).

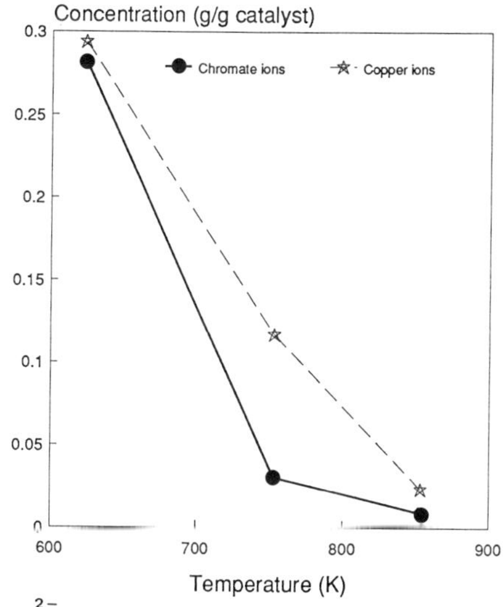

Fig. 3 Amounts of chromates and of copper ions extracted by a NH_4OH/NH_4NO_3 solution as a function of the calcination temperature for a Cu/Cr= 50:50 (atom. ratio) catalyst.

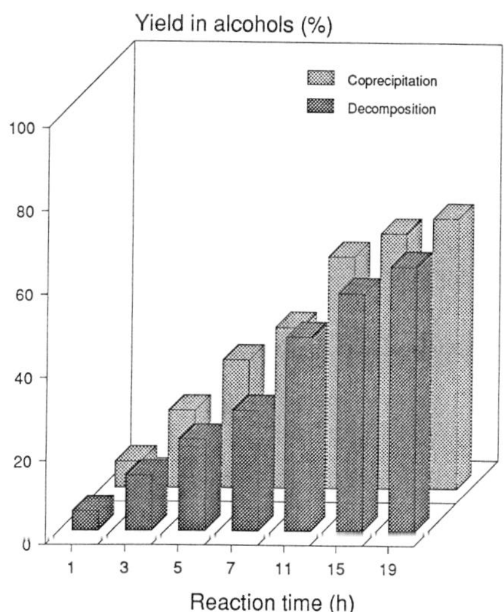

Fig. 4. Catalytic activity in the hydrogenation of an oxo-aldehyde mixture for a Cu/Cr= 50:50 (atom. ratio) catalyst prepared by different methods. T= 410K; P= 7.0MPa.

Different is the behavior of the Co/Cr system, for which the non-stoichiometry is mainly "apparent", the formation of monophasic samples being associated with the oxidation of a part of the Co^{2+} ions to Co^{3+} ions. In fact, normal spinel structures with tetrahedral site/octahedral site occupancy ratios near 0.5 were obtained by calcination. Furthermore, (i) the linear decrease in the lattice parameter a with increasing cobalt content (Fig. 5), which reflects the smaller size of octahedrally coordinated Co^{3+} cations with respect to octahedrally coordinated Cr^{3+} cations and

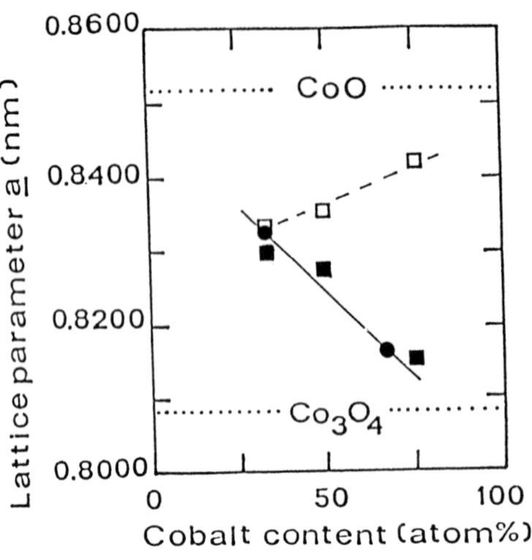

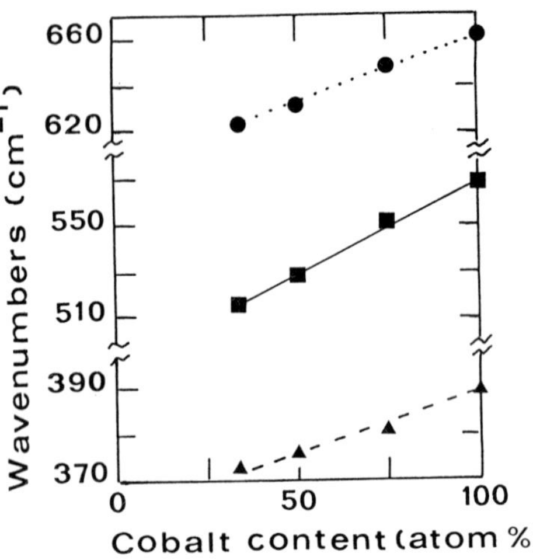

Fig. 5 Dependence of the lattice parameter *a* on cobalt content for calcined (■) and spent catalysts (□). Circles: literature data for $CoCr_2O_4$ and Co_2CrO_4.

Fig. 6 Dependence of the IR frequencies of the spinel phase on cobalt content for calcined catalysts. Co= 100% : data for Co_3O_4.

(ii) the continuous shift of the frequencies of the IR bands (Fig. 6) are both indicative of solid solution spinels arising from isomorphous Cr^{3+}/Co^{3+} cation substitution. A full characterization of the bulk and surface properties of these solids has been reported elsewhere (36).

When activated up to 623K, these samples remain monophasic, showing an evolution towards rock-salt type phases, related to the reduction of the Co^{3+} ions to Co^{2+} ions and structural rearrangement. As a consequence the structure becomes highly defective and this gives rise to an increase in the *a* parameter as well as in lattice disorder. However, the samples activated in these conditions show very low activity in CO hydrogenation, which increases only slightly as cobalt content increases (Table 1).

Table 1 Total productivity in hydrogenated compounds and selectivity in methane for the catalysts activated for 24h at 623K (A) or 773K (B). *P*= 1.2MPa; *GHSV*= $3600h^{-1}$.

Catalyst	React. Temp. (K)	Productivity ($mol\ h^{-1}\ kg^{-1}$)	Selectivity (%)
(A)			
Co 33	563	0.3	25
	583	0.5	22
Co 50	563	0.4	23
	583	1.4	20
Co 75	563	1.2	10
	583	4.0	15
(B)			
Co 33	563	2.0	27
	583	5.0	34
Co 50	563	10.1	29
	583	34.4	42
Co 75	563	60.9	41
	583	81.8	55

Selectivity calculated on carbon atom basis.

On the contrary, when the activation temperature is increased up to 773K, only the sample with the stoichiometric Co/Cr ratio (33:67) maintains a stable structure. In the cobalt-rich samples a considerable part of the cobalt segregates as well crystallized metallic particles. In agreement with the XRD data, the IR spectra of the samples with the highest cobalt content (Co/Cr>33:67) show a clear shift of the two more intense bands, now detected near 623 cm^{-1} and 518 cm^{-1}. This is clear evidence that, although considerably modified, a spinel-type structure is retained even after catalytic tests. Furthermore, the absence of a broad absorption band in the region 800-500 cm^{-1} allows the presence of CoO to be excluded (37). Activation at the highest temperature investigated gives rise to a drammatic increase in activity, especially for the samples with the highest cobalt content (Table 1), in which a considerable increase in selectivity for methane has been observed.

These data illustrate the role of the different species of cobalt (ionic or metallic) on the stability of the structure and on the catalytic activity, strictly correlated to the different activation conditions adopted.

Ternary and quaternary systems

The presence of different elements inside the same structure may have a considerable influence on both the stability and the catalytic activity of the non-stoichiometric samples. For example, a 20% substitution of Cu^{2+} ions for Zn^{2+} ions or the contrary, does not modify the structure and the mechanisms of formation of non-stoichiometric phases for the Zn/Cr and Cu/Cr systems, respectively, but does increase their stability. For both systems, this increase in stability corresponds to an increase of about 100K in the calcination temperature. Furthermore, the nature of the second bivalent element also plays an important role. For example, Co^{2+} ions are better physical promoters than Zn^{2+} ions in the Cu/Cr system, probably because of their partial oxidation to Co^{3+}. Therefore, in the binary non-stoichiometric structures previously discussed part of the ions may be substituted with different ions in order to modify the physical and/or catalytic properties.

The progressive substitution of zinc ions with copper ions gives rise to considerable differences in the catalytic activity, as a function of the copper content (Fig. 7). However, two general behaviours are found: 1) Up to a Cu/Cu+Zn ratio$\leq$ 0.5, the presence of copper considerably increases the activity in methanol synthesis. 2) For the highest ratios, a drammatic deactivation is observed, accompanied also by a considerable change in selectivity (35). It should be pointed out that the main increase in catalytic activity takes place for the catalysts in which copper ions substitute for only 20% of the zinc ions. The productivity of this catalyst is similar to the best value reported in the literature, if based on kg of catalyst (38-40), but clearly better if calculated on the basis of kg of copper, thus indicating the formation of very active copper-containing centers. We attribute this increase in activity to the presence of copper ions inside the Zn/Cr non-stoichiometric spinel-type structure, which favours the development of Cu-Zn synergic effects (38,41). Furthermore, it should be noted that a rough correlation holds between the catalytic activity in methanol synthesis and the whole chemisorption activity towards CO (Fig. 7). In spite of some small discrepancies it seems clear that the samples with a low copper content have remarkable catalytic activity and a corresponding high chemisorption capacity, whereas the Cu-rich samples exibit lower values of both.

However, Cu-rich catalysts show a surprisingly high activity in the hydrogenation of oxo-aldehyde mixtures associated with the presence of another element, such as zinc or cobalt, inside the Cu/Cr structure (Fig. 8). This effect can not be attributed to differences in the physical properties (all the catalysts have similar values of surface area and pore volume), but must be associated with a synergic effect between the copper ions and the zinc and/or cobalt ions, probably both located in octahedral sites (42). Furthermore, the ternary systems show a higher stability than the classic barium-promoted chromite catalysts towards the usual byproducts (formic esters, acetals and hemiacetals, higher esters, ketones and aldols) always present in crude oxo-aldehyde mixtures (12,43).

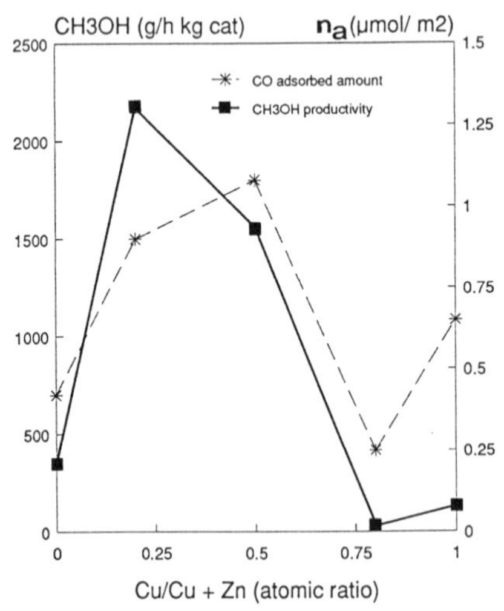

Fig. 7 Comparison of the catalytic activity in methanol synthesis and CO adsorption capacity as a function of the copper content. *T*= 555K; *P*= 6.0MPa; *GHSV*= $15000h^{-1}$. n_a: *T*= 300K; *P*= 5.3kPa.

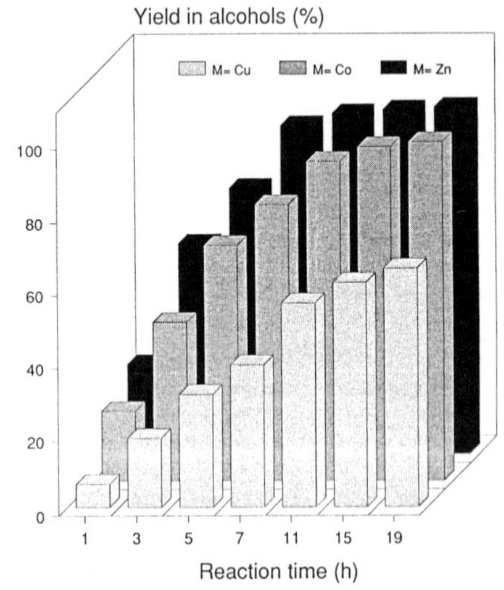

Fig. 8 Catalytic activity in the hydrogenation of an oxo-aldehyde mixture for Cu:M:Cr (40:10:50, atom. ratio) catalysts. *T*= 410K; *P*= 7.0MPa.

We have investigated a large number of ternary and quaternary cobalt, copper, zinc, and chromium mixed oxide systems for a wide range of compositions (10,44). Calcination of compositions in these systems results essentially in the formation of spinel-type phases, characterized by M^{2+}/M^{3+} (M= metal) ratios much higher than the value required for the stoichiometric spinel, also taking into account partial oxidation of Co^{2+} ions to Co^{3+} ions. In anology to that previously discussed for the Zn/Cr system , also in these cases the formation of non-stoichiometric spinel-type phases may be hypothesized, where the M^{2+} ions can partially occupy octahedral positions. Most of these phases were also stable after both activation up to 623K and catalytic tests of hydrocarbon synthesis, without any evidence of phase sintering phenomena, and the formation of metallic cobalt and/or cobalt oxides was not observed (even though it is not possible to exclude their presence in very small amounts).

Unlike the Co/Cr samples, catalysts containing both copper and cobalt are very active in CO

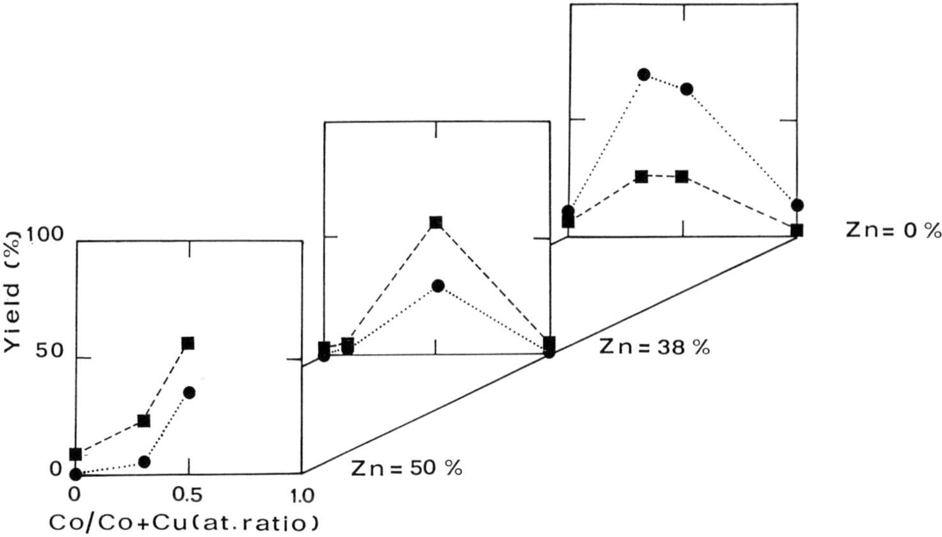

Fig. 9 Yields in hydrogenated compounds (■) and in CO_2 (●) as a function of the Co/Co+Cu ratio for catalysts with 24% chromium (atom. ratio) and three different zinc contents. T= 563K; P= 1.2MPa; $GHSV$= $3600h^{-1}$.

hydrogenation, also when activated at low temperatures (up to 623K), and form mainly hydrocarbons with typical Schulz-Flory distributions. Moreover, a maximum is observed for both Fischer-Tropsch and gas shift reactions for catalysts containing comparable amounts of cobalt and copper (Fig. 9). On the other hand, the depletion of one of them gives rise to a considerable change in selectivity and/or activity of the catalysts (10). This effect cannot be attributed to surface area changes, since the differences in surface area of the catalysts or of metallic copper are smaller than the differences in activity.

Therefore, a synergic effect must exist between cobalt and copper, with the formation of centers that are very active and selective for the synthesis of hydrocarbons. This effect seems to be related to the presence of a non-stoichiometric spinel-type phase, in which copper, cobalt, and zinc can be found in the octahedral positions, or to an

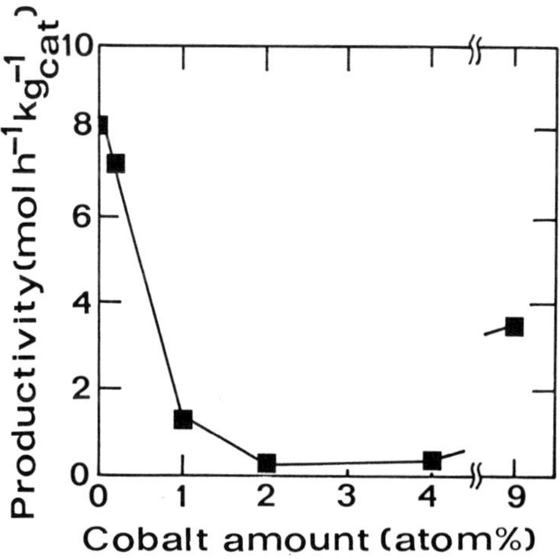

Fig. 10 Total productivity in hydrogenated compounds as a function of the cobalt content for Co/Cu/Zn/Cr catalysts containing 24% chromium and 38% zinc (atom. ratio). T=533K; P= 1.2MPa; $GHSV$= $15000h^{-1}$.

interaction between this phase and well-dispersed metallic copper formed in reducing conditions.

On the other hand, the presence of small amounts of Co^{2+} ions (up to 4%) inside the structure of typical methanol catalysts (44,45) has a drammatic poisoning effect on the activity without changing the selectivity, whereas at the highest cobalt contents a further increase in the activity takes place, along with a change in selectivity towards hydrocarbon synthesis (Fig. 10).

CONCLUSIONS

Non-stoichiometric phases may be useful precursors of hydrogenation catalysts with unusual physical and catalytic properties. These phases may be obtained by low-temperature methods and have different mechanisms of formation as a function of the nature of the elements present, which also regulates the physical and catalytic properties of these solids. The spinel structure is fairly empty and flexible in the accomodation of non-stoichiometry, which is characterized by an excess of bivalent cations, located in both tetrahedral and octahedral sites. The presence of Cr^{6+} ions is compatible with this model, considering that $ZnCrO_4$ has a crystal structure closely related to the spinel structure and has the same oxygen lattice.

The cations in the spinel structure can be substituted by many other cations to give multicomponents systems, whose properties can be regulated by proper selection of the cations and appropriate adjustament of the composition. In this way, specific interactions between the different cations present in the same structure may be obtained, which affect on the catalytic activity. The strong increase in activity in methanol or hydrocarbon synthesis from syngas, or the high activity and stability in the hydrogenation of oxo- aldehyde mixture may be attributed to the existence of specific synergic effects.

ACKNOWLEDGMENT

This work is a part of a scientific programm carried out within the framework of the "Progetto Finalizzato Energetica 2, CNR-ENEA".

REFERENCES

1. J.J. Burton and R.L. Garten (Ed.s), "Advanced Materials in Catalysis", Academic Press, N.Y., 1977.
2. O.T. Sorensen (Ed.), "Non-stoichiometric Oxides", Academic Press, N.Y., 1981.
3. R.K. Grasselli and J.F. Bradzil (Ed.s), "Solid State Chemistry in Catalysis", ACS Symp. Series 279, ACS Publ., Washington, 1985.
4. H. Yanagida, *Angew. Chem. (Engl. Ed.)* **27**, 1389 (1988).
5. G. Centi, F. Trifiro' and A. Vaccari, *Chim. Ind. (Milan)* **71**, 5 (1989).
6. A. Jones, P. Moseley and B. Tofield, *Chem. Brit.* **8**, 749 (1987).
7. E. Errani, F. Trifiro', A. Vaccari, M. Richter and G. Del Piero, *Catal. Lett.* **3**, 65 (1989).
8. P. Courty, D. Durand, E. Freund and A. Sugier, *J. Mol. Catal.* **17**, 241 (1982).
9. A. Riva, F. Trifiro', A. Vaccari, G. Busca, L. Mintchev, D. Sanfilippo and W. Manzatti, *J. Chem. Soc., Faraday Trans. 1* **83**, 2213 (1987).
10. G. Fornasari, S. Gusi, F. Trifiro' and A. Vaccari, *I&EC Res.* **26**, 1500 (1987).
11. R. Bechara, G. Wrobel, M. Daage and J.P. Bonnelle, *Appl. Catal.* **16**, 15 (1985).

12. G. Braca, A.M. Raspolli Galletti, F. Trifiro' and A. Vaccari, Italian Patent n. 21831A (1989).
13. R.J. Tilley, in "Surface Properties and Catalysis by non-Metals" (J.P. Bonnelle, B. Delmon and E. Derouane, Ed.s), Reidel Pub., Dordrecht (NL), 1982, p. 83.
14. G. Del Piero, F. Trifiro' and A. Vaccari, *J. Chem. Soc., Chem. Commun.*, 656 (1984).
15. M. Di Conca, A. Riva, F. Trifiro', A. Vaccari, G. Del Piero, V. Fattore and F. Pincolini, in "Proc. 8th Int. Congress Catalysis", Dechema, Frankfurt a.M., 1984, vol. 2, p. 173.
16. H.P. Klug and L.E. Alexander, "X-Ray Diffraction Procedures", Wiley, N.Y., 1974, p. 531.
17. M. Bertoldi, B. Fubini, E. Giamello, G. Busca, F. Trifiro' and A. Vaccari, *J. Chem. Soc., Faraday Trans. 1* **84.**, 1405 (1988).
18. E. Giamello, B. Fubini, M. Bertoldi, G. Busca and A.Vaccari, *J. Chem. Soc., Faraday Trans. 1* **85**, 237 (1989).
19. A. Riva, F. Trifiro', A. Vaccari, L. Mintchev and G. Busca, *J. Chem. Soc., Faraday Trans. 1* **84.**, 1423 (1988).
20. B. Fubini, E. Giamello, F. Trifiro' and A. Vaccari, *Therm. Acta* **133**, *155 (1988).*
21. J. Escard, I. Mantin and R. Sibut-Pinote, *Bull. Soc. Chim. France*, 3403 (1970).
22. P.E. Hojlund Nielsen, *Nature (London)* **267**, 822 (1977).
23. N. Von Laqua, S. Dudda and B. Revtlec, *Z. Anorg. Allg. Chem.* **428**, 151 (1977).
24. G. Del Piero, M. Di Conca, F. Trifiro' and A. Vaccari, in "Reactivity of Solids" (P. Barret and L.C. Dufour, Ed.s), Elsevier, Amsterdam, 1985, p. 1029.
25. G. Busca and A. Vaccari, *J. Catal.* **108**, 491 (1987).
26. G. Wrobel, J. Arsene, M. Lenglet, A. D' Huysser and J.P. Bonnelle, *Materials Chem.* **6**, 19 (1981).
27. A. D' Huysser, G. Wrobel and J.P. Bonnelle, *Nouv. J. Chimie* **6**, 437 (1982).
28. L. Jalowiecki, G. Wrobel, M. Daage and J.P. Bonnelle, *J. Catal.* **107**, 375 (1987).
29. N.T. McDevitt and W.L. Baun, *Spectrochim. Acta* **20**, 799 (1964).
30. R.A. Nyquist and R.O. Kagel, "Infrared Spectra of Inorganic Compounds", Academic Press, N.Y., 1971.
31. G. Wrobel, A. D' Huysser and J.P. Bonnelle, *Nouv. J. Chimie* **8**, 291 (1984).
32. A. Iimura, Y. Inoue and I. Yasumori, *Bull. Soc. Chim. Jpn* **56**, 2203 (1983).
33. H. Adkins, R. Connor and K. Folker, *J. Am. Chem. Soc.* **54**, 1138 (1932).
34. J.R. Monnier, M.J. Hanrahan and G. Apai, *J. Catal.* **92**, 119 (1985).
35. M. Piemontese, F. Trifiro', A. Vaccari, B. Fubini, E. Giamello, I. Rumori, in "Actas 12 Simp. Iberoamericano de Catalise", BTP/CAT, Rio de Janeiro, 1990, Vol. 2, p. 356.
36. G. Busca, F. Trifiro' and A. Vaccari, *Langmuir*, in press.
37. Z.T. Fattakova, A.A. Ukharskii, P.A. Shiryaev and A.D. Berman, *Kinet. Katal.* **27**, 884 (1986).
38. K. Klier, in "Advances in Catalysis" (D.D. Eley, H. Pines and P.B. Weisz, Ed.s), Academic Press, N.Y., Vol. 31, .1982, p. 243.
39. G.C. Chinchen, P.J. Denny, J.R. Jennings, M.S. Spencer and K.C. Waugh, *Appl. Catal.* **36**, 1 (1988).
40. E.B.M. Doesburg, R.H. Höppener, B. de Koning, X. Xiaoding and J.J.F. Scholten, in "Preparation of Catalysts IV" (B. Delmon, P. Grange, P.A. Jacobs and G. Poncelet, Ed.s), Elsevier, Amsterdam, 1987, p. 767.
41. S. Gusi, F. Trifiro', A. Vaccari and G. Del Piero, *J. Catal.* **94**, 120 (1985).
42. G. Wrobel, L. Jalowiecki and J.P. Bonnelle, *New J. Chem.* **11**, 715 (1987).
43. F. Trifiro', A. Vaccari, G. Braca and A.M. Raspolli Galletti, to be published.
44. E. Errani, G. Fornasari, T.M.G. La Torretta, F. Trifiro' and A. Vaccari, in "Actas XI Simposio Iberoamericano de Catalisis" (F. Cossio, O. Bermúdez, G. del Angel and R. Gómez, Ed.s), Instituto Mexicano del Petroleo, Mexico D.F., 1988, Vol. 3, p. 1239.
45. F.N. Lin and F. Pennella, in "Catalytic Conversion of Synthesis Gas and Alcohols to Chemicals" (R.G. Herman Ed.), Plenum Press, N.Y., 1984, p. 53.

R.K. Grasselli and A.W. Sleight (Editors), *Structure-Activity and Selectivity Relationships in Heterogeneous Catalysis*
© 1991 Elsevier Science Publishers B.V., Amsterdam

SUPPORTED METAL CATALYSTS PREPARED FROM AMORPHOUS METAL ALLOYS

A. Baiker[1], J. De Pietro[1], M. Maciejewski[1] and B. Walz[2]

[1]Department of Industrial and Engineering Chemistry, Swiss Federal Institute of Technology, ETH-Zentrum, CH-8092 Zürich, Switzerland

[2]University of Basel, Institute of Physics, CH-4056 Basel, Switzerland

SUMMARY

Ni/ZrO_2 and Pd/ZrO_2 catalysts have been prepared by controlled oxidation of amorphous $Ni_{64}Zr_{36}$ and $Pd_{33}Zr_{67}$ metal alloys in oxygen containing atmospheres. The oxidation which largely influences the morphological, structural and chemical properties of as-prepared catalysts has been studied using thermoanalytical methods (TG,DTA), XRD, XPS, gas adsorption and electron microscopy. The catalysts derived from the metallic glasses exhibit some unique structural and chemical properties which are discussed. Their potential for the liquid phase hydrogenation of organic compounds is illustrated using the hydrogenation of trans-2-hexene-1-al as an example. Hydrogenation over Ni/ZrO_2 yielded hexane-1-ol, whereas over Pd/ZrO_2 hexane-1-al could be produced selectively.

INTRODUCTION

Amorphous metal alloys have gained interest in catalysis research due to their potential as model catalysts and as catalyst precursors. Progress in this field has been discussed in two recent reviews [1,2]. Here we report the preparation of zirconia supported nickel and palladium catalysts from corresponding metallic glass precursors. The major aim was to learn more about the chemical and structural changes the metallic glass precursors undergo during their transformation to the active catalysts and about the suitability of as-prepared catalysts for liquid phase hydrogenations of organic compounds.

EXPERIMENTAL

The metallic glass precursors, $Ni_{64}Zr_{36}$ and $Pd_{33}Zr_{67}$, were prepared from the pure metals using the technique of melt spinning. Before use as precursor materials the ribbons were ground in liquid nitrogen to flakes of about 0.5 - 1mm size. Catalysts were prepared by oxidizing the precursor materials in an oxygen containing atmosphere under appropriate conditions and subsequent reduction in hydrogen at 600 K. The structural and chemical changes the metallic glass precursors underwent during their transformation to the active catalysts have been studied using powder X-ray diffraction (XRD), thermal analysis (TG,DTA), X-ray photoelectron spectroscopy (XPS), gas adsorption and scanning and transmission electron microscopy. Catalytic tests were performed in a 500 ml autoclave under constant hydrogen pressure using an agitator speed of 1500 rpm. Products were analyzed by gas chromatography using a HP 5890A GC equipped with a HP-FFAP capillary column (30m x 0.53mm x 1μm).

RESULTS AND DISCUSSION

Genesis of active catalysts from metallic glasses

Nickel on zirconia from $Ni_{64}Zr_{36}$

Nickel on zirconia was prepared by controlled oxidation of the amorphous Ni-Zr alloy in air and subsequent reduction. The oxidation in air in the temperature range 570-750K resulted in solids containing ZrO_2 and metallic nickel besides unreacted amorphous metal alloy. Significant Oxidation of Ni to NiO was only observed after almost complete oxidation of the zirconium in the alloy. Figure 1 depicts the XRD patterns of the amorphous $Ni_{64}Zr_{36}$ alloy corresponding to different degrees of oxidation (α) of the amorphous metal alloy. α was measured gravimetrically and denotes the fraction oxygen consumed divided by the amount of oxygen required to convert Zr to ZrO_2. The XRD patterns indicate the built-up of small crystalline particles of tetragonal and monoclinic ZrO_2 and metallic nickel upon oxidation.

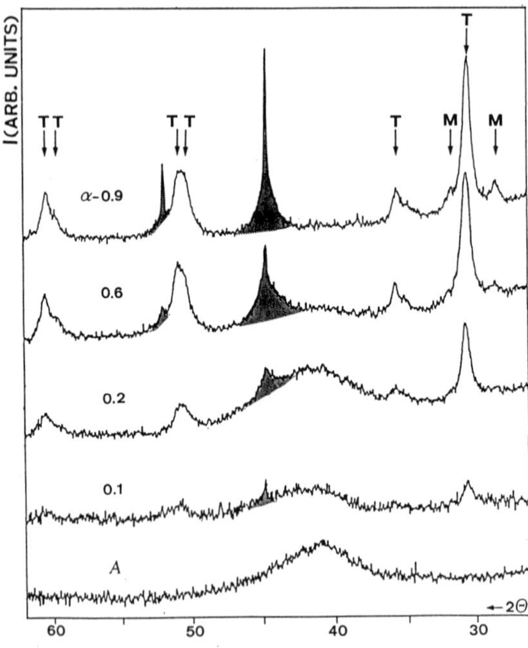

Fig. 1 XRD patterns of amorphous $Ni_{64}Zr_{36}$ alloys of different degree of oxidation α A - as quenched alloy. Reflections of Ni are shaded, arrows indicate positions of main reflections of tetragonal (T) and monoclinic (M) ZrO_2.

The crystallization behavior of as-prepared samples investigated by DSC measurements under an inert gas atmosphere is shown in Fig. 2. Note that the temperature range of crystallization did not depend significantly on the degree of oxidation of the alloy. This behavior was further supported by the observation that the specific heat of crystallization referred to the unreacted core of the alloy was constant ca. 40 J/g, regardless of the degree of oxidation of the alloy sample (3). Thus the presence of zirconia in the oxidized alloy did virtually not influence the crystallization behavior of the unreacted part of the alloy.

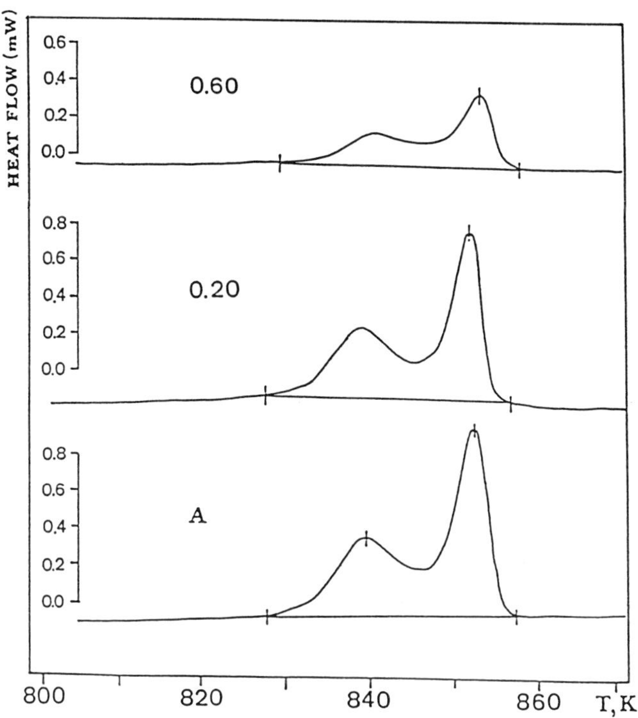

Fig. 2 Crystallization behavior of amorphous and partially oxidized $Ni_{64}Zr_{36}$ alloy investigated by DSC measurements. (A) corresponds to as-quenched amorphous alloy, the degree of oxidation α is indicated on curves. Heating rate 5K/min.

The chemical and structural changes of the bulk were accompanied by similar drastic changes in the textural properties of the alloy. The BET surface area of the precursor material (0.02 m^2/g) increased to 10 - 25 m^2/g depending on the oxidation conditions used.

The surface oxidation behavior of the amorphous precursor alloy was investigated by means of XPS and UPS (4). Oxygen doses up to 2000L were used to study the initial stages of the oxidation of the clean surfaces in the temperature range from room temperature to 570K.

Fig. 3 compares the XPS Zr 3d spectra of the fresh amorphous $Ni_{64}Zr_{36}$ alloy, the sample after exposure to 80 L O_2, and a ZrO_2 reference sample. The Zr 3d levels of the alloy cleaned by argon ion bombardment are located at E_b = 179.4 eV. The shift compared to clean metallic Zr (E_b= 179.0 eV) is due to alloying (5). After exposure to 80 L an additional doublet can be seen which is attributed to Zr in an oxidized state shifted by 3.1 eV with respect to pure Zr. The different shifts in the Zr 3d core levels of the ZrO_2 reference sample and the sample obtained by exposure to 80 L O_2 indicate a different stoichiometry of these zirconium oxides. Comparison of the observed shifts with literature data indicated that the zirconia formed upon oxygen exposure was deficient in oxygen $ZrO_{1<x<2}$. It appears that oxygen deficient zirconia is formed predominantly in the initial stage of the oxidation, i.e. when metallic zirconium is still present in the sample. With higher degree of oxidation (i.e. when the bulk of the alloy is oxidized) stoichiometric zirconia (ZrO_2) becomes prevalent, as the XRD patterns (Fig. 1) indicate.

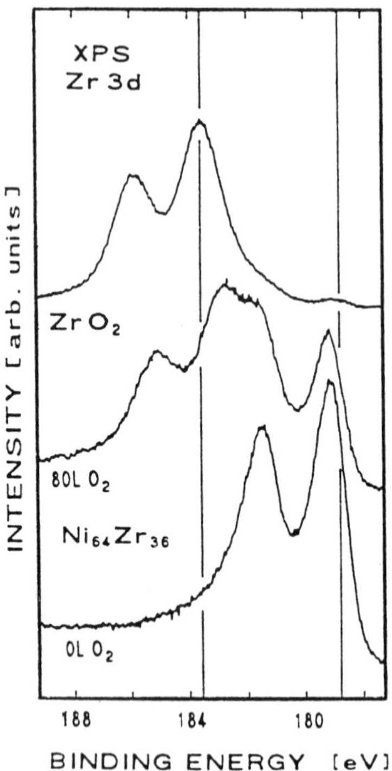

Fig. 3 XPS core level spectra of Zr 3d levels of $Ni_{64}Zr_{36}$ after exposure to 80 L O_2 at 420K. Reference spectra of ZrO_2 (obtained after exposure of Zr to 1000 L O_2) and clean metallic alloy (0 L O_2) are shown for comparison. Zr $3d_{5/2}$ core level positions of Zr and ZrO_2 are indicated by vertical lines.

Palladium on zirconia from $Pd_{33}Zr_{67}$

Figure 4 shows the XRD patterns of the amorphous $Pd_{33}Zr_{67}$ alloy after oxidation in air at 590K for different times. In contrast to the behavior observed with the Ni-Zr alloy significant oxidation of the group VIII transition metal occurs already at relatively low degree of oxidation α. Note that the degree of oxidation α was defined here as the fraction oxygen consumed divided by the amount of oxygen required to convert the alloy to PdO and ZrO_2. The bulk concentration of metallic Pd first increases and then decreases with increasing degree of oxidation α. The fully oxidized sample contained ZrO_2, PdO and a little Pd. Although monoclinic and tetragonal ZrO_2 exist in the oxidized samples, the monoclinic phase is dominant independent of the degree of oxidation.

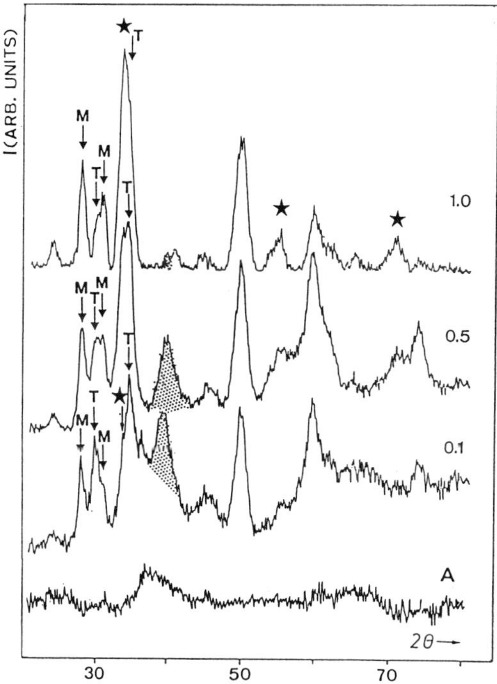

Fig. 4 XRD patterns of amorphous $Pd_{33}Zr_{67}$ after different degree of oxidation α. Oxidation carried out in air at 590K. A - as-quenched alloy; reflections of Pd are shaded, asterisks indicate reflections due to PdO, arrows indicate positions of main reflections of monoclinic (M) and tetragonal (T) ZrO_2.

The crystallization behavior of the oxidized samples (Fig. 4) has been investigated using DSC (Fig. 5.)

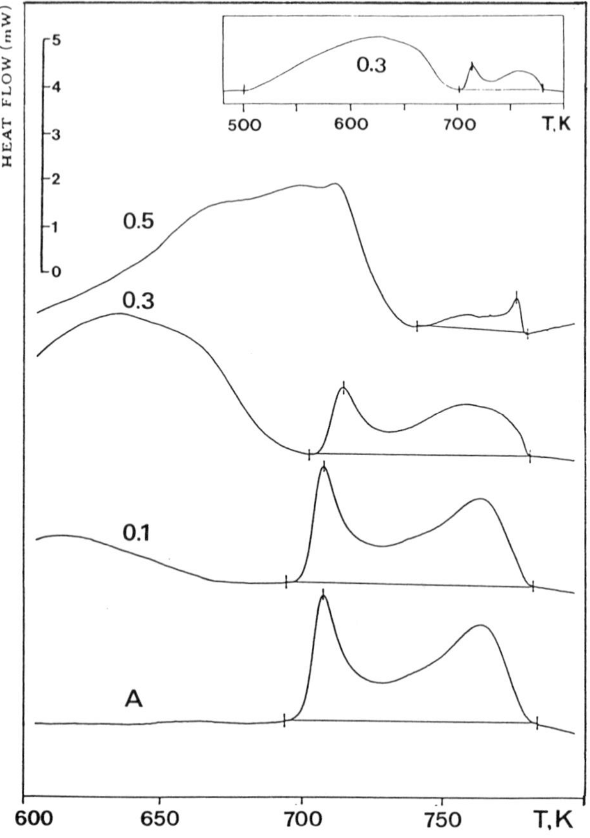

Fig. 5 Crystallization behavior of the $Pd_{33}Zr_{67}$ alloy after oxidation in air at 590K investigated by DSC under inert gas atmosphere. The degree of oxidation α is marked on the curves. Trace (A) corresponds to as-quenched amorphous alloy. Inset in upper right corner represents overview of DSC curve for $\alpha = 0.3$ and illustrates the occurrence of the solid state reduction 2 PdO + Zr $\rightarrow$ 2 Pd + ZrO_2 (broad signal at 500-700K) previous to crystallization of the unreacted alloy at higher temperature. Heating rate 5K/min.

The DSC curves shown in Fig. 5 indicate that the as-quenched sample starts to crystallize around 700K under the conditions used. Most interesting is that the partially oxidized samples show an additional exothermal process at significantly lower temperature. Simultaneous TG measurements revealed that during both thermal events the sample weight did not change. The exothermal process occurring at lower temperature is attributed to the solid state reduction of PdO by metallic Zr: $2\ PdO + Zr \rightarrow 2\ Pd + ZrO_2$ which occurred in the partially oxidized samples. For the sample with $\alpha = 0.3$ this reaction was not complete before the crystallization of the unreacted amorphous alloy occurred. The superposition of the two thermal events is even more pronounced for the sample with $\alpha = 0.5$. Note that the larger the degree of oxidation (α) of the sample was, the higher was the temperature of the reduction of PdO by metallic zirconium.

As a result of the drastic chemical and structural changes of the bulk material the BET surface area of the amorphous precursor increased from 0.025 to about 60 m^2/g depending on the oxidation conditions used. Pore size distribution measurements using nitrogen capillary condensation indicated that the material contained mainly pores of 2 - 4 nm size besides some larger pores.

The morphological changes are illustrated by the scanning electron micrograph presented in Fig. 6. The initially flat surface of the precursor alloy changed to a rough surface which was built up of small agglomerates containing zirconia and palladium as evidenced by electron dispersive X-ray analysis. High resolution electron microscopy as well as electron diffraction showed that the agglomerates were made up of intimately mixed intergrown small crystallites of zirconia and palladium. Due to this particular structural property as-prepared catalysts exhibit an extremely large interfacial area between the active metal species and the oxidic support material, as has been demonstrated in detail elsewhere (6). Similar large interfacial areas between the metal and the oxidic support are generally not observed in conventionally prepared supported metal catalysts.

Figure 7 depicts the XPS core level spectra of Zr 3d levels of $Pd_{33}Zr_{67}$ measured after different exposures of the alloy to oxygen at 420K. Similar characteristics were observed as with the Ni-Zr alloy. With increasing oxygen exposure the minimum between the two Zr 3d peaks decreases and a shoulder grows at the higher binding energy side of the Zr $3d_{3/2}$ peak indicating the formation of zirconium oxide. Again the zirconium oxide formed is non-stoichiometric, i.e. deficient in oxygen, as emerges from the observed shifts of the Zr 3d peaks (compare with reference in Fig. 3).

The XRD and XPS investigations indicate that oxygen deficient ZrO_2 is prevalent in the initial stage of the oxidation (surface and subsurface region), whereas in completely oxidized samples (bulk oxidation) stoichiometric ZrO_2 prevails. This behavior was found to be characteristic for $Ni_{64}Zr_{36}$ as well as $Pd_{33}Zr_{67}$ alloys.

Fig. 6 Scanning electron micrographs illustrating the morphology of the amorphous $Pd_{33}Zr_{67}$ alloy after oxidation in oxygen atmosphere (0.9 bar O_2) during 5 hours. White bare on left bottom side corresponds to 1 μm.

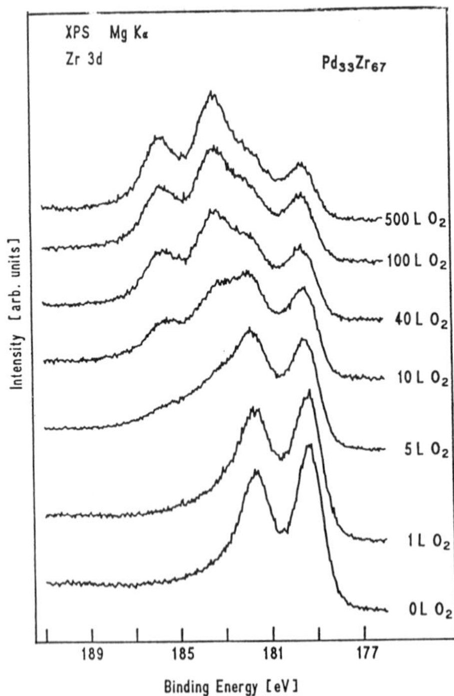

Fig. 7 XPS core level spectra of Zr 3d levels of $Pd_{33}Zr_{67}$ after different exposure to oxygen at 420K (4).

Catalytic properties of catalysts

Figure 8 compares the catalytic properties of the metal/zirconia catalysts derived from the amorphous Ni-Zr and Pd-Zr alloys for the liquid phase hydrogenation of trans-2-hexene-1-al. Note that over Ni/ZrO_2 the hydrogenation occurred in two consecutive steps, finally producing hexane-1-ol (C). In contrast, over Pd/ZrO_2 the consecutive hydrogenation step is not occurring and hexane-1-al (B) is formed with high selectivity.

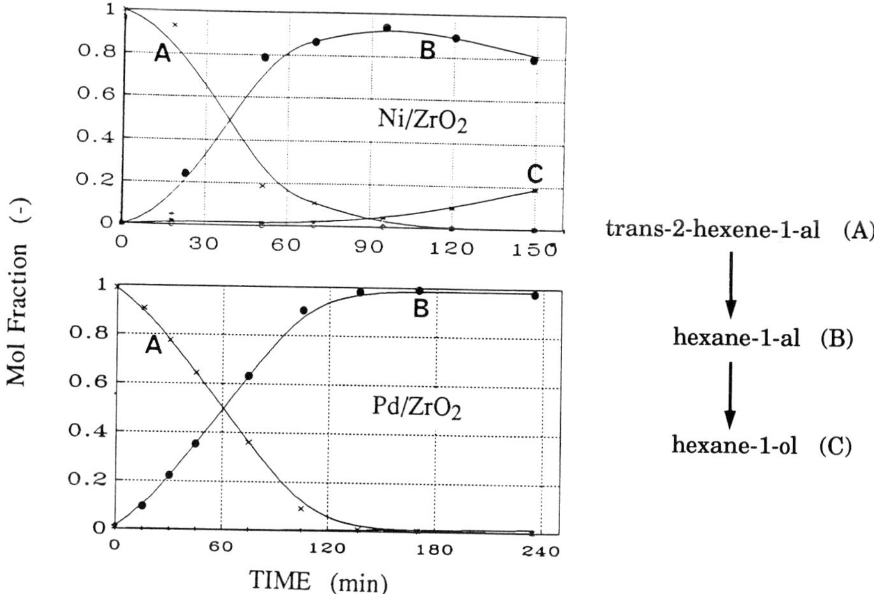

Fig. 8 Hydrogenation of trans-2-hexene-1-al. Composition of reactant mixture versus time. Conditions: Ni/ZrO_2 (45g/l), 400 K, 10 bar H_2; Pd/ZrO_2 (60g/l). 370 K, H_2 1.2 bar.

CONCLUSIONS

Ni/ZrO_2 and Pd/ZrO_2 catalysts with interesting chemical and structural properties were prepared from corresponding metal-zirconium alloys by controlled oxidation in an oxygen containing atmosphere and subsequent reduction in hydrogen. The interplay between the simultaneously occurring oxidation and crystallization processes during the oxidation of the precursor alloy was found to be crucial for the development of the structural and chemical properties of the final catalysts. Ni-Zr and Pd-Zr alloys exhibit significantly different oxidation behavior. With Ni-Zr alloys the zirconium is almost selectively oxidized and a solid containing metallic nickel and zirconia is formed. The oxidation of nickel is only observed after depletion of the metallic zirconium in the alloy or when the oxidation is carried out at high temperatures. In contrast, with Pd-Zr alloys both components are oxidized simultaneously and the resulting solid contains PdO, Pd and ZrO_2. In the presence of metallic zirconium (i.e. with partially oxidized samples) the solid state reduction $2\ PdO + Zr \rightarrow Pd + ZrO_2$ was found to occur when samples were heated in an inert gas atmosphere. This phenomenon indicates that the metallic zirconium present in partially oxidized alloys can act as oxygen scavenger which suppresses deactivation of the active metal species by oxygen contamination.

The active catalysts are porous and exhibit BET surface areas in the range of 10-60 m^2/g depending on the precursor alloy and the oxidation conditions used. A characteristic structural feature of as-prepared catalysts is that the active metal particles are intimately associated with the zirconia phase resulting in unusually large interfacial areas between the metal and the oxidic phases. This characteristic structural property is likely to be of importance for all phenomena where the interfacial area plays a role such as metal-support interaction, and adsorption and spillover of hydrogen.

Another peculiarity of the catalysts derived from the metallic glasses is the fact that ZrO_2 is present in monoclinic and tetragonal form. Furthermore, the XPS investigations indicated that part of the zirconia (surface and subsurface region) exists as non-stoichiometric, i.e. oxygen deficient $ZrO_{1<x<2}$.

The catalysts derived from the metallic alloys are suitable for liquid phase hydrogenations in stirred autoclaves as has been demonstrated using the example of the hydrogenation of trans-2-hexene-1-al. With Ni/ZrO_2 the hydrogenation to hexane-1-ol occurs, whereas with Pd/ZrO_2 hexane-1-al can be produced selectively.

Acknowledgement

Thanks are due to Lonza AG, Switzerland for financial support of this work.

REFERENCES

1 A. Molnar, G.V. Smith, and M. Bartok, Adv. Catal. 36 (1989) 329-383.

2 A. Baiker, Faraday Discuss. Chem. Soc. 87 (1989) 239-251; A. Baiker, in H. Beck and H.J. Güntherodt (Eds.),Topics in Applied Physics, Glassy Metals III, Springer, Berlin,1991, in press.

3 M. Maciejewski and A. Baiker, J. Chem. Soc. Faraday. Trans. 86 (1990) 843-848.

4 B. Walz, PhD. Thesis, University of Basel, 1989; B. Walz, P. Oelhafen, H.J. Güntherodt, and A. Baiker, Appl. Surf. Sci., 37 (1989) 337-352.

5 P.H. Oelhafen, in H. Beck and H.J. Güntherodt (Eds.),Topics in Applied Physics, Vol. 53, Glassy Metals II, Springer, Berlin, 1983, p. 283.

6 A. Baiker, D. Gasser, J. Lenzner, A. Reller and R. Schlögl, J. Catal., in press.

R.K. Grasselli and A.W. Sleight (Editors), *Structure-Activity and Selectivity Relationships in Heterogeneous Catalysis*
1991 Elsevier Science Publishers B.V., Amsterdam

STRUCTURE SENSITIVITY IN ZEOLITE CATALYSTS

F. G. DWYER
Mobil Research and Development Corporation, Paulsboro Research Laboratory, Paulsboro, NJ 08066

ABSTRACT

Almost from its beginning, zeolite catalysis has had a strong dependency upon the catalyst structure. Shape selective catalysis using zeolites has controlled selectivity by constraining the reactants that can be admitted to the catalytic sites, the products that can emit from the zeolite pores and the products that can be formed within the zeolite cavity. Although advances continue to be made in this area, they are mainly subtle refinements of the original concepts. A major challenge is and has been to combine the shape selective constraints with homogeneous and/or enzymatic catalysis. The major obstacle has been the limited size of zeolite pores and cavities to accommodate the host homogeneous or enzymatic agents. With the recent discovery of the 18 membered $AlPO_4$, VPI-5/MCM-9, hope for the synthesis of zeolites to accommodate such systems has been renewed.

INTRODUCTION

When we speak of structure sensitivity in zeolite catalysis, we are really referring to what is more commonly referred to as shape selectivity. This term, originally coined by Weisz and Frilette in 1960 (1), embodies several structure sensitive phenomena. Size exclusion, in which reactants are prohibited in reaching the catalytic surface because of their size or products are barred from returning from the catalytic surface to the bulk, was the first phenomenon studied. Subsequently spatiospecific selectivity, where reactions are inhibited by the inability of reaction intermediates to form in the confines of the zeolite cavity, and diffusive selectivity were observed and applications developed.

This unique ability to control the selectivity and essentially direct chemical reactions by purely physical constraints has always led to speculation of how we can combine these effects with catalytic surfaces or sites not found in zeolites. Although multifunctional catalysts have been prepared by incorporating metals into zeolites, the metal functions have complemented the zeolite catalysis rather than been enhanced by the zeolite's shape selective constraints.

For the sake of this discussion catalytic cracking has not been included. It is generally accepted that in catalytic cracking the large, high molecular weight hydrocarbons in the gas oil feed are first cracked either thermally or catalytically on the amorphous matrix component of the catalyst and it is the cracked products that can now diffuse into the zeolite portion of the catalyst reacting further. It is speculative whether the selectivity would be further enhanced, more gasoline/less coke, if the zeolite pores were large enough to admit the large hydrocarbon feed molecules. Nevertheless if we were to extend our definition of shape selectivity, even catalytic cracking with zeolite catalysts would be included.

ZEOLITE FUNDAMENTALS

For the purpose of this discussion, I will define zeolites as porous crystalline materials formed by the connection of numerous elements, usually tetrahedrally coordinated, through oxygen bridges. They usually have ion exchange capacity and can be readily converted into a variety of catalytic forms. The pore systems are regular and well-defined and can afford access to the interior of the crystal in a unidirectional mode, such as a tube, or by dual pore systems, intersecting and nonintersecting, or even a three-dimensional access. To demonstrate some of these characteristics, Figure 1 depicts the structure of faujasite, a large pore zeolite with three-dimensional access. In Figure 2, the two-dimensional pore system and the 010 projection of ZSM-5 are shown. Zeolites of interest in catalysis are usually classified according to the number of tetrahedral units forming the pore opening: 12 - large, 10 - intermediate, and 8 - small. Examples of zeolites classified in this manner are shown in Figure 3.

TYPES OF SHAPE SELECTIVITY

As has been stated before, shape selectivity in zeolite catalysis can be divided into three general categories: size exclusion, spatiospecific and diffusive. We will discuss each type separately citing examples and, where applicable, processes built on these concepts.

Size Exclusion

Early examples of size exclusive selectivity were the selective dehydration of n-butanol in the presence of i-butanol over the zeolite CaA (2) and the selective cracking of n-hexane in the presence of 3-methylpentane (3), Tables 1 and 2. Commercial applications of this characteristic were demonstrated in the post-catalytic reforming processes, Selectoforming and M-forming, in which the low octane n-paraffins in the reformate were selectively cracked to lighter hydrocarbon fragments resulting in higher octane number product. Selectoforming employed erionite as the zeolitic catalytic component which had an 8-membered ring pore structure and M-forming utilized ZSM-5 having a 10-membered ring pore structure. The selective cracking phenomenon is illustrated in Tables 3 and 4 showing the compositional changes brought about by the process with the corresponding octane number increase.

By far the broadest application of size exclusion is in catalytic dewaxing processes. In these processes the low temperature rheology is changed to a more fluid product by cracking normal and slightly branched paraffins in the feed to lighter products. Mobil Distillate Dewaxing, MDDW, is the process for improving the low temperature fluidity of diesel fuel, jet fuel and heating oils while Mobil Lube Dewaxing, MLDW, is the process for lube stocks. Both processes

Figure 1

Faujasite

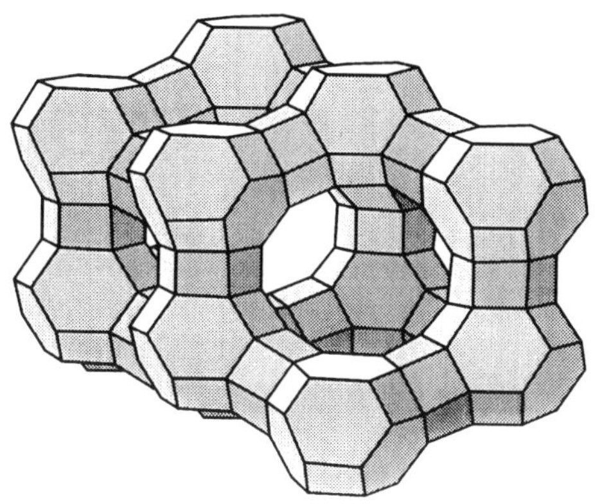

Figure 2

ZSM - 5

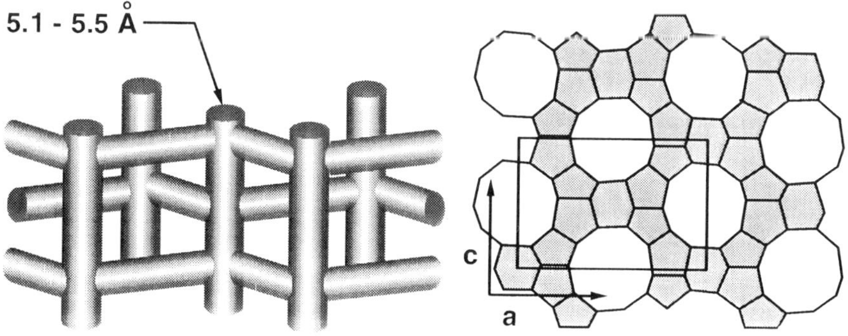

Figure 3

Pore Openings of Zeolite Structures

Phase	Tetrahedral Units	Dimensions (Å)*
Faujasite, X,Y	12	7.4
Mazzite, ZSM-4	12	7.4
Mordenite	12	6.7 x 7.0
Offretite	12	6.4
Cancrinite	12	6.2
Heulandite	10	4.4 x 7.2
Stilbite	10	4.1 x 6.2
Ferrierite	10	4.3 x 5.5
ZSM-5	10	5.4 x 5.6
Zeolite A	8	4.1
Erionite	8	3.6 x 5.2

* Based on X-Ray Structure Determination Using an Oxygen Radius of 1.35Å

Table 1

Dehydration of Primary Butyl Alcohols

1 Atm. Pressure

Temperature °C	130	230	260
Over Linde CaA Molecular Sieve			
Isobutyl Alcohol, Wt%		<2	<2
n-Butyl Alcohol, Wt%		18	60
sec-Butyl Alcohol, Wt%	~ 0		
Over Faujasite-Type CaX Molecular Sieve			
Isobutyl Alcohol, Wt%		46	85
n-Butyl Alcohol, Wt%		9	64
sec-Butyl Alcohol, Wt%	82		

Table 2

Molecular-Shape Selective Cracking

Catalyst	Hydrocarbon Charge	Time on Stream (Min)	Temperature (°C)	Conversion (%)
H Gmelinite	n-Hexane	10 to 33	370	47 to 30
H Gmelinite	2-Methylpentane	10 to 20	320 to 540	0 to 0.7
H Gmelinite	Methycyclopentane	10 to 20	510 to 540	0.4 to 1.9
H Erionite	n-Hexane	26	320	52.1
H Erionite	2-Methylpentane	26	430	1.0
H Erionite	2-Methylpentane	26	540	4.7
H Chabazite	n-Hexane	30	260	10.0
H Chabazite	2-Methylpentane	10	540	1.5

Table 3

Selectoforming a C_5 -82 °C Light Naphtha

Composition	Feed	Product at 395°C	Product at 425°C
Methane + Ethane		1.5	5.0
Propane		21.9	24.7
Isobutane		0.4	1.3
n-Butane		6.0	7.7
Isopentane	22.2	22.1	21.7
n-Pentane	23.4	11.1	2.8
C_6^+	54.4	37.0	31.5
C_5^+ Octane, R+O	67.4	79.6	84.0

at 28 Atm., 1.6 LHSV

employ a ZSM-5 containing catalyst tailored to the specific process. Figure 4 shows the relative effect of hydrocarbon components of petroleum stocks on the pour point, the property we use to measure low temperature fluidity. Another interesting aspect of dewaxing is the effect the shape of the pore opening can have on the effectiveness of the process. ZSM-23 is a zeolite which like ZSM-5 has a 10-membered pore opening but with a tear-shaped opening. Figure 5 shows the projections of the pore openings for ZSM-5 and ZSM-23 with their corresponding effective and crystallographic dimensions. When dewaxing a distillate stock to the same pore point, ZSM-23 performs less conversion than ZSM-5 and gives a product with a higher viscosity index, a measure of the temperature/viscosity response (Table 5). This result is interpreted as due to the further restriction of the pores of ZSM-23 reducing the amount of branched paraffins that are cracked in the dewaxing process.

Finally, this size exclusive property of ZSM-5 is also exhibited when used as an octane enhancing additive catalyst in catalytic cracking. In this operation, low octane n-paraffins and olefins are cracked to light gases while some olefin isomerization also occurs. Table 6 compares operation with and without ZSM-5 for both FCC and TCC.

Spatiospecificity

When both the reactant molecule and the product molecule are small enough to diffuse through the zeolite pore channels, but the reaction intermediates are larger than either the reactants or products and are spatially constrained either by their size or orientation, we refer to this as spatiospecific selectivity. Spatioselectivity or transition state selectivity is independent of crystal size and activity, but depends on the pore diameter and zeolite structure. This type of selectivity was first proposed by Csicsery (4) in 1971.

Spatioselectivity plays a major role in the selective cracking of paraffins in medium pore zeolites. For example, n-hexane and 3-methylpentane are readily sorbed by ZSM-5, yet the singly-branched molecule cracks at a significantly slower rate than the straight chain molecule. 3-methylpentane, being a bulkier molecule than n-hexane, apparently requires more space than n-hexane to form the reaction intermediate as shown in Figure 6.

A very significant commercial application of spatioselectivity is in the isomerization of xylenes using ZSM-5 containing catalysts. Xylenes can undergo isomerization via intramolecular 1,2-methyl shift as well as a disproportionation via diphenylmethanes, Figure 7. The intracrystalline cavity of ZSM-5 cannot readily accommodate the disproportionation reaction intermediate, hence shifting the selectivity dramatically towards isomerization. Figure 8 shows this selectivity effect in comparing the relative amounts of disproportionation and isomerization for ZSM-5 vs faujasite, a large pore zeolite with a much larger intracrystalline cavity.

Table 4

Compositional Change at Different M-Forming Severities

C_6-80°C Mid-Continent Naphtha, 28 Atm.

	Feed	Product		Product	
C_5^+, R+O	84.5	89.6		92.7	
Aromatics					
B	18.5	16.9	-1.6	16.3	-2.2
T	23.4	22.5	-0.9	21.7	-1.7
X	0.6	1.1	+0.5	1.5	+0.9
C_9	0.2	2.9	+2.7	3.7	+3.5
C_{10}	0	2.0	+2.0	3.3	+3.3
C_{11}	0	0.3	+0.3	0.4	+0.4
Total	42.7	45.8	+3.0	46.8	+4.1
Paraffins					
C_6	30.9	26.2	-4.7	23.4	-7.5
C_7	12.9	9.7	-3.2	8.7	-4.2
C_8	0.2	0	-0.2	0.3	+0.1
C_9	0	0	0	0	0
Total	44.0	35.9	-8.1	32.4	-11.6
Naphthenes	1.4	1.6	+0.2	1.1	-0.3
Pentanes	8.0	8.6	+0.6	9.0	+1.0
Butanes and Lighter	3.9	8.1	+4.2	10.7	+6.8

Figure 4

Conversion of Alkane and Alkylbenzene Molecular Classes During Dewaxing Over ZSM-5

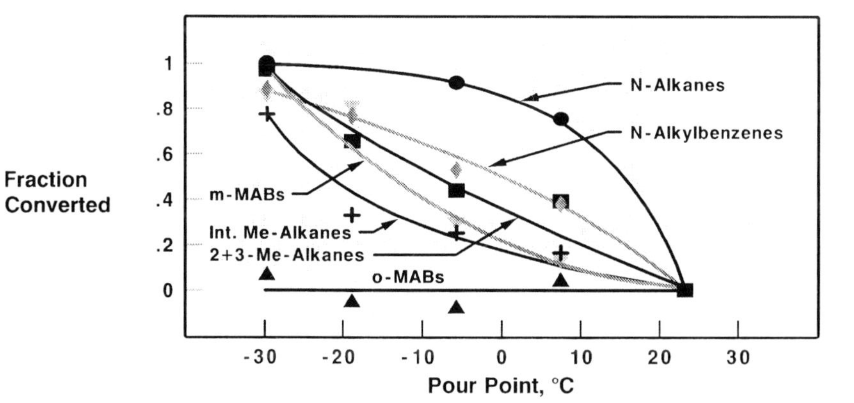

Figure 5

Projections of ZSM - 5 and 23 Structures

Crystallographic Pore Size (Å)

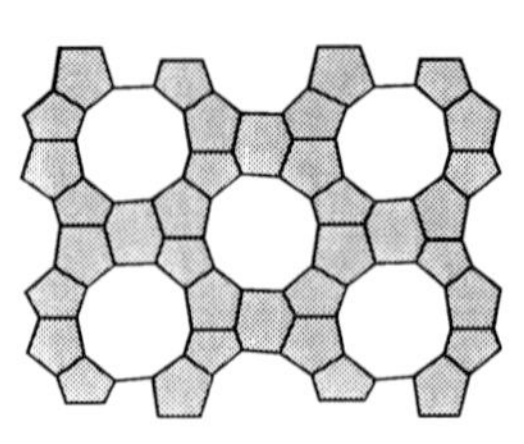

5.1 x 5.5

ZSM - 5

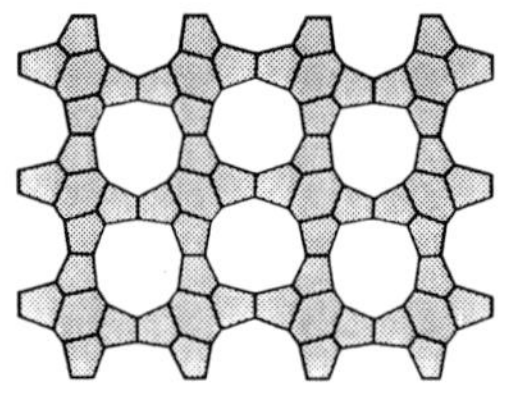

4.5 x 5.6

ZSM - 23

Table 5

Dewaxing Over ZSM-5 and ZSM-23

Catalyst	ZSM-5	ZSM-23
Conversion, Wt%	15	11
Product Pour Point, °C	-12	-12
Viscosity Index	101.0	108.7

Table 6

Commercial Test of ZSM-5

	TCC			FCC	
	Base Catalyst	Catalyst Containing ZSM-5		Base Catalyst	Catalyst Containing ZSM-5
Time on Stream, Days	0	72	106	0	37
Conversion, Vol%	53.0	53.0	53.0	73.2	73.2
C_5^+ Gasoline, Vol%	42.3	40.9	41.0	53.9	51.6
Butenes, Vol%	3.8	4.6	4.2	7.9	9.0
Propene, Vol%	3.7	4.7	4.2	6.2	7.0
Light Fuel Oil, Vol%	29.9	29.1	26.4	26.5	26.5
Coke, Wt%	--	--	--	6.7	6.7
Research Octane Number R+O	86.0	90.2	91.2	87.3	89.0
Motor Octane Number M+O	77.4	79.2	79.5	78.1	78.7
Potential Alkylate, Vol%	12.7	15.7	14.2	24.9	28.3
Total Gasoline, Vol%	55.0	56.6	55.2	78.8	79.9

Figure 6

Mechanism of Paraffin Cracking

n - hexane

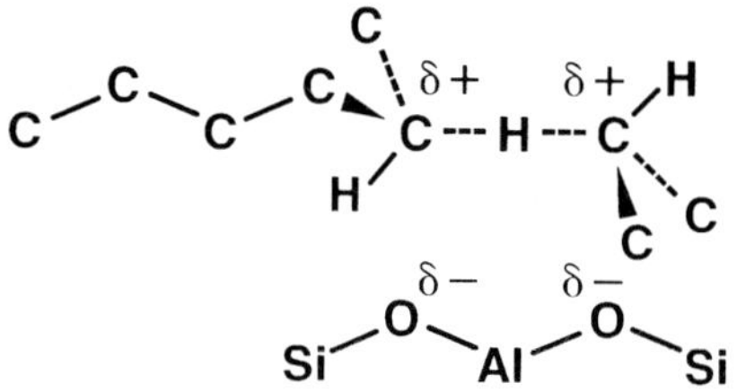

Cross - Section

4.9 x 6 Å

3 - methylpentane

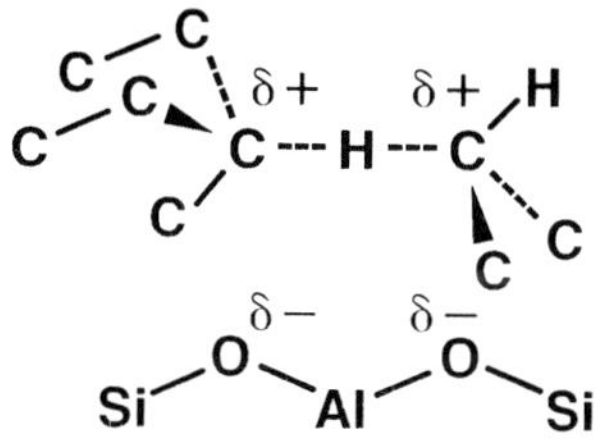

6 x 7 Å

Figure 7

Xylene Isomerization

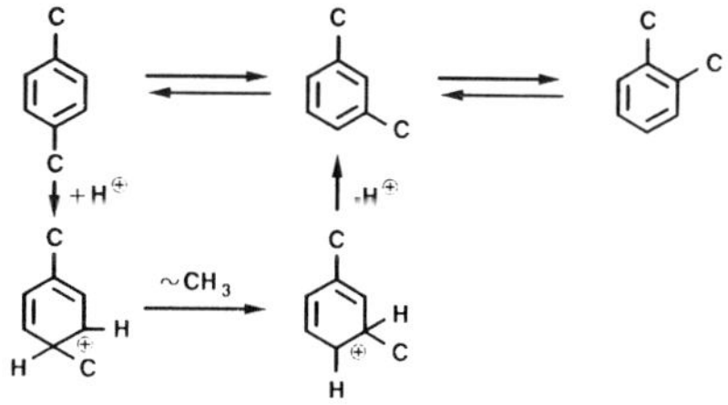

$+H^{\oplus}$

$\sim CH_3$

$-H^{\oplus}$

Disproportionation

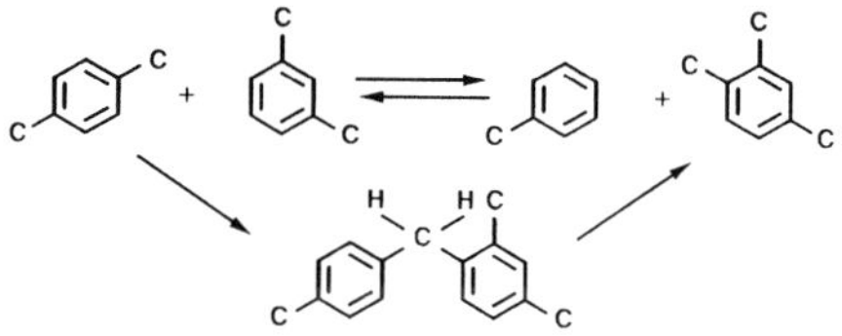

Figure 8

Disproportionation vs. Isomerization

m-Xylene Feed, 573 K (300°C)
Comparison of ZSM-5 with Faujasite*

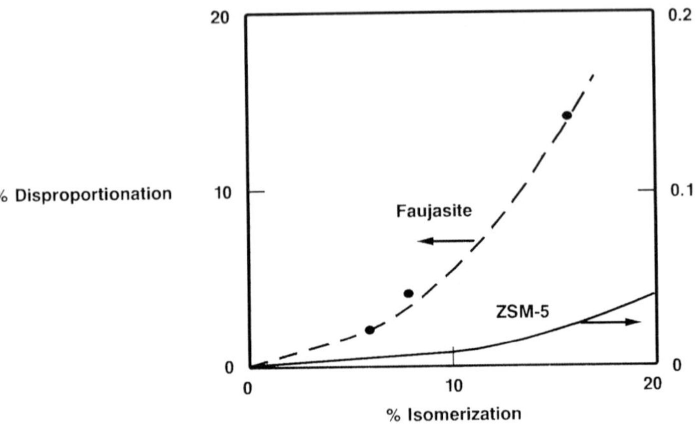

*Lanewalda & Bolten, J. Org. Chem. 34,3107 (1969)

Figure 9

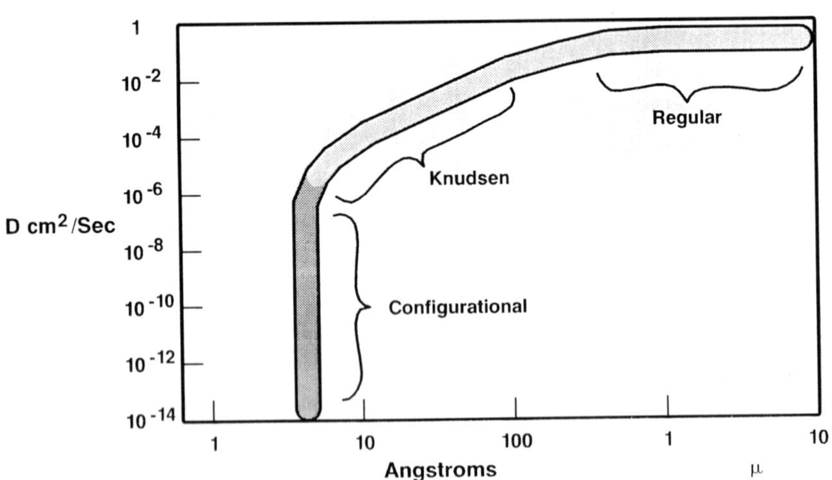

Diffusivity and Size of Aperture (Pore); the Classical Regions of Regular and Knudsen and the New Regime of Configurational Diffusion

Table 7

Hydrogenation of a Mixture of Trans- and Cis- Butene-2

Temp., °C	Initial Composition Trans.	Cis.	Final Composition Trans.	Cis.	n-Butane	Conversion ϵ, Wt% Trans.	Cis.	$\frac{k_{Trans}}{k_{Cis}}$
120	78.7	21.3	37.1	17.0	45.9	52.9	20.8	3.3
103	78.7	21.3	57.3	19.8	22.9	27.2	7.1	4.3
98	78.7	21.3	69.4	20.9	9.7	11.8	1.8	7.0

$k_{Trans}/k_{Cis} = \ln(1 - \epsilon_{Cis})/\ln(1 - \epsilon_{Trans})$.

Figure 10

Kinetic Model for p-Xylene Preference with ZSM-5 Catalyst

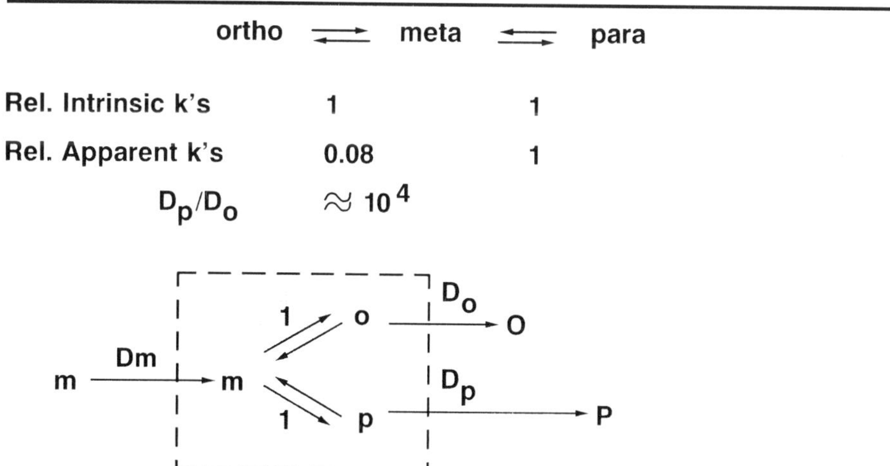

Table 8
Toluene Disproportionation

	Large Crystal ZSM-5	Mg ZSM-5	Thermodynamic Equilibrium
Temperature, °C	550	550	
WHSV	30	3.5	
Feedstock	Toluene	Toluene	
Conversion, Wt%			
Toluene	13.2	10.9	
% Xylenes			
Para	35	88	23
Meta	46	10	51
Ortho	19	2	26

Diffusive Selectivity

Diffusion in zeolites does not conform to the principles we usually apply in diffusive processes. The diffusive process in zeolites has been described by Weisz (5) as configurational diffusion. In this diffusion regime, even a subtle change in the dimensions of a molecule can result in a large change in its diffusivity as shown in Figure 9. This type of diffusion and its effect on catalytic selectivity was shown (Chen and Weisz) (6) in the hydrogenation of trans- and cis-butene where, although the two molecules differ in size by only 0.2Å, the much higher diffusion rate of trans-butene results in a much higher reaction rate (Table 7). Commercial processes utilizing this structure sensitive property include several aromatics processing applications such as xylene isomerization, selective toluene disproportionation (STDP), para-ethyl toluene synthesis and p-diethylbenzene synthesis. In xylene isomerization the slight difference in the size of the para-isomer but large difference in diffusivity leads to a para-selective product as shown in Figure 10.

In the case of toluene disproportionation, an equilibrium mixture of xylenes is obtained using a standard ZSM-5 catalyst containing ZSM-5 crystals <0.5 μm. Increasing the crystal size to >3 μm or chemically modifying the crystal with Mg to reduce pore size results in a dramatic shift to para-selectivity, Table 8.

Chemical modification has also been shown to be effective in the alkylation of toluene with ethylene to make p-ethyltoluene, Table 9, and ethylbenzene alkylation with ethylene to make p-diethylbenzene, Table 10.

It should be pointed out that these structure sensitive phenomena in zeolites are not mutually exclusive and in many reactions the phenomena occur simultaneously.

Table 9
Toluene-Ethylene Alkylation

		Unmodified HZSM-5	Si-ZSM-5
Temperature, °C		350	350
WHSV			
Toluene		6.9	6.9
C_2H_4		0.5	0.5
Toluene/C_2H_4 (Mole)		4.5/1	4.3/1
Conversion, Wt%			
Toluene		20.7	15.3
C_2H_4		94.4	71.2
Selectivity to Products, %			
Benzene		0.8	0.5
Ethylbenzene		1.4	0.8
Xylenes		1.4	0.8
p-Ethyltoluene		23.3	90.6
m-Ethyltoluene		53.3	2.8
o-Ethyltoluene		8.3	0
Normalized Ethyltoluene	Equilibrium		
Para	33.7	27.4	97.0
Meta	49.9	62.8	3.0
Ortho	16.3	9.8	0

Table 10
Ethylbenzene Disproportionation

	Large Crystal HZSM-5	Mg-P-ZSM-5
Temperature, °C	500	525
WHSV		
EB	3.5	30.2
Conversion, %		
EB	89.6	22.5
Selectivity to Products, Wt%		
Benzene	60.8	62.4
Toluene	10.4	1.3
Xylenes, EB	3.9	0
Ethyltoluene	1.1	0.6
Diethylbenzene	1.8	15.4
Other Aromatics	4.1	2.7
Light Gas	--	17.6
C_5-C_9	1.9	--
Gas	16.0	--
Total	100.0	100.0
Diethylbenzene		
Para	33.3	99.3
Meta	65.0	0.7
Ortho	1.7	0

FUTURE OPPORTUNITIES

The application of shape selective catalysis of zeolites is far from being exhausted. Extension to other petroleum based and petroleum related feedstocks has shown potential with existing zeolitic materials. Applications of shape selective catalysis in fine chemical manufacture seem to be relatively unexplored.

Combining homogeneous catalysis with the shape selectivity of zeolites has found no success because of size incompatibility. With the discovery of larger pore zeolitic like materials such as VPI-5, we might be able to overcome the barrier.

LITERATURE CITED

1 P. B. Weisz and V. J. Frilette, J. Phys. Chem., 64 (1960) 382.
2 P. B. Weisz, V. J. Frilette, R. W. Maatman and E. B. Mower, J. Cat., 1 (1962) 307.
3 J. N. Miale, N. Y. Chen and P. B. Weisz, J. Cat., 6 (1966) 278.
4 S. M. Csicsery, J. Cat., 23 (1971) 124.
5 P. B. Weisz, Chemtech, 3 (1973) 498.

R.K. Grasselli and A.W. Sleight (Editors), *Structure-Activity and Selectivity Relationships in Heterogeneous Catalysis*
© 1991 Elsevier Science Publishers B.V., Amsterdam

CONFORMATIONAL EFFECTS IN HETEROGENEOUS CATALYSIS

Dan Fărcaşiu

Department of Chemical and Petroleum Engineering, University of Pittsburgh, 1249 Benedum Hall, Pittsburgh, PA 15261

ABSTRACT

The importance of conformation of a molecule for its interaction with solid catalysts is discussed. The contributions of conformational enthalpy and entropy are examined. For molecules having a large number of conformations the entropy term can dominate the reaction. Cracking of paraffins on small and medium-pore zeolites is entropy-controlled; a simple molecular sieve effect (shape selectivity) cannot explain the higher reactivity of normal isomers.

1. INTRODUCTION

Any correlation of structure with activity is tied to the question of reaction mechanism and the relationship between structure and mechanism.

A full description of a reaction mechanism has to address several questions. Thus, it must establish the reaction steps and intermediates. Then, it needs to determine the energy barriers (both heights and shapes), that is the reaction kinetics. Next, the mechanistic picture has to give an account of the movement of atoms in each reaction step. Ultimately, the mechanistic representation should give a description of the flow of electrons in all reaction steps, experimentally or by calculations and inferences.

The ways in which structural modifications influence the reaction mechanism are electronic and steric; these patterns of influence are called "effects" [1]. A broader list of effects proposed for catalysis [1] can be reduced to the two fundamental patterns mentioned above. Likewise, conformational effects, which are the subject of this chapter, belong to the broad class of steric effects.

Steric effects are described by a quantity named strain [2]. Strain is the measure, in energy terms, of the distortions from the ideal geometry of the molecule. There are several types of strain, identified by the geometrical parameter which is distorted [3]: bond strain, angle strain, torsional strain, and non-bonded (van der Waals) strain. Strain affects the stability of the molecule and also the rates of reactions for which there is an increase or decrease in strain at the transition state relative to the initial state, resulting in rate retardation and rate acceleration, respectively. This influence upon the reaction kinetics can also be observed in a catalytic reaction, if strain is manifested in the interaction between the catalyst and the substrate.

2. STERIC EFFECTS IN CATALYSIS

Reactions dominated by steric effects in solution [4] show the same type of sensitivity when conducted in heterogeneous catalysis [5]. More important for catalysis, however, are the steric effects determined by the interaction of the substrate with the solid catalyst. The existence of such steric effects, especially nonbonded interactions, in heterogeneous catalysis is well known.

The consequences of catalyst-substrate steric interactions are changes in reaction rates and in selectivities concerning the product or the substrate. The sterically controlled selectivity in reactions on solid catalysts has been referred to as shape selectivity [6]. Some pertinent examples are given below.

A. Product selectivity [7] is encountered when the specific interaction with the catalyst leads to preferential formation of one of the possible products from a starting material. The steric effects can be manifested either in the chemical reaction, or in a subsequent step, in which case they affect the separation of reaction products existing in rapid equilibrium.

An example of the product selectivity induced by steric interactions at the transition state of a reaction is shown in Fig. 1 for the alkylation of toluene with ethylene [8]. If the reaction occurs in a constricted space, the transition states along the reaction pathways forming ortho- and meta-ethyltoluene suffer from nonbonded interaction strain that raises their energy relative to the barrier for para attack, which does not suffer from steric interactions.

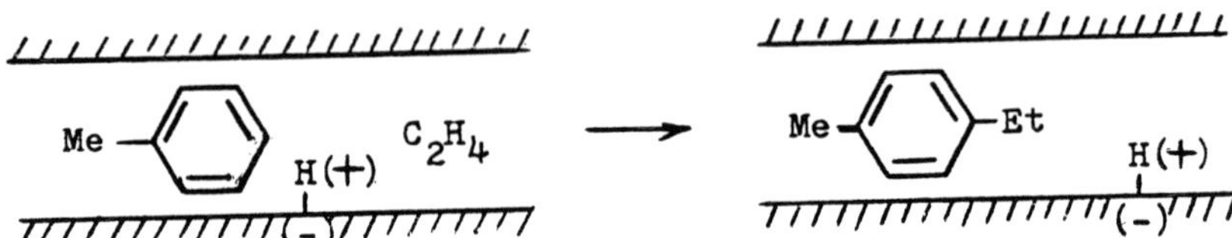

Fig. 1. Selectivity Induced Sterically at the Transition State

The selectivity observed is influenced by features of the reaction mechanism. Thus, ethyltoluenes interconvert by disproportionation [9] and isomerization [10]. For larger groups, dealkylation [11] is seen. Even if these reactions do not occur inside the channel, they can take place on the external surface, after the product exits the pore. Therefore, high selectivity requires large crystals and deactivated external surface of the catalyst [12].

Selectivity determined by steric interactions involving the reaction products is seen in the interconversion of xylenes (Fig. 2) [13]. If diffusion from the crystal is much slower than reaction inside, the isolated product ratio is determined by isomer diffusivities and concentrations at equilibrium [14].

If diffusion is much faster for *p*- than for *o*- and *m*-xylene, the former is obtained selectively.

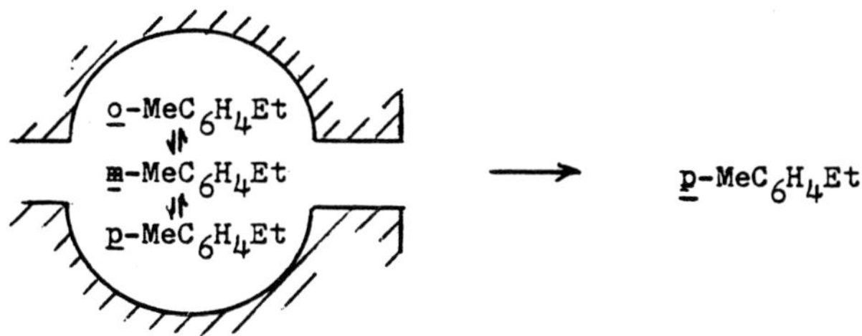

Fig. 2. Selectivity Induced by Steric Hindrance of Diffusion

Again, large crystals or selective poisoning of their exterior are required to minimize the re-equilibration of isomers on the external surface [12,13]. In any case, the maximum *p/m* ratio attainable starting from the meta isomer is the equilibrium ratio inside the cavity [15]: when the concentration ratios inside and outside are equal, the ratio of the flux of *p* and of *m* in the two directions is the same. Thus, *p*-xylene is isomerized on HZSM-5 selectively poisoned on the external surface [16]. Therefore, a *p/m* ratio higher than the equilibrium value [17] is not due to diffusion differences, but to side reac- tions of the isomers [18], that is the loss of *m* and *o* isomers to side products, including coke and gases. Para selective catalysts form coke easily [12a].

B. Substrate selectivity. Differences in steric interactions with the cata- lyst may result in the preferential conversion of a compound present in a mix- ture of reactive compounds. An example is the discrimination by size known as molecular sieve effect, exhibited by zeolites and used for drying organic mat- erials [19] and separation of straight-chain from branched hydrocarbons [20]. Both processes are physical equilibria conducted at or near room temperature.

Substrate selectivity induced by different chemical reaction rates is seen in the preferential cracking of linear alkanes in mixtures with their branched isomers, on small- [21] or medium-pore zeolites [22] (catalytic dewaxing). Shape selectivity was invoked to explain this finding; it was assumed that the locus of catalysis is the interior of the zeolite channels, and selectivity is observed when the size of the channels matches the size (van der Waals diameter) of a linear chain [6] or the size of the activated complex for its conversion [6b]. The branched hydrocarbons, too large to penetrate the pores, are not converted efficiently. Thus, the role of external acid sites is considered negligible.

The role of conformational effects in heterogeneous catalysis has been most often overlooked. The selective cracking of alkanes is a reaction in which conformational effects play an essential role, as it will be shown below.

3. CONFORMATIONAL ANALYSIS OF STRAIGHT CHAINS.

Conformations are structures that differ by rotation about one or more bonds, separated by relatively low energy barriers.

Butane is the smallest *n*-alkane that has distinguishable conformations. Of its three carbon-carbon bonds, one is conformationally relevant; its rotation produces two types of conformations, *anti* (lower in enthalpy) and *gauche*. The gauche form, chiral, exists in two enantiomers, *g*(+) and *g*(-).

Likewise, pentane has two conformationally relevant carbon-carbon bonds and six distinguishable conformations. The relative enthalpies and entropies of the conformers (rotamers) of pentane are given below. The entropy is determined both by count and by molecular symmetry, because the existence of a twofold symmetry axis introduces an entropy penalty of Rln2 [23].

	aa	*ag(+) ag(-)*	*g(+)g(+) g(-)g(-)*	*g(+)g(-)*
ΔH^o	0.00	0.90	1.63	3.22
ΔS^o count	0.00	Rln2	Rln2	0.00
ΔS^o symmetry	-Rln2	0.00	-Rln2	0.00

The conformations of very long chains had been discussed in literature [24], but the simplifications introduced make the treatment unsatisfactory [25]. To undertake a conformational analysis of linear chains we developed a method to calculate the number of rotamers and the number of rotamers with a given number of *gauche* bonds in the chain as a function of the number of relev- ant bonds in the chain, *n* (equal to the number of carbons, minus 3). For the method of calculation the reader is referred to the original articles [25,27].

From our equations we found, for instance, that a chain with 10 carbons ($n = 7$) has 1134 rotamers and a chain with 25 carbon atoms ($n = 22$) has 1.569×10^{10} conformers. Likewise, we found that decane, with a maximum number of gauche bonds possible $i = n = 7$, has 286 conformers with $i = 4$ and 348 conformers with $i = 5$, that eicosane ($n = 17$) has 1,244,672 conformers with $i = 7$ and 9.958 million conformers with $i = 10$, etc. The actual list of rotamers for each normal chain was generated by a simple procedure [25].

Each linear chain has only one all-*anti* (extended) conformation; all others are coiled, mostly heavily coiled. The entropy-determined preference for the coiled forms is such that even at room temperature coiled forms predominate in normal alkanes higher than pentane [26].

4. ALKANE CONFORMATIONS AND SHAPE SELECTIVITY

The origin of selectivities observed in the separation of alkanes and in alkane cracking has been considered to be discrimination by size among substrate molecules by the zeolite channel [6b]. Cracking and separation differ, however, in two ways: First, cracking reactions are run at high temperature. Second, cracking is kinetically controlled: its selectivity is determined by differences in rates, rather than by relative stabilities of the starting material in two states. Shape selectivity requires that *n*-alkanes react at the process temperature faster inside the channels than on the external surface.

We noted that the match of the van der Waals radius of the chain and the pore diameter holds only for the extended alkane conformation. A *gauche* bond along a normal chain is equivalent in space requirements to a methyl substituent [25]. Two or more *gauche* bonds have greater effect than the same number of methyl substituents, since the chain folds upon itself. Therefore, the *gauche* conformations should uncoil before entering to react within the channels.

It follows that the mechanism of *n*-alkane reactions on small- and medium-pore zeolites changes with the location of the catalytic site [27]:

(a) Inside the channels, the chemical reaction is preceded by the conformational equilibrium (eq. 1) and by the diffusion into the pore (eq. 2):

RH (coiled) $\rightleftharpoons$ RH (all-*anti*) (K) (1)

RH (all-*anti*) $\longrightarrow$ RH (confined) (2)

RH (confined) $\longrightarrow$ Products (k'_{int}) (3)

The conformational equilibrium of eq. 1 is illustrated below for one of the coiled forms with 7 *gauche* bonds of eicosane ($C_{20}H_{42}$, n = 17).

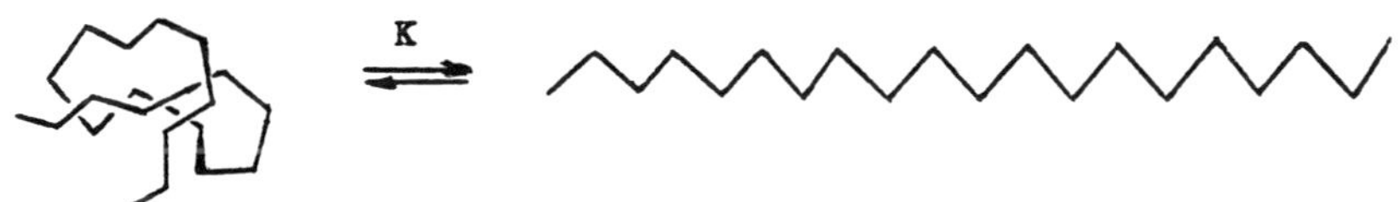

The measured (effective) rate constant for the reaction (k_{int}) will be the product of the intrinsic rate constant on the internal catalytic site (k'_{int}) and the equilibrium constant for the conformational preequilibrium:

$$k_{int} = k'_{int} K \quad (4)$$

For simplification it was assumed that the diffusion step (eq. 3) is fast and irreversible; otherwise, the overall rate constant, k_{int}, would be lower.

(b) On the external surface of the catalyst, the reaction (eq. 5) does not require readjustment of conformation or diffusion into the crystal. The coiled chain can interact sideways with the active centers, and the effective rate con stant is the same as the intrinsic rate constant on the site (eq. 6):

RH (coiled) $\longrightarrow$ Products (k'_{ext}) (5)

$$k_{ext} = k'_{ext} \quad (6)$$

The equilibrium constant *K*, needed to compare the rates for the linear alkane within the pores and on the external surface, is obtained from the differences in enthalpy and entropy between conformers.

$\ln K = -\Delta G^\circ/RT$; $\Delta G^\circ = \Delta H^\circ - T\Delta S^\circ$

The relative enthalpies of formation of conformers can be calculated as the differences between their steric energies obtained by molecular mechanics calculations [28]. We had used before this method for the elucidation of a number of problems of reaction mechanism [29]. The heats of formation calculated for alkanes are about as good as those obtained by combustion calorimetry.

The relative entropy is determined by counting rotamers with the same number of gauche bonds and the same steric energy, and applying the symmetry correction to the forms that have a twofold symmetry axis [25].

We performed extensive calculations of conformational enthalpies and conformer counting. To convert the results into values for the conformational equilibrium constant *K*, we had to make a number of choices:

1. The value of *K* is temperature dependent. The temperatures indicated in disclosures of selective *n*-alkane cracking were 450-950°F (505-783 K) [21] or 550-1100°F (560-866 K) [22c]. We chose for our evaluation a temperature of 550 K.
2. It was reported that a medium-pore zeolite (ZSM-5) accepts and cracks within its channels a normal or monomethyl-substituted alkane, but it excludes a dimethyl-substituted chain [6b]. Therefore, the equilibrium between the forms with two *gauche* bonds or more, and the all-*anti* forms should be considered. We chose, however, to include in the calculation of *K* only the rotamers with four *gauche* bonds and higher.
3. For certain conformers four energy minima are found upon rotation of some bonds [24,27]. We used for the calculation of entropy only three conformations for each bond: *a*, *g*(+), and *g*(-).

These choices assured that our estimate of the rate constant for conversion of *n*-alkanes inside the zeolite channels (k_{int}) was, if anything, too high. The free energy differences between the conformers with $i > 4$ and the all-*anti* conformers for the *n*-alkanes from heptane to undecane are plotted in Fig. 3. It is seen that already for undecane ΔG° is 4.4 Kcal/mol and it varies monotonically with the chain length. For the C_{20} chain ΔG° should be in excess of 10 Kcal/mol ($K = 10^{-4}$ [27].)

The boiling point of eicosane ($C_{20}H_{42}$) matches the low end (343°C, 650°F) of the boiling range of the oil fraction used in selective cracking experiments [22c]. For the longer straight chains contained in that feed the retardation of reaction inside the channels because of the conformation preequilibrium

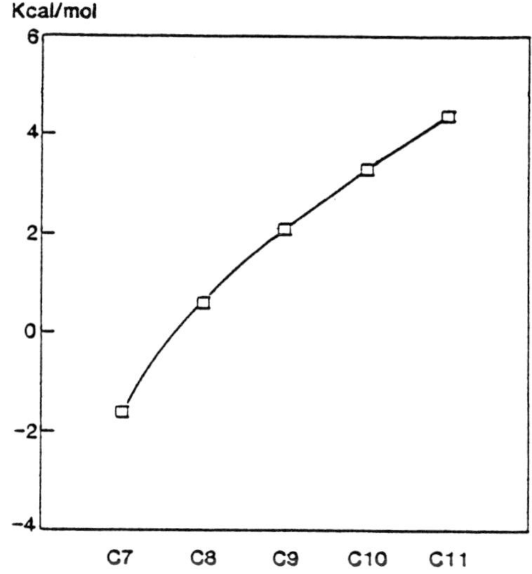

Fig. 3. Conformational energy for n-alkanes

should be enormous. The reaction is, therefore, controlled by conformational effects.

Experiments that substantiate our representation have been published. In a study of distribution of alkanes between gas phase and zeolite ZSM-5 [30], the solid was equilibrated with a flow of a 1:1 mixture of *n*-hexane and 2-methylpentane vapors diluted with He. The linear isomer predominated in the adsorbed phase by a ratio of 28:1 at 293 K, but only 4:1 at 403 K. As predicted by our model, the change was due mostly to the decrease in adsorption of *n*-hexane; the branched isomer has fewer conformations available and its conformational energy is less temperature-dependent. The results [30] indicate that at the temperature of the catalytic cracking the amount of linear isomer present in the pores should be essentially zero [27].

Another study looked at the energy barriers for the passage of alkyl groups through the molecular cavity of rigid cyclic ethers (cyclophanes) of 2,5-disubstituted hydroquinones [31]. The substituents were carboxyl groups esterified with alcohols of varying lengths (e=COOR, Fig. 4). The compounds in which both rings carried substituents were also studied. The macrocycle was rigid, only the benzene rings could rotate around their respective O-C(1)-C(4)-O axes. A variable temperature NMR study provided the activation parameters for the rotation, from the coalescence temperature (T_c) of the signals for H_a and H_b, and from a complete line-shape analysis [31].

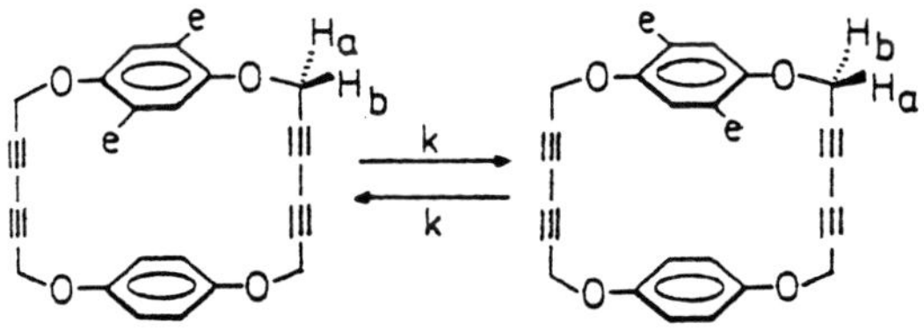

Fig. 4. Cyclophanes with rigid molecular cavity

As shown in Fig. 5, the space requirement of the properly oriented ester group upon passage through the molecular cavity during rotation does not change with the length of the alkyl chain R. Nonetheless, the energy barriers for the process were found to increase with the size of a linear alkyl chain [31]:

R =	T_c =	$\Delta G^{\ddagger}$ =
Ethyl	42.4°	15.3 kcal/mol
Propyl	85.9°	17.2
Decyl	(>100°)	..could not be measured

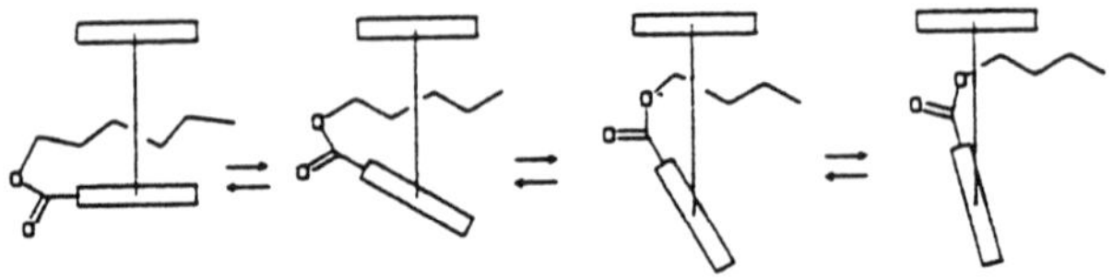

Fig. 5. Passage of a linear chain through a molecular cavity.

5. THE EFFECT OF CATALYST-SUBSTRATE ORIENTATION ON SELECTIVITY.

Whether the alkane chain is extended or coiled, its penetration into the channel requires that the end of the chain meets the opening of the channel when the molecule collides with the zeolite crystal [27,32]. If the orientation is not favorable, the hydrocarbon molecule will sit on the catalyst crystal until it achieves the proper position for penetration, or it reacts at an acid site on the external surface. The orientation requirement brings about another retarding factor for the reaction inside the channels, which depends upon the zeolite geometry and length of the hydrocarbon. For eicosane on ZSM-5 this factor is about 1/36 [27].

6. ALTERNATIVE ORIGINS OF SHAPE SELECTIVITY IN ALKANE CRACKING

We have shown that simple size discrimination (high-temperature molecular sieve effect) cannot be the reason for the selective cracking of *n*-alkanes on small-pore and medium-pore zeolites. Another mechanistic rationalization for this reactivity pattern is needed. We have considered four possibilities:

(1) Selectivity is determined by the ratio of internal to external surface of the zeolite, that is the internal to external area ratio is significantly greater than the retardation of the reaction inside the pores. This ratio is determined by the crystal structure and size [34,35]. A 10^5 ratio is achieved for 3-5 mm single crystals, a size just not available for ZSM-5 [36a].

(2) Chain uncoiling and chain penetration are concerted. A molecule that is properly oriented and finds a pore opening can be visualized as slowly penetrating the channels as it unwinds. The exothermic interaction of the hydrocarbon with the channel walls, which helps adsorption at low temperature, makes the molecule resist entering the pore at the high temperature of the

cracking process (Le Chatelier principle). The adsorption heat evolved when the end of the chain penetrates the channel is transferred to the rest of the chain and increases its movements, thus slowing down further penetration. Most of the reaction would, therefore, occur in a layer 5-7 Å thick or on the external surface of the crystal, around pore openings. The model also requires that the catalyst be not more selective for long straight chain molecules than for branched alkanes with linear ends or side chains of five or six carbon atoms, which is not substantiated by experiment [6b].

(3) The intrinsic activity of a site inside the channel is much higher than that of a site on the external surface. Such a relationship between the catalytic activity of the two type of sites ($k'_{int} >> k'_{ext}$) would then be unique for alkane cracking, because for other types of reactions previous authors have shown that reactivity of internal and external sites are not much different [16,34,36]. There is no result or theory that substantiates such a dramatic change in catalyst properties with the type of reaction.

(4) At the temperature used for cracking the reactions of both branched and linear alkanes occur largely on the external surface. This model requires that linear chains, reacting in coiled form, have a higher intrinsic reactivity than the branched chains on the medium-pore zeolites at the reaction temperature. The reasons for such a reactivity difference between alkane isomers are not clear.

Experiments are needed to determine the true origin of the selectivity. It is clear, however, that the application of a simple molecular sieving mechanism to a high temperature reaction such as cracking is unsatisfactory.

ACKNOWLEDGMENT

The American Chemical Society allowed us to reproduce Figures 4 and 5.

REFERENCES

1. R.S. Weber, *J. Catal.* **122** (1990) 198.
2. E.L. Eliel, N.L. Allinger, S.J. Angyal, and G.A. Morrison, Conformational Analysis, Ch. 1, Interscience, New York, 1965.
3. (a) M.S. Newman (ed.), Steric Effects in Organic Chemistry, John Wiley, New York, 1956. (b) See also: J.E. Williams, P.J. Stang, and P.v.R. Schleyer, *Annu. Rev. Phys. Chem.*, **19** (1968) 531, and references therein.
4. (a) K. Kindler, *Liebigs Ann. Chem.* **464** (1928) 278. (b) C.K. Ingold, *J. Chem. Soc.*, **1930**, 1032.
5. (a) G.C. Thomas and C.W. Davies, *Nature*, **159** (1927) 373. (b) S. Affrossman and J.P. Murray, J. Chem. Soc. (B), 1968, 579.
6. (a) P.B. Weisz, V.J. Frilette, R.W. Maatman, and E.B. Mower, *J. Catal.*, **1**, (1962) 307. (b) W.O. Haag, R.M. Lago, and P.B. Weisz, *Disc. Faraday Soc.*, **72** (1982) 317. (c) N.Y. Chen and W.E. Garwood, *Catal. Rev. Sci. Eng.*, **28** (1986) 185, and references therein.

7. Alternative classification: E. Derouane, in: F. Ramoa Ribeiro, A.E. Rodrigues, L. Rollman, and C. Naccache (Eds), Zeolites, Science nad Technology, NATO ASI Series E: Applied Sciences No. 80, Martinus Nijhoff, The Hague, 1984, p. 347.
8. W.W. Kaeding, I.B. Young, and C.-C. Chu, *J. Catal.*, **189** (1984) 267.
9. (a) G. Baddeley, G. Holt, and D. Voss, *J. Chem. Soc.*, **1952**, 100. (b) D.A. McCaulay and A.P. Lien, *J. Am. Chem.* Soc., **74** (1952) 6246.
10. R. Heise, *Ber. Dtsch. Chem. Ges.*, **24** (1891) 768.
11. V.N. Ipatieff and H. Pines, *J. Am. Chem. Soc.* **59**, (1937) 56.
12. (a) W.W. Kaeding, *J. Catal.*, **95** (1985) 512. (b) A.G. Ashton, S. Batmanian, I.S. Elliott, and F. Fitch, *J. Molec. Catal.*, **34** (1986) 73.
13. L.B. Young, S.A. Butter, and W.W. Kaeding, *J. Catal.*, **76** (1982) 418.
14. C.N. Satterfield, Mass Transfer in Heterogeneous Catalysis, M.I.T. Press, Cambridge, MA, 1970.
15. D. Seddon, *J. Catal.*, **98** (1986) 1.
16. S. Namba, S. Nakanishi, and T. Yashima, *J. Catal.*, **88** (1984) 505.
17. F.A. Smith, A.B. Schwartz, and L.L. Breckenridge, Eur. Pat. 432, quoted in ref. 15.
18. D. Fărcaşiu, *Rev. Roumaine Chim.*, **12** (1967) 795.
19. D.R. Burfield, G.-H. Gan, and R.H. Smithers, *J. Appl. Chem. Biotechnol.*, **28** (1978) 23.
20. R.M. Barrer, *Quart. Revs.*, **3** (1949) 293.
21. (a) J. Eng., U.S. Pat. 3039953, 1962; (b) E.M. Gladrow, R.B. Mason, and G.P. Hamner, U.S. Pat., 3395096, 1968.
22. (a) R. Argauer and G.R. Landolt, U.S. Pat. 3702886, 1972. (b) R.H. Daniels, G.T. Kerr, and L.D. Rollman, *J. Am. Chem. Soc.*, **100** (1978) 3097. (c) N.Y. Chen, S.J. Lucki, and W.E. Garwood, U.S. Pat. 3700585, 1972.
23. S.W. Benson, Thermochemical Kinetics, Wiley, New York, 1968.
24. (a) P.J. Flory, Statistical Mechanics of Chain Molecules, Interscience, New York, 1969.
25. D. Fărcaşiu, P. Walter, and K. Sheils, *J. Comput. Chem.*, **10** (1989) 520.
26. P.v.R. Schleyer, J.E. Williams, and K.R. Blanchard, *J. Am. Chem. Soc.*, **92** (1970), 2377, and references therein.
27. D. Fărcaşiu, J. Hutchison, and L. Li, *J. Catal.*, **122** (1990) 34.
28. (a) U. Burkert and N.L. Allinger, Molecular Mechanics, American Chemical Society, Washington, D.C., 1982.
29. D. Fărcaşiu, E. Seppo, M. Kizirian, D.B. Ledlie, and A. Sevin, *J. Am. Chem. Soc.*, **111** (1989) 8466, and references therein.
30. D.S. Santilli, *J. Catal.*, **99 (1986) 335.**
31. **B.J. Whitlock and H.W. Whitlock,** *J. Am. Chem. Soc.*, **105** (1983), 838.
32. E. Ruckenstein and D.B. Dadyburjor, *Chem. Eng. Commun.*, **14** (1982) 59.
33. (a) S. Ramdas, J.M. Thomas, P.W. Betteridge, A.K. Cheetham, and E.K. Davies, *Angew. Chem. Int. Ed. Engl.*, **23** (1984) 671. (b) W.M. Meier and D.H. Olson, Atlas of Zeolite Structure Types, The Structure Commission of the International Zeolite Association, 1987.
34. J.P. Gilson and E.G. Derouane, *J. Catal.*, **88** (1984) 538.
35. M. Farcasiu and T.F. Degnan, *Ind. Eng. Chem. Res.*, **27** (1988) 45.
36. (a) D. Fraenkel and M. Levy, *J. Catal.*, **118** (1989) 10. (b) G. Paparatto, E. Moretti, G. Leofanti, and F. Gatti, *J. Catal.*, **105** (1987) 227.

R.K. Grasselli and A.W. Sleight (Editors), *Structure-Activity and Selectivity Relationships in Heterogeneous Catalysis*

© 1991 Elsevier Science Publishers B.V., Amsterdam

EFFECTS OF MORPHOLOGY AND ELECTRONIC STRUCTURE ON THE CATALYSIS OF ZEOLITE ENCAGED PALLADIUM PARTICLES

Z. KARPIŃSKI[1], S.T. HOMEYER, W.M.H. SACHTLER
V.N. IPATIEFF LABORATORY, CENTER FOR CATALYSIS AND SURFACE SCIENCE, NORTHWESTERN UNIVERSITY, EVANSTON, IL 60208

ABSTRACT

After calcination of ion exchanged Pd/Nay at 500°C, the Pd^{2+} ions have lost their ligands and are located in sodalite cages, but after calcination at 250° $Pd(NHY_3)^{2+2}$ ions are present in supercages. Reduction leads to Pd atoms in sodalite cages or to Pd particles in supercages. An important difference between these cases is due to the co-product of hydrogen reduction: NH_4^+ ions or protons. Probing these catalysts with the conversion of neopentane shows that Pd particle morphology has only a minor influence on the rate of conversion, but the presence of protons has a marked effect, increasing the activity per site by two orders of magnitude. This result is tentatively ascribed to electron deficient $[Pd_n-H_x]^{x+}$ clusters as superactive sites.

1. INTRODUCTION

Catalysts containing transition metal particles in zeolite cages are of great potential importance for the petroleum and chemical industries, as they combine a high catalytic activity with a number of qualities which are unique to these systems:

1. The metal particles tend to be more uniform in size than metals on amorphous supports;
2. The geometric constraints of the zeolite matrix impose stereospecificity upon the catalytic reactions. Only molecules which pass through the channel system will reach the metal particles; the space between metal particle surface and cage wall imposes strict limitations on the transition states for chemical reaction(12),(2),(3).
3. Reduction of bare metal ions with H_2 is accompanied by formation of protons

---

[1] Institute of Physical Chemistry, Polish Academy of Sciences, Warsaw, Poland

of high Brønsted acidity, rendering the catalyst bifunctional. The acidity can be controlled by the calcination conditions(4-8).

4. Catalyst promotion by the electronic "ligand effect" is facilitated for small metal particles which share the same zeolite cage with the promoting ion(9).

Recent work has shown that the critical morphological and electronic parameters of the zeolite encaged metal particles can be controlled by the preparation conditions, in particular the temperatures where ion exchanged zeolites are calcined and reduced. This paper summarizes these findings.

2. CALCINATION

A schematic description of the important processes during calcination and reduction of metal ions in zeolite cages is shown in Figure 1. If metal

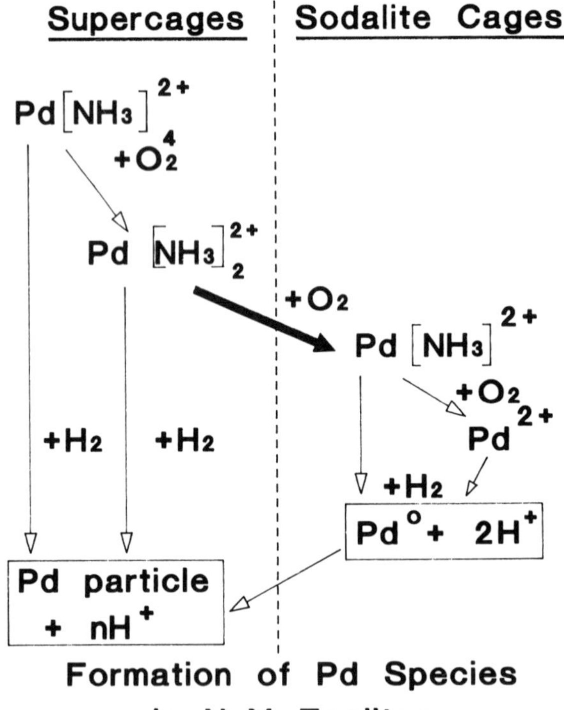

Schematic description of elementary processes during calcination and reduction of ion exchanged $Pd(NH_3)_4^{2+}$ in NaY

precursors are introduced into zeolite cages by ion exchange, the pH of the solution must be compatible with zeolite stability and ionicity. For platinum and palladium the tetrammine cations are appropriate precursors. After ion exchange with zeolite NaY, these complexed ions are located inside the supercage network of the zeolite. For the relatively low metal loadings used in our research one Pd ion per supercage is introduced; i.e. per unit cell there are 40 Na^{+} ions and 8 $Pd(NH_3)_4^{2+}$ ions compensating 56 negative charges. Calcination is necessary to remove adsorbed water and to destroy the ammine ligands. To obtain high metal dispersion it is essential that autoreduction of Pd by decomposing ammine ligands is prevented; therefore calcination is preferentially done at 1 atm of pure oxygen with a very low heating program, e.g. 0.5°C/min.

It has been found that the decomposition of the ammine ligands occurs in discrete steps.(4) If calcination is limited to 250°C, e.g. by holding the sample at this temperature for 2 hrs., the ratio of ammine ligands to Pd ions is 2:1; most Pd ions are thus present as diammine complexes. Diffuse reflectance and EXAFS results have shown that these Pd ions have a square planar coordination with two NH_3 groups in cis-position and two oxygen ions, O_z, of the zeolite cages completing the coordination.(10) These complexed ions are still located in the supercages. At higher calcination temperatures the remaining ammine ligands are destroyed; the Pd^{2+} ions migrate into smaller cages. The driving force for this rearrangement of the charge compensation ions is thought to be the the high density of negative charge in the sodalite cages and hexagonal prisms; Coulomb interaction therefore favors a distribution with dipositive Pd ions in these cages, but monopositive Na ions in supercages. Indeed, we recently observed by TPR that no migration of Pd ions into smaller cages takes place if the remaining charge compensating ions are also dipositive., e.g. Mg^{2+} or Ca^{2+}(11). For the subsequent reduction of Pd/NaY two major scenarios are of interest:(12).

Scenario #1: Calcination at 250°C; the Pd ions are almost exclusively located inside supercages;

Scenario #2: Calcination at 500°C: the Pd ions are located predominantly in sodalite cages, a small fraction is present in hexagonal prisms.

In Scenario # 1 Pd nuclei are formed in supercages, Pd ions adhere to these nuclei and are subsequently reduced. The size of the resulting Pd particles depends on the reduction temperature, T_R. At low T_R primary particles (diameter < 7.5 Å) of very few atoms are formed; they migrate and coalesce with increasing T_R. The resulting decrease of the metal dispersion with T_R is steep for small particles which can traverse the windows between cages, but much less steep for larger particles which can only grow by Ostwald ripening or via local collapse of the zeolite lattice(5),(6). When Pd^{2+} ions are adsorbed and reduced on window facing parts of the Pd_n particles growth extends into adjacent cages, i.e. clusters of contiguous particles, filling adjacent supercages, also called "grape-shaped" particles, are formed(13).

The Pd particle morphology and its dependence on T_R is different in Scenario #2 (calcination at elevated temperatures has positioned the Pd ions into small zeolite cages). In this case, the primary reduction product at low T_R consists of isolated Pd atoms which are located inside these cages. EXAFS shows that their coordination number with other Pd atoms is zero and the "metal dispersion" expressed by the ratio of chemisorbed hydrogen to reduced Pd atoms is initially very small, because isolated Pd atoms are unable of dissociatively chemisorbing H_2 molecules under the conditions used. With increasing T_R, the metal dispersion increases because Pd atoms escape from the small cages and form Pd_n nuclei in supercages, capable of chemisorbing H_2. For T_R = 350°C our EXAFS data show that the Pd atoms in such particles have four Pd atoms as first neighbors. The Pd–Pd distance is compressed in comparison to bulk Pd(11). These primary Pd particles migrate and coalesce with each other at higher temperature. A plot of the metal dispersion versus T_R thus displays a maximum which is typical for Scenario #2 of Pd/NaY.(5)

The activation energy for the escape of isolated metal atoms from small cages into larger cages is higher for Pt than for Pd. This explains some important differences in the reduction behavior of Pt/NaY and Pd/NaY. Isolated Pt atoms in sodalite cages are reoxidized by protons at elevated temperature (e.g. 500°C):

$$Pt^{o} + 2\ H^{+} \Longrightarrow Pt^{2+} + H_2$$

in reversal of the ion reduction process(14). We never observed this type of re-oxidation with Pd/NaY, because at the high temperature where it would occur, the Pd atoms have left the small cages (e.g. at 350°C) and agglomerated to multiatomic clusters which do not react with protons in this way for either metal. Also the high dispersion which can be achieved for Pd after positioning the precursor into small cages, cannot be obtained for Pt, because Pt atoms require a very high temperature to leave these cages.

The calcination of $Pd(NH_3)_4$/NaY, besides affecting the morphology of the particles after reduction, also determines the acidity of the zeolite and the extent of "electron deficiency" of the metal particles, because ammine ligands, which survive calcination, are able to neutralize the protons generated during metal reduction.

In Scenario #1, most Pd ions are present as $Pd(NH_3)_2^{2+}$ ions, their reduction with H_2 thus proceeds according to:

$$Pd(NH_3)_2^{2+} + H_2 \Longrightarrow Pd^{o} + 2\ NH_4^{+}$$

In scenario #2, however, the Pd ions are bare, reduction proceeds according to:

$$Pd^{2+} + H_2 \Longrightarrow Pd^{o} + 2\ H^{+}.$$

The formation of protons in a zeolite implies that acid sites of high Brønsted acidity are formed. The resulting catalyst is therefore "bifunctional" i.e. metal sites and acid sites catalyze reaction steps including metal adsorbed alkyl groups and carbenium ions respectively

This is very clearly demonstrated with methylcyclopentane (MCP) conversion as the probe reaction. If the Pd/NaY catalyst has been prepared according to Scenario #1, the metal catalyzed ring opening of MCP to the isomeric hexanes, n-hexane, 2-methylpentane and 3-methylpentane, is the prevailing reaction. If the catalyst with the same metal loading has, however, been prepared using the calcination scenario # 2, probing with MCP at precisely the same conditions leads to a different result: in this case the carbenium ion intermediated ring enlargement of the five membered ring compound to six membered compounds prevails. As the primary product is easily (de-) hydrogenated over the Pd particles, the reaction product consists of a mixture of benzene and cyclohexane(15).

3. OXIDATIVE REDISPERSION

Aged supported metal catalysts contain larger particles than virgin catalysts. In the case of Pd/NaY it is possible to exploit the high tendency of Pd^{2+} ions to escape into small zeolite cages for a catalyst rejuvenation strategy. Basically it consists of two steps: the aged catalyst is oxidized under conditions where PdO particles will react with zeolite protons to form Pd^{2+} ions + H_2O. Ideally, this process brings the sample into the same state as the freshly calcined catalyst in scenario #2. In a second step the catalyst is gently reduced and small Pd particles are formed.(16) This strategy is particularly effective for zeolites which contain a large concentration of protons. If the environment of an oxidized particle is depleted of protons, additional "tricks" can be used to favor proton migration into that region, so that even very large particles can be effectively redispersed(17).

4. ELECTRON DEFICIENT METAL PARTICLES

It has been reported by various authors that Pt in zeolites can acquire a positive electric charge, as evidenced by various physical techniques(18),(19). These"electron deficient" particles display an unusually high catalytic activity for metal catalyzed reactions such as the conversion of neopentane (= 2,2-dimethylpropane)(20). The nature of the electron acceptor was unclear, but it has been reported that "electron deficiency" of the zeolite supported metal particles increases with the acidity of the zeolite.(21) We recently found evidence strongly suggesting that an electron deficient Pd particle is actually an adduct of a Pd particle with one or several protons. If e.g. one proton is interacting with a Pd particle consisting of n Pd atoms, the positive charge in the resulting $[Pd_n-H]^+$ complex is no longer localized on the proton, but it is smeared out over the particle surface which, apart from its positive charge, is indistinguishable from a Pd particle with an adsorbed H atom. This positive charge on Pd has been detected by several physical and chemical methods, including the FTIR spectrum of adsorbed CO, for which the typical stretching frequencies are shifted to higher wave numbers Previously we had shown that Pd carbonyl complexes in zeolite supercages react with zeolite protons in an analoguous manner, resulting in a hydrido carbonyl palladium cluster cation(22),(23). The identification of a metal–proton adduct is not only of relevance to the reported "superactivity" of metals in acidic zeolites, this adduct might also act as a "bifunctional site" in reactions where classical "bifunctional catalysts" require that intermediates shuttle between separated metal and acid sites(24).

Further, a proton which is simultaneously interacting with the zeolite cage and a metal particles can also function as an "anchor", delaying migration of small metal particles through the system of supercage channels. These possibilities are schematically depicted in Figure 2.

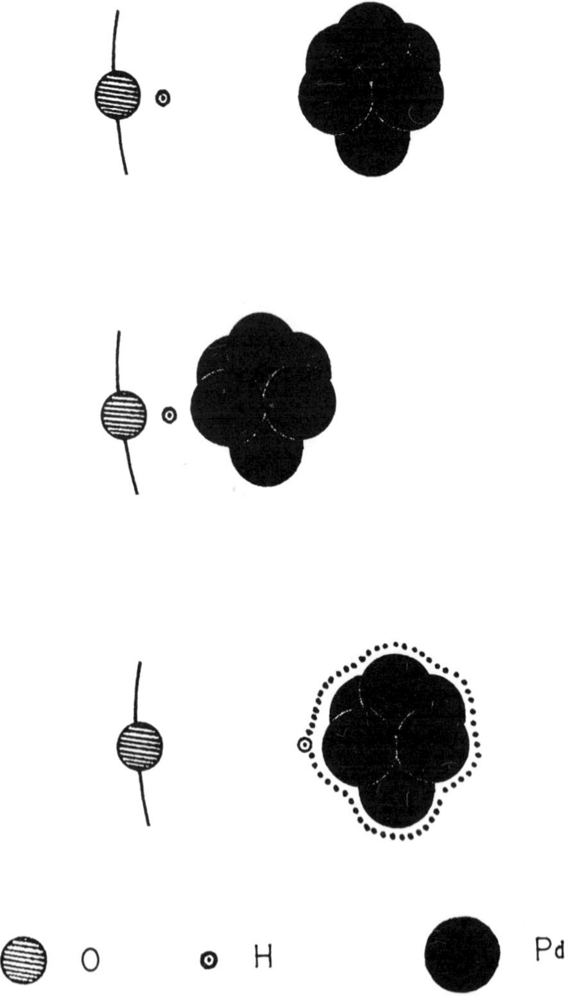

Schematic description of interactions between Pd_n particle inside a supercage with a zeolite proton in the same cage
a. No direct interaction (= classical site pair for bifunctional

catalysis
b. Hydrogen bridge anchoring metal particle to zeolite cage
c. $[Pd_n-H]^+$ Adduct

5. CONCLUSIONS

The combination of EXAFS, diffuse reflectance, electron microscopic and dynamic studies, including TPR, TPO and catalytic tests, has provided novel information on the factors which determine size, shape and charge of zeolite supported metal particles. The enhanced catalytic activity of palladium particles in zeolite cages containing protons of high Brønsted acidity is ascribed to adducts such as $[Pd_n-H]^+$ which possess all qualities of "electron deficient" metal particles for which "catalytic superactivity" has been reported earlier. The hypothesis that such adducts also act as bifunctional sites in hydrocarbon conversion reactions deserves further study.

6. ACKNOWLEDGEMENT

Financial support by the National Science Foundation (contract CTS 8911184) and a grant-in-aid by the Mobil Foundation are gratefully acknowledged.

REFERENCES

1 W.M.H. Sachtler, Ultramicroscopy 20 (1986) 135
2 M.S. Tzou, H.J. Jiang, and W.M.H. Sachtler, Appl. Catal. 20 (1986) 231
3 G. Moretti, W.M.H. Sachtler. J. Catal. 116 (1989) 350.
4 S.T. Homeyer and W.M. H. Sachtler, J. Catalysis 117 (1989) 91
5 S.T. Homeyer and W.M.H. Sachtler J. Catal. 118 1989, 266
6 S.T. Homeyer and W.M.H. Sachtler in Zeolites: Facts, Figures, Future, P.A. Jacobs, and R.A. van Santen Eds., Elsevier, Amsterdam 1989, p. 975-84
7 S. T. Homeyer, Z. Karpiński, W.M.H. Sachtler, Recl. Trav. Chim. Pays- Bas (J. Roy. Neth. Chem Soc.) 109 (1990) 81
8 S.T. Homeyer, Z. Karpiński and W.M.H. Sachtler, J. Catal. 123 1990, 60
9 R.A. van Santen, Chemisch Weekblad 1989 439
10 Z. Zhang, H. Chen and W.M.H. Sachtler, Zeolites (in press)
11 Z. Zhang, T. Wong and W.M.H. Sachtler, J. Catal. (submitted)
12 W.M.H. Sachtler, in Proc. Int. Summer Inst. Surface Science, (Springer Verlag, Berlin, 1990) p. 60-85 (in press)
13 G. Bergeret, P. Gallezot and B. Imelik, J. Phys. Chem. 85 411 (1981)
14 M.S. Tzou, B.K. Teo and W.M.H. Sachtler, J. Catal. 113 220 (1988)
15 S.T. Homeyer, Z. Karpiński and W.M.H. Sachtler, J. Catal. 123 1990, 60
16 S.T. Homeyer and W.M.H. Sachtler, Applied Catalysis 1989 54 189
17 O. Feeley and W.M.H. Sachtler, to be submitted to Applied Catalysis
18 K. Foger, J.R. Anderson, J. Catal. 54, 1978, 318
19 P. Gallezot in Metal Clusters, M. Moskovits Ed. (J. Wiley & Sons 1986) p. 219-247
20 R.A. Dalla Betta, M. Boudart: in Proc. 5th Int. Congr. Catalysis, H. Hightower Ed. (North Holland Publ., Amsterdam 1973) p. 1329
21 P. Gallezot, J. Dataka, J. Massardier, M. Primet, and B. Imelik in Proc. 6th Intern. Congre. Catalysis G. C. Bond et al. Eds. (Chem. Soc., London, 1976) Vol. 2 p. 696
22 L.L. Sheu, H. Knözinger and W.M.H. Sachtler, J. Am. Chem. Soc. 111 (1989) 8125
23 L.L. Sheu, H. Knözinger and W.M.H. Sachtler, J. Molec. Catal. 57, 62 (1989)
24 X.L. Bai and W.M.H. Sachtler, paper in preparation

R.K. Grasselli and A.W. Sleight (Editors), *Structure-Activity and Selectivity Relationships in Heterogeneous Catalysis*
© 1991 Elsevier Science Publishers B.V., Amsterdam

FUNDAMENTAL CHARACTERISTICS OF THE CATALYST SYSTEM PLATINUM–LOADED ZEOLITE L

J. M. NEWSAM#, B. G. SILBERNAGEL, A. R. GARCIA, M. T. MELCHIOR and S. C. FUNG

Exxon Research and Engineering Company, Route 22 East, Annandale, NJ 08801

ABSTRACT
A variety of characterization techniques have been applied to probing the structure – selectivity interrelationship in the selective dehydrocyclization catalyst platinum–loaded zeolite L. Powder X-ray (PXD) and neutron diffraction (PND) have yielded precise data on framework geometries, aluminum partitionings (and, at Si:Al = 1.0, evidence for long range Si:Al ordering), and the non-framework cation configurations. Hydrocarbon location and motional characteristics have been probed by PND and ^{2}H nmr respectively, complemented by computer simulations. The preferred Pt atom sites in the zeolite loaded with Pt at high dispersion remain unmeasured, although insight has been gleaned from simple hard-sphere modeling. Aggregated Pt clusters can be detected by electron microscopy and high resolution PXD.

INTRODUCTION

In 1980 Bernard demonstrated that platinum-loaded zeolite L is an active deydrocyclization catalyst, with excellent selectivity in the conversion of n-hexane to benzene [1-3]. To translate these early laboratory data into a viable commercial process has involved addressing a number of issues such as relating to catalyst lifetime, regenerability, scale-up, and sensitivity to poisoning by other components (such as sulfur-containing species) in the real feed. Additionally, these early data posed a number of interesting fundamental questions, notably about the particular characteristics of the zeolite L system that give rise to the exceptional benzene selectivity. We outline here some of the strategies that we have pursued in attempting to develop answers to these various questions.

Although most of the issues involved are common to other zeolite catalyst systems, our ability to apply sophisticated characterization tools to the real catalyst under process conditions remains limited. Most progress has evolved from studying better defined, monophasic materials at temperatures at or below ambient, or from examining the state of catalyst composites after exposure to oil.

It should be noted that in this, as in many other zeolite systems, our focus has been on exploring *structure – selectivity* relationships rather than, as traditionally phrased, the coupling between structure and *activity*.

Present address: BIOSYM Technologies Inc, 10065 Barnes Canyon Rd. San Diego, CA 92121

STRUCTURAL CHARACTERISTICS OF THE 'NAKED' ZEOLITE

The basic features of the zeolite L structure were determined in 1969, based on powder X-ray diffraction (PXD) data measured from a hydrated Na, K - form of the zeolite [4]. The framework structure is hexagonal, space group P6/mmm with a = 18.4Å and c = 7.5Å, and with a unidimensional 12-ring pore system (Figure 1). The framework has two crystallographically distinct T-sites (T = Tetrahedral species, Si or Al). From the average T–O bond lengths determined in the PXD study preferential aluminum atom placement on the hexagonal prism (Si2) sites was inferred. The framework aluminum distribution is of interest in the context of catalysis, and was one characteristic that we sought early to define. Four non-framework cation sites (**A**, **B**, **C** and **D**) were found to be occupied, together with the suggestion that an additional site (**E**) might be populated in the dehydrated material. A PXD study of hydrated (K)Ba-G(L) was described in 1972 [5].

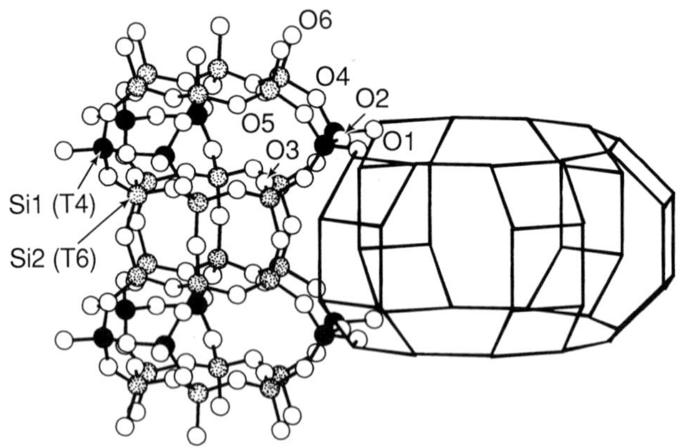

Figure 1. The construction of the LTL-framework structure of zeolite L

Framework aluminum placement

Powder X-ray diffraction has only limited direct sensitivity to Si:Al ordering or partitioning phenomena because the atomic scattering powers of Si^{2+} and Al^{1+} are similar. The neutron scattering lengths of Si (0.41491 fm) and Al (0.3449 fm), however, differ sufficiently for direct measurement of the aluminum partitioning in powder neutron diffraction (PND) experiments [6]. In a series of dehydrated materials spanning a range of framework compositions, $2.4 \leq$ Si:Al ≤ 3.1, the direct measure obtained via Rietveld refinements demonstrates a significant aluminum preference for the Si1 (12-ring window) sites. The optimized coordinates yield average T–O distances for the two T-sites that are consistent with the direct measures of site Al content. The contrary indication of the earlier PXD results is attributed to imprecision (and underestimated standard deviations). To enhance sensitivity in the diffraction experiment (with both X-rays and neutrons) to the character of the T-atom partitioning, the structure of a dehydrated material with complete Ga substitution for Al was determined by Rietveld analysis of PND data. The synthesis chemistry here, however, proves subtly different, as gallium is found to be randomly partitioned between the two T-sites [7, 8]. We have noted similarly reduced site preference tendencies in a number of other gallosilicate zeolites [9].

Although local Si:Al order is indicated by the ^{29}Si nmr data (below), direct diffraction evidence for Al - O - Al linkage avoidance is not obtained, for at Si:Al = 2.8 the alternation does not develop long-range coherence. At lower Si:Al ratios (below a percolation threshold of approximately Si:Al = 1.4) long range coherence is expected. Although the earlier PXD study (K)Ba-G(L) with Si:Al = 1.0 indicated no evidence of ordering, the associated superlattice (that involves a doubling of the hexagonal **c**-axis) is subtle. The improved resolution (reduced peak widths give better peak separations, combined with peakier features which then stand out well above a smooth background) and increased intensity configured in an experiment on beamline X10B at the National Synchrotron Light Source, Brookhaven, however, enabled direct observation of the superlattice 131 reflection [10]. The peak width is comparable to those of the adjacent (subcell) reflections demonstrating long-range Si:Al ordering coherence. The development of a superlattice in LTL-framework materials at Si:Al ~ 1 is directly analogous with the superlattice ordering in zeolite A that results in a doubling of the cubic unit cell constant from 12.3Å to 24.6Å [11].

Solid state ^{29}Si nmr proves of limited direct value for probing framework aluminum placement in zeolite L materials. The mean geometries about the Si1 and Si2 sites are found to be almost identical implying that the ^{29}Si (and ^{27}Al) chemical shifts for the two sites will be similar. A typical ^{29}Si pattern (Figure 2) does indeed give the appearance of having only a single set of Si-nAl; n=0-4 contributors (such as generally observed for zeolites X, Y or ZK-4). However, recognizing that the Si-nAl peaks for the two sites are effectively coincident, the measured intensity can be appropriately apportioned to yield framework compositions. Full profile analysis [12] assuming only a single contributing set of Si:nAl peaks gives a reasonable fit (Figure 2), and a framework composition that agrees well with that measured by Inductively Coupled Plasma Emission Spectroscopy (ICPES).

If we could estimate the ^{29}Si chemical shifts, δ_n, with sufficient accuracy, this type of full profile analysis procedure could be applied generally, such as to the more complex zeolites, permitting extraction of framework compositions and local site environment populations even for highly overlapped spectra. For this reason we have been attempting steadily to improve quantitation of the geometrical and compositional influences on the ^{29}Si chemical shifts. Quantitative analysis of ^{29}Si chemical shift data for a series of LTA and FAU-framework materials [13] yielded (by extrapolation to Si:Al = ∞), a mean change per first shell Al atom of 6.2ppm, notably much larger than the value of ~4.5ppm that is still frequently quoted. The discrepancy reflects the significant contributions from second neighbor aluminum atoms, which result in substantial changes in the Si-nAl; n < 4 peak positions with framework composition. The form of these changes could initially be approximated by assigning a change of 0.6ppm to each second shell Al atom [13].

Partly as a result of this (largely unappreciated) compositional sensitivity, correlations between Si–T distance or functions of the 4 Si–O–T angles described earlier offer only limited accuracy. Circumvention of the problem associated with this compositional dependence is possible by studying purely siliceous materials. Well-resolved ^{29}Si spectra are usually obtained for SiO_2 compositions (there is less variability in the local site environments when framework Al and associated non-framework species are completely absent), although the range of T-O-T angles observed is concomitantly more limited. We have therefore focussed on the Si-4Al environment [14] (for which, by Loewenstein's rule, second shell sites are occupied only by Si), extending the useful range of T-O-T angles by non-framework cation substitution (verifying simultaneously that the direct effect of these cations is small).

To estimate ^{29}Si chemical shifts for zeolite L, we use the mean T–O–T angles calculated from the refined structural data (for the dehydrated zeolite – we have not yet completed studes of hydrated materials), the correlation between mean T–O–T angle and $\delta_{n=4}$ [14], and assume changes of b_1 = 6.2 ppm and b_2 = 0.6 ppm from each Al substitution in the first and second shells respectively [13]. Agreement with the measured values is encouraging, but inexact. Use of a single deshielding influence for each first and second shell site is known to be simplistic [15]. If the first shell influence, b_1, is taken as a variable (which may be rationalized in terms of the different constraints on the local site geometries for zeolite L and zeolites X and Y) it can be deduced from the measured spectrum [16]. For the zeolite L data illustrated in Figure 2, a value b_1 = 5.95 ppm gives the calculated peak positions given in Figure 2. The agreement with the least-squares optimized values is now reasonable. This same approach can clearly also be applied to other complicated systems [16], including cases in which reliable structural data are unavailable (via parameter calibrations based on measured peak positions).

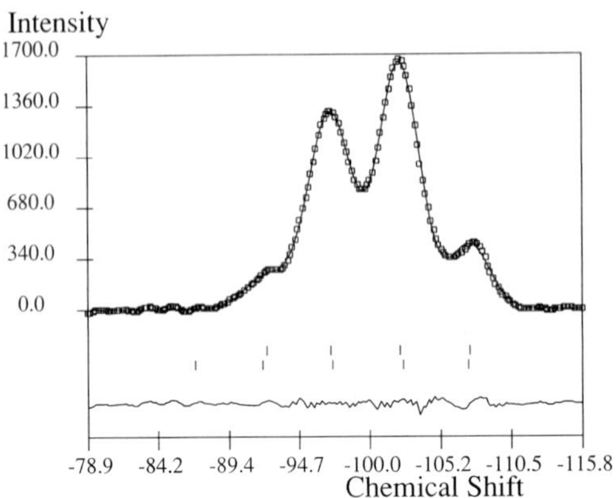

Figure 2. Observed (squares), calculated (continuous line) and difference (lower) ^{29}Si nmr spectra for zeolite L at a typical framework Si:Al ratio of 2.8. The optimized peak positions are indicated by the upper set of 4 vertical bars. The lower 5 bars indicate the Si-nAl; n=0-4 peak positions calculated based on an empirical treatment of the geometrical and compositional dependences of the ^{29}Si chemical shifts.

Non-framework cation configuration

The non-framework cation sites described in 1969 [4] continue to provide the basis on which the configurations in all materials studied to date can be described. The hexagonal prism site (**A**) observed populated to an incomplete extent by Na^+ in the hydrated Na,K-form [4] has at most minimal K^+ occupancy in the pure K^+ form of zeolite L. The cancrinite cage site (**B**) and the coplanar bridging site (**C**) are typically fully occupied. The remainder of the K^+ cations occupy type **D** channel wall sites. The **D** site occupancy (on a unit cell basis) is found to be close to 4.5 out of 6 available sites at a typical Si:Al of 2.9. Lower and higher **D** site occupancies are measured for materials synthesized with respectively higher and lower Si:Al ratios. This pattern leads to a limit of Si:Al = 2.27 for full **B**, **C** and **D** sites, below which occupation of a further site must commence or, as has actually been effected, divalent cations (Ba^{2+}) must be introduced in partial replacement for K^+ {The mineral perlialite reported [17] to have a framework Si:Al = 1.86, also contains the divalent cations Ca^{2+}, Ba^{2+} and Mg^{2+} although at a level apparently insufficient for occupancy of sites other than **B**, **C**, and **D** (and **A** by Na^+) not to be necessary. Structural details have yet to be published}. Similarly, at Si:Al = 6.2 the type **D** sites will presumably be completely depopulated. Still higher ratios would necessarily require some attrition of the **C** and/or **B** site(s). We have not yet studied a material with a Si:Al ratio sufficiently high to observe such depopulation. Presumably in part because of the difficulty in extracting K^+ cations from the **B** and **C** sites, dealumination of zeolite L to Si:Al > ~6 is difficult.

Barrer and Villiger recognized the potential for a migration of cations at elevated temperatures from type D sites to sites **E** within the channel walls. We find evidence for, at most, marginal occupancy of site **E** by K^+ in dehydrated materials for compositions Si:Al > 2.4. Insufficient **D** – **E** site separations likely prohibit simultaneous occupation of a type **D** site and the adjacent type **E** site. However, at Si:Al = 2.8 on average 1.5 of the 6 type **D** sites in any one unit cell are vacant, probably allowing in aggregate near full occupancy of the 3 type **E** sites in each unit cell. The observed absence of substantial **E** site occupancy thus indicates that other factors (such as the 3.7Å separations to type **C** sites) are also at work. Evidence has been presented for the locations and migrations at elevated temperatures of various cations in zeolite L based on ion-exchange, elution and Szilard-Chalmers cation recoil studies [18, 19] and, for Fe^{2+}, Mössbauer spectroscopy [20, 21].

BEHAVIOR OF SORBED HYDROCARBONS

Location

To understand the properties of hydrocarbons in the active platinum-loaded zeolite under catalytic conditions requires prior knowledge of their behavior under more controlled, sorptive conditions. A first question posed inquires as to the preferred (energy minimum) hydrocarbon configuration(s) under relatively dilute loadings (where direct sorbate-sorbate interactions are likely to be small). This question can be addressed quite readily by computer simulation (below), but to develop confidence in such results (and, indeed, to test and improve methods, potential functions and parameters) direct measurements are essential. Although single crystal diffraction measurements on, for example, benzene-loaded zeolite X at low temperature [22] (necessary to minimize population of higher energy hydrocarbon configurations) have yielded reasonably precise benzene molecule locations, suitably large single crystals of zeolite L have not yet proven synthesizable. Low temperature powder X-ray diffraction has, as yet, been applied little to zeolite - sorbate complexes and the precision of results remains modest [23].

Powder neutron diffraction gives good sensitivity to the scattering from C and ^2H (deuterium is preferred as incoherent scattering from ^1H contributes a large background) and measurements can straightforwardly be made on sealed samples over a wide range of temperatures. Results for perdeuterobenzene (~1.0 molecules per channel lobe) [24] and perdeuteropyridine (~1.2) [8] reveal that both adopt sites indicative of substantial interaction between the ring π-electron density and a type **D** K^+ cation. In contrast to the symmetrical arrangement observed for benzene (in which all C-atoms are (approximately) equidistant from the K^+ cation), pyridine is apparently displaced towards the adjacent type **D** K^+ cation site. Structural results for other hydrocarbons sorbed within zeolite L have not yet been reported.

Motional properties

One proposal made to account for the excellent benzene selectivity of the Pt K–L system argues for the channel system imposing a 'collimating' influence on diffusing n-hexane molecules such that terminal adsorption on Pt centers is promoted [3]. Little data on the diffusional properties of hydrocarbons such as n-hexane in zeolite L are available, preventing further direct evaluation of this proposed mechanism. Measurements that yield diffusion constants [25, 26] generally provide little insight into the character of the motion on a molecular scale. Quasielastic neutron scattering can measure diffusion constants (in the approximate range 10^{-7} to 10^{-5} cm^2 s^{-1}) and data on the geometry of the diffusional process can also be extracted, although often with limited definition, particularly when more than one type of motion is present (see, e.g. [27]).

The form of a hydrocarbon ^2H (deuterium) nmr spectrum depends on the degree (and rate) to which molecular reorientations occur. ^2H nmr experiments on the single set of C_6 hydrocarbons benzene, n-hexane and cyclohexane in zeolite L over a range of temperatures and loading levels have provided a rich picture of the hydrocarbon motional characteristics [28, 29]. The ^2H nmr spectrum of perdeuterobenzene at 150K at a loading level of 1 molecule per channel lobe (or per unit cell) shows a reduction in the effective quadrupolar coupling constant from the static value of almost exactly one half. The reduction scales as $[3\cos^2\beta - 1]/2$, and the data are thus consistent with benzene molecule reorientations about the unique axis (which makes an angle, β, with the C–D bond of 90°) on a time scale shorter than 10^{-6} s. At temperatures above 250K, the form of the quadrupolar spectrum and the spin-lattice relaxational (T_1) characteristics of the ^2H nuclei indicate the commencement of activated hopping between preferred sites, with a deduced activation energy of 4.8 kcal mol^{-1} [28].

For n-hexane, the ^2H nmr spectra demonstrate that isotropic reorientations do not occur on the sub micro-second scale for temperatures below 300K. Rather, the data are fully explicable based on C – C torsional rotations. The C-D vectors for the 8 methylene deuterons are reoriented by rotations about two distinct C–C bond vectors (the further C-C vectors are parallel to one or other of these two directions). The 6 methyl C-D vectors are, additionally, reoriented by rotations about the respective terminal C-C bond. At the tetrahedral angle of 109.47° to the C-D bond each of these rotations gives a 1/3 reduction factor, leading to total reductions of $(1/3)^2 = 9$ and $(1/3)^3 = 27$ for the methylene and methyl deuterons respectively, almost exactly the values observed at 298K (the relative intensities of the two components in the measured spectrum are in the approximate intensity ratio 8:6) [28]. This reptative mode of motion is consistent with a substantial hydrocarbon – zeolite interaction, such that even at room temperature the n-hexane dwells mainly close to the walls of the zeolite channel.

PLATINUM PLACEMENT

To begin to probe even in a simplistic fashion the constraints on the reaction profile imposed by the enclosing zeolite cage we are missing a key piece of information. Comparison between the catalytic performance of Pt loaded zeolite L and that for Pt loaded on other supports (such as silica) argues strongly that the (most) active Pt component is housed within the zeolite. However, we do not yet know the preferred sites for isolated Pt atoms or for small (≤ 5 atoms) Pt clusters within the zeolite. Additionally, although evidence indicates that the most active catalysts are those with the Pt in its most dispersed state, data to indicate the character of the most active Pt centers are unavailable. Thus although we propose that the benzene selectivity displayed by Pt-loaded zeolite L reflects at least in part the steric constraints imposed on the reaction profile by the surrounding zeolite cage, a quantitative exploration of these steric factors hinges on the character and location assumed for the Pt species within the zeolite [30]. Encouraging X-ray diffraction studies of metal atoms in zeolite X were reported some years ago [31], but the Pt dilution in the present system (the Pt loading corresponds to on average one Pt atom per 9 zeolite unit cells at the hypothetical atomic dispersion limit) has so far prevented definitive diffraction studies.

Although computer simulations (below) can in principle address this question, reliable parameters, particularly those for describing the interaction between a Pt atom and its environment are unavailable. A simple picture based on effective (hard-sphere) radii [30] suggests that for atomic radii of 1.4Å for both Pt and O, the most favorable site in potassium zeolite L will be close to a type **D** cation site such that contact distances to framework oxygen atoms are similar to those for K^+ - O (i.e. close to 2.8Å). At typical compositions on average 1.5 of the 6 such sites in any given channel lobe are not populated by K^+ cations, 13x more than the maximum number of Pt atoms to be housed. A representation of a Pt atom at such a site (together with a bonded n-hexane molecule) is given in Figure 3.

Figure 3. Stereoview representation of an isolated Pt atom occupying a type D cation site in the zeolite L channel, to which a n-hexane molecule has been bonded in a carbinoid configuration

Larger clusters and discrete particles

Electron microscopy is a first choice technique for observing small metal particles. However, it proves experimentally taxing to correlate the occurrence of supported small metal particles with the structure of the support. The most definitive data on the zeolite framework structure is obtained using conventional Transmission Electron Microscopy (TEM), although even in this mode the problem of damage in the high intensity electron beam makes detailed studies troublesome. Further, conventional TEM has limited value for studying very small dispersed particles. Much better performance, for heavier atoms, is achieved by using a high angle (annular) detector in a scanning transmission electron microscope (STEM) which records electrons scattered with large momentum change [32]. The relative scattering by the atoms with larger atomic number then dominates, greatly improving the sensitivity to small metal clusters (even, in very favorable cases, permitting detection of single atoms such as, for example, uranium atoms/ions on an organic substrate). Concomitantly however, the moderated intensity recorded from the substrate reduces the degree to which the highlighted metal particles can be placed with respect to the supporting zeolite framework. Encouraging work on Pt-loaded zeolite L has been reported recently [32].

The mechanism of Pt migration and aggregation within the zeolite remains little understood. The spatial extent of Pt clusters within the zeolite is limited by the constraints of the zeolite cage. Clusters up to ~Pt_{15} [33] can be contained within a single channel lobe. Further growth is then restricted to development along the channel direction, unless framework degradation were to occur (for which there is no evidence under normal conditions). No evidence for extended Pt whiskers within the zeolite has been reported and the preferred mechanism for further Pt cluster growth apparently involves migration to exterior regions. External Pt particles can be observed in the electron microscope or, for particle sizes larger than some 25Å, by conventional powder X-ray diffraction. At the low loading levels involved, the Pt (111) reflection is relatively weak, and further, at d = 2.2655Å, occurs in close proximity to the (070/350) pair of reflections (d=2.2734Å; the (033) at d=2.2676Å is very weak) making direct observation difficult. Here, again, the higher intensity and better resolution available in PXD scans using a synchrotron X-ray source [10] can facilitate characterization.

A DEVELOPING ROLE FOR COMPUTER SIMULATIONS

For regimes that tax current measurement methods, computer simulations offer considerable attractions. The picture of the Pt-loaded zeolite L system that we have steadily been building still leaves largely unexplored the influences of temperature (all of the results discussed above refer to temperatures at close to ambient or below), atmosphere and pressure, and time scale (intermediates or transition states involving Pt have yet to be detected). Particularly in the last category, no viable analytical tools yet even promise to provide the structural and dynamical infromation that we seek. Although computer simulations can in principle be applied at varying degrees of sophistication, over a wide range of conditions, detailed comparisons against observation under conditions that are experimentally tractable are vital in order to test models, methods and parameters.

In common with other zeolite systems, we are still at an early stage in this process for Pt-loaded zeolite L. Nevertheless, relatively simple modeling treatments have provided some insight. Atom-atom potential modeling of pyridine in zeolite L apparently gave a computed global minimum position close to that measured in consort experimentally [8]. For benzene, however, using the same methods and parameters, the predicted location places the molecule near-perpendicular to the channel wall [34]. At low loading levels no evidence is seen for occupation of such a site [24], or for the 12-ring window site considered earlier as a possible benzene site in the (hypothetical) purely siliceous material [35] (and substantiated by the subsequent observation of benzene occupation of the 12-ring window in sodium zeolite Y [36]). As above, direct observation determines that the π-electron density K^+ cation interaction dominates [24]. Given this measured location atom-atom potential summations can be used to ask the simpler question as to the rotational mobility about the unique axis. Such an analysis (in which K^+ interactions were ignored) yields a calculated activation energy of less than 0.5 kcal mol^{-1} for reorientation, not inconsistent with the ^2H nmr results. An intriguing aspect of potassium zeolite L is its ability to take up more than two molecules of benzene per unit cell. Three sites for accomodating the additional benzene molecules might be considered. The 12-ring window site, although superficially attractive, may be destabilized when benzene molecules in adjacent cells are occupying the π-complex sites. A benzene location near-perpendicular to the channel wall [34] places it parallel with the two molecules occupying the π-complex sites. However, the amount of free space between these two molecules appears insufficient to accept an additional benzene molecule. There may, however, be sufficient space to accomodate 3 benzene molecules on π-complex sites in any one channel lobe (Figure 4 – based on the low-temperature structure results, without any adjustment of coordinates). If this is actually the arrangement adopted, there would appear to be only a kinetic restriction on the zeolite taking up a full 3 molecules of benzene per unit cell.

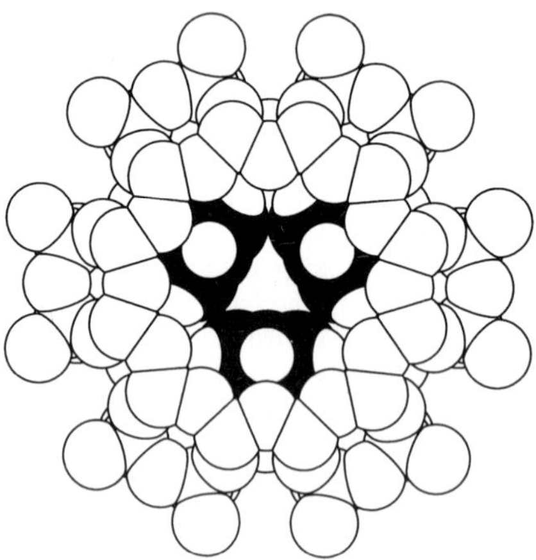

Figure 4. Representation of benzene molecules occupying three of capping sites (π-complex) above type D K^+ cations based on the structural results at 78K for a loading level of 1 molecule per unit cell (see text – drawing radii: Si 1.4, K 1.4, O 1.35, C 1.77, H 1.17Å).

Molecular graphics utilities prove extremely helpful during the generally protracted process of zeolite structure determination. Additionally, as above, they permit simple exploration of possible sites for non-framework cations, metal atoms or clusters (Figure 3), and hydrocarbon sorbates (Figure 4). Simple aspects of the reaction chemistry can also be considered. Thus molecular mechanics studies suggest that the steric constraints imposed by the surrounding cage may play a significant role in promoting those conformations of a n-hexane molecule terminally bonded to a Pt atom (at a type **D** site) that are likely to give rise to 1-6 ring closure, with subsequent dehydrogenation to benzene [30]. Both algorithm and computer hardware developments promise soon to advance substantially our simulations capabilities. Thus, framework flexibility and a proper treatment of the Coulombic (charge - charge) terms can now be incorporated [37], enabling exploration of the energetic and dynamical properties of a range of hydrocarbons in systems such as potassium zeolite L.

CONCLUSION

Contributions to our current understanding of the interrelationship between structure and selectivity in the dehydrocyclization catalyst Pt–loaded zeolite L have been made by a range of techniques. Powder X-ray (PXD) and neutron diffraction (PND) have afforded data on zeolite framework geometries, aluminum partitionings, and the non-framework cation configurations. Hydrocarbon location and motional characteristics have been probed by PND and ^2H nmr respectively, complemented by computer simulations. The preferred Pt atom sites at high dispersion remain unmeasured, but insight has been gleaned from simple hard-sphere modeling. Detection of aggregated Pt clusters is possible in the electron microscope, and measurements of still larger Pt particles (external to the zeolite) are feasible using high resolution PXD methods.

ACKNOWLEDGEMENTS

We thank A. J. Jacobson, A. E. Schweizer, J. J. Steger, S. J. Tauster, T. H. Vanderspurt, D. E. W. Vaughan and J. P Verduijn for numerous helpful discussions and input during the course of these various studies.

REFERENCES

1. J. R. Bernard, in L. V. C. Rees (Ed.), Proc. Fifth Int. Conf. Zeolites, Heyden, London, 1980, pp. 686-695.
2. T. R. Hughes, W. C. Buss, P. W. Tamm and R. L. Jacobson, in Y. Murakami, A. Iijima and J. W. Ward (Eds.), New Developments in Zeolite Science and Technology, Kodansha and Elsevier, Tokyo and Amsterdam, 1986, pp. 725-732.
3. S. J. Tauster and J. J. Steger, in M. M. J. Treacy, J. M. White and J. M. Thomas (Eds.), Microstructure and Properties of Catalysts (MRS Symp. Proc. Vol. 111), Materials Research Society, Pittsburgh, PA, 1988, pp. 419-423.
4. R. M. Barrer and H. Villiger, Z. Kristallogr., Kristallgeom., Kristallphys., Kristallchem., 128 (1969) 352-370.
5. C. Baerlocher and R. M. Barrer, Z. Kristallogr., Kristallgeom., Kristallphys., Kristallchem., 136 (1972) 245-254.
6. J. M. Newsam, J. Chem. Soc. Chem. Comm., (1987) 123-124.
7. J. M. Newsam, Mater. Res. Bull, 21 (1986) 661-672.
8. P. A. Wright, J. M. Thomas, A. K. Cheetham and A. K. Nowak, Nature (London), 318 (1986) 611-614.
9. J. M. Newsam and D. E. W. Vaughan, in Y. Murakami, A. Iijima and J. W. Ward (Eds.), New Developments in Zeolite Science and Technology, Kodansha and Elsevier, Tokyo and Amsterdam, 1986, pp. 457-464.
10. J. M. Newsam and K. S. Liang, Int. Rev. Phys. Chem., 8 (1989) 289-338.
11. W. M. Meier and D. H. Olson, Atlas of Zeolite Structure Types, Butterworths, Guildford, UK, 1987.
12. J. M. Newsam, M. T. Melchior and H. Malone, Solid State Ionics, 26 (1988) 125-131.
13. J. M. Newsam, J. Phys. Chem., 89 (1985) 2002-2005.
14. J. M. Newsam, J. Phys. Chem., 91 (1987) 1259-1262.
15. M. T. Melchior and J. M. Newsam, in P. A. Jacobs and R. A. van Santen (Eds.), Zeolites: Facts, Figures, Future (Stud. Surf. Sci. Cat. No. 49), Elsevier, Amsterdam, 1989, pp. 805-814.
16. J. M. Newsam, M. T. Melchior and R. A. Beyerlein, in M. M. J. Treacy, J. M. White and J. M. Thomas (Eds.), Microstructure and Properties of Catalysts (MRS Symp. Proc. Vol. 111), Materials Research Society, Pittsburgh, PA, 1988, pp. 125-134.
17. Y. P. Menshikov, Zap. Vses. Mineral. O-va, 113 (1984) 607.
18. P. A. Newell and L. V. C. Rees, Zeolites, 3 (1983) 22-27.
19. P. A. Newell and L. V. C. Rees, Zeolites, 3 (1983) 28-36.
20. F. R. Fitch and L. V. C. Rees, Zeolites, 2 (1982) 33.
21. F. R. Fitch and L. V. C. Rees, Zeolites, 2 (1982) 279-289.
22. Y. F. Shepelev, A. A. Anderson and Y. I. Smolin, Kristallografiya, 33 (1988) 359-364.
23. B. F. Mentzen, Mater. Res. Bull., 22 (1987) 337-343.
24. J. M. Newsam, J. Phys. Chem., 93 (1989) 7689-7694.

25. D. M. Ruthven, Principles of Adsorption and Adsorption Processes, Wiley-Interscience, New York, 1984.
26. J. Karger and D. M. Ruthven, Zeolites, 9 (1989) 267-281.
27. J. M. Newsam, T. O. Brun, F. Trouw, L. E. Iton and L. A. Curtiss, in R. T. K. Baker (Ed.), New Catalytic Materials and Techniques (ACS Symp. Ser.), American Chemical Society, Washington, DC, 1990, in press.
28. B. G. Silbernagel, A. R. Garcia, J. M. Newsam and R. Hulme, J. Phys. Chem., 93 (1989) 6506-6511.
29. B. G. Silbernagel, A. R. Garcia, R. Hulme and J. M. Newsam, in P. A. Jacobs and R. A. van Santen (Eds.), Zeolites: Facts, Figures, Future (Stud. Surf. Sci. Cat. No. 49A), Elsevier, Amsterdam, 1989, pp. 615-622.
30. S. T. Weidman, J. M. Newsam, J. J. Steger and J. L. Larsen, (1990) in preparation.
31. G. Bergeret, Tran Manh Tri and P. Gallezot, J. Phys. Chem., 87 (1983) 1160-1165.
32. S. B. Rice, J. Y. Koo, M. M. Disko and M. M. J. Treacy, (1990) submitted.
33. E. G. Derouane and D. J. Vanderveken, Appl. Catal., 45 (1988) L15-L22.
34. A. K. Nowak and A. K. Cheetham, in Y. Murakami, A. Iijima and J. W. Ward (Eds.), New Developments in Zeolite Science and Technology, Kodansha and Elsevier, Tokyo and Amsterdam, 1988, pp. 475-479.
35. J. M. Newsam, Materials Science Forum, 27/28 (1987) 385-396.
36. A. N. Fitch, H. Jobic and A. Renouprez, J. Phys. Chem., 90 (1986) 1311-1318.
37. M. W. Deem and J. M. Newsam, (1990) in preparation.

R.K. Grasselli and A.W. Sleight (Editors), *Structure-Activity and Selectivity Relationships in Heterogeneous Catalysis*
1991 Elsevier Science Publishers B.V., Amsterdam

CATION INDUCED CHANGES IN CHEMICAL REACTIVITY OF SMALL METAL PARTICLES

A.P.J. JANSEN, and R.A. van SANTEN

Schuit Institute of Catalysis, Eindhoven University of Technology, P.O. Box 513, 5600 MB Eindhoven (The Netherlands)

ABSTRACT

The chemical reactivity of small metal clusters dispersed in microcavities of zeolites is shown to be significantly influenced by charge-compensating cations. Adsorption properties of CO or Lewis bases are also found to be changed.

Here we report HFS-LCAO calculations of CO, H_2, and H chemisorbed onto small Ir_4 clusters in the presence of a Mg^{2+} ion. Analysis of our results shows that the chemical reactivity of CO, and also its infrared spectra, should be a sensitive function of the relative positions of adsorbate, and cation. This appears also to be the case for H_2 chemisorption. But the effects are different. Whereas the CO bond appears to be strengthened when the Ir_4 cluster is in between the Mg^{2+} ions and CO, the reverse is found for H_2. The bond strength of chemisorbed H_2 appears to be significantly weakened due to the presence of the Mg^{2+} ion. It appears to be mainly due to the reduced repulsive interactions between H_2 and Ir_4.

The result of these calculations will be analysed with respect to their catalytic implications.

1. INTRODUCTION

The chemical reactivity of small metal clusters is of considerable fundamental as well as practical interest.[1] When dispersed in microcavities of zeolites they are active hydrocarbon conversion catalysts. In the work reported here we present a quantum chemical study of the interaction of small molecules (CO, H_2, and H) with a tetrahedral Ir_4 cluster, and investigate how this interaction is changed when a Mg^{2+} ion is co-adsorbed on the Ir_4 cluster.[2,3] This model is chosen to explore how a cation in close contact with a noble metal cluster affects its chemical reactivity. This is a question of particular interest for zeolite catalysis.

Large changes in chemical reactivity of metal particles in zeolites have been reported, especially for the zeolites X and L, when cations of low charge were replaced by cations of high charge.[4,5] In order to study this more closely studies have been made in which these changes were probed by chemisorption of CO,[6,7] or basic molecules.[8]

2. METHOD AND CLUSTER GEOMETRIES

We used Local Density Functional theory with the $X\alpha$-potential for the exchange-correlation potential ($\alpha = 0.7$), and the LCAO approach as implemented by the group of Baerends.[9-11] The adsorption energies were calculated using the Ziegler transition state method.[12] The adsorption energies were decomposed into steric repulsion and interaction energy. The steric repulsion is defined as the energy of the metal cluster, plus the energy of the admolecule, minus the energy of a Slater-determinant in which the non-relaxed molecular orbitals of the metal cluster and the admolecule are put. The interaction energy is the energy decrease that is found when these molecular orbitals are allowed to relax. Basis and fit sets can be found in Refs. 2 and 3.

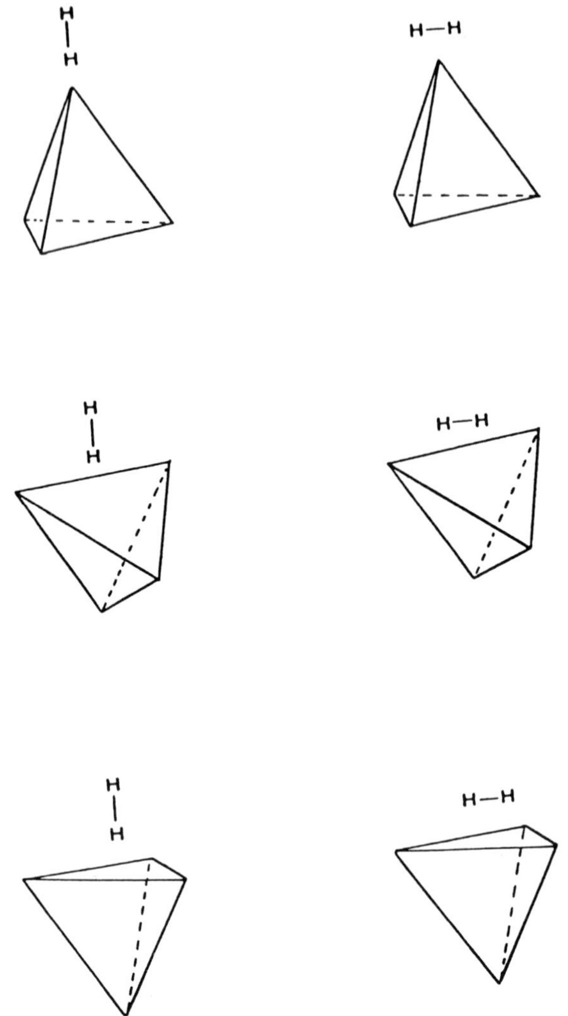

Fig. 1. Structures of the H_2-Ir_4 complexes. For the CO molecule only the three structures on the left, with the lower hydrogen atom replaced by carbon, and the upper by oxygen were investigated. For the hydrogen atom only the three structures on the left, with the top atom left out, were investigated.

We used in all our calculations a tetrahedral Ir_4 cluster. The distance between the iridium atoms was the same as the nearest-neighbour distance in bulk iridium; i.e. 2.72 Å. The admolecules were placed in 1-, 2-, and 3-fold adsorption sites. For CO only end-on adsorption with carbon closest to the Ir_4 cluster was studied. For H_2 end-on as well as side-on adsorption was studied (see Fig. 1). In all cases the distance of the admolecule to the Ir_4 cluster was optimized. For H_2 we also optimized the H–H distance. We fixed the C–O distance to the experimental gas phase value of 1.13 Å. The Mg^{2+} ion was placed at the other side of the Ir_4 cluster as the admolecule. Its coordination being 1-fold for 3-fold, 2-fold for 2-fold, and 3-fold for 1-fold admolecule adsorption (see Fig. 2). The distance of the Mg^{2+} ion

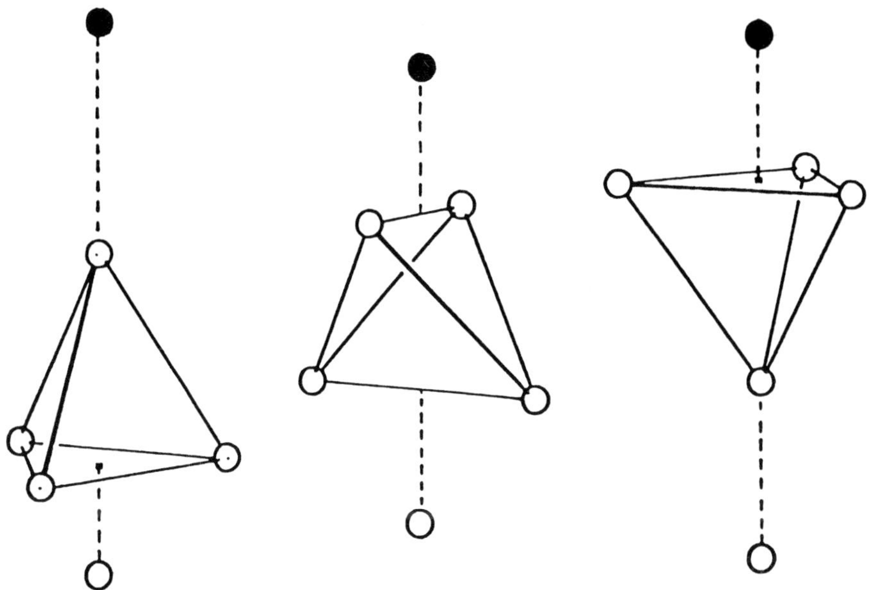

Fig. 2. The relative positions of the admolecule, depicted by the dark sphere at the top, and the Mg^{2+} ion, depicted by the light sphere at the bottom.

Table 1. Adsorption energies and their decomposition (in eV) of CO on a tetrahedral Ir_4 cluster.

geometry	adsorption energy	steric repulsion	interaction energy
1-fold	0.93	1.71	2.64
2-fold	2.42	7.48	9.89
3-fold	2.39	7.94	10.30

to the nearest iridium atom was fixed to the ionic radius (0.66 Å) plus half the Ir–Ir distance.

3. ADSORPTION IN THE ABSENCE OF THE CATION

Before evaluating the effect of the cation on the adsorption we start with a discussion of the adsorption in the absence of the cation. Table 1 shows the various adsorption energies for CO, and their decomposition. The adsorption of the CO molecule is usually explained in terms of the Blyholder mechanism.[13–15] Various calculations however have shown that the interaction of the 5σ orbital of CO with the metal orbitals is not simply a matter of electron donation.[2,16,17] In general a strong repulsion is found between the 5σ orbital and occupied metal orbitals; high coordination sites showing a higher steric repulsion than low coordination sites. This repulsion results in a change in the electronic configuration upon CO approach. The 5σ orbital of the CO molecule and metals orbitals interact to form an anti-bonding orbital that is pushed above the Fermi-level. Transferring the electrons of the orbital to non-σ orbitals reduces the repulsion. If we look at the electron density difference (Fig. 3), we see a decrease in the region between the carbon and the nearest metal atom. An increase is observed at the sides of the same metal atoms. The CO molecule seems to push electrons on the metal cluster aside.

Fig. 3. Contour plot of the electron density difference $\rho(Ir_4CO) - \rho(Ir_4) - \rho(CO)$ for 1-fold (top-left), 2-fold (top-right), and 3-fold adsorption (bottom). Dashed lines show a decrease, solid lines an increase of the electron density, except for the solid lines next to dashed lines which depict nodal surfaces. Subsequent contours correspond to ± 0.001, ± 0.003, ± 0.007, ± 0.013, ± 0.027, ± 0.054, ± 0.108, ± 0.218, and ± 0.439 electrons per $Å^3$.

Although the CO molecule feels a stronger repulsion at the high coordination sites the adsorption energy is nevertheless larger. Table 1 shows that this is due to a larger interaction energy. A decomposition of the interaction energy into different symmetries shows that, apart from the σ relaxation to reduce the repulsion, it is largely the interaction of metal d orbitals with the CO $2\pi^*$ orbitals that yields this interaction energy. In the 1-fold adsorption geometry the overlaps, and hence the interactions, of the $2\pi^*$ orbitals with the metal orbitals are much smaller than in the 2- and 3-fold adsorption geometries. There is also a smaller number of metals orbitals that interact with the $2\pi^*$ orbitals. This can also clearly be seen in the electron density difference plots (see Fig. 3). A much large increase is observed in the $2\pi^*$ region for the 2- and 3-fold adsorption geometries.

The preference for the 2- and 3-fold adsorption seems to be at variance with experimental data that indicate that CO adsorbs 1-fold.[18-21] We think that this is a cluster size effect. Recent slab calculations of CO on Cu yield a 1-fold adsorption, whereas the same method applied to Cu clusters show that higher coordination is preferred.[22,23]

The hydrogen molecule yields only very small adsorption energies (see Table 2). The relevant orbitals here are the bonding σ_g, and the anti-bonding σ_u^*. The interaction of the σ_g orbital with occupied metal orbitals seems to dominate. The mechanism in which an anti-bonding orbital is pushed above the Fermi-level does not work for the hydrogen molecule. Moreover, the σ_u^* orbital is too high in energy to interact appreciably with occupied metal orbitals, so that the steric repulsion due to the σ_g orbital is not compensated.

Table 2. Adsorption energies and their decomposition (in eV) of H_2 on a tetrahedral Ir_4 cluster.

geometry		adsorption energy	steric repulsion	interaction energy
1-fold	end-on	0.24	0.35	0.59
	side-on	0.30	0.47	0.77
2-fold	end-on	0.17	0.32	0.49
	side-on	0.03	0.07	0.11
3-fold	end-on	0.10	0.09	0.20
	side-on	0.02	0.00	0.03

In contradistinction to the CO molecule the preferred adsorption site is the 1-fold site. As for the CO molecule, for the 2- and 3-fold adsorption the steric repulsion is stronger, and here it determines the adsorption. This is especially so for the side-on 2- and 3-fold adsorption. The steric repulsion may seem to be small in Table 2, but we have to realize that for these geometries the metal cluster-hydrogen molecule distance is much larger as for all other geometries. (We found for this distance for end-on 2.03, 2.03, and 2.48 Å, and for side-on 2.08, 2.87, and 3.44 Å for 1-fold, 2-fold, and 3-fold adsorption, respectively.) If we change for example the distance for the 3-fold side-on adsorption from 2.54 to 2.01 Å, then the repulsion increases 0.80 eV

For the hydrogen atom we find again an appreciable adsorption energy (see Table 3). The repulsion in this case is relatively small compared to CO. Indeed, in terms of the electronic energies we only get a repulsion if there is a large overlap of the $1s$ orbital of the hydrogen atom and a metal orbital, whereas for larger distances (and thus smaller overlap) the electronic interactions are attractive. Only the nuclear repulsion counteracts these electronic interactions. Just as for the CO molecule the 2- and 3-fold adsorption sites are preferred. This can simply be ascribed to the larger number of metal orbitals with which the $1s$ orbital of the hydrogen atom can interact.

4. ADSORPTION IN THE PRESENCE OF A CATION

Table 4 shows the adsorption energies and their decomposition, and Table 5 shows the changes upon introduction of the Mg^{2+} ion onto Ir_4CO. We find the very remarkable fact that the adsorption energy does not change (we estimate the numerical error in our calculations to be about 0.1 eV); but that the steric repulsion and the interaction energies on the other hand change a lot. The latter changes seem to cancel each other, however. The steric repulsion is reduced for all adsorption geometries. This can be explained using Fig. 4). This figure

Table 3. Adsorption energies and their decomposition (in eV) of H on a tetrahedral Ir_4 cluster.

geometry	adsorption energy	steric repulsion	interaction energy
1-fold	2.87	2.02	4.88
2-fold	4.02	3.94	7.93
3-fold	3.92	5.17	9.06

shows the polarization of the Ir_4 cluster (without the CO molecule) due to the Mg^{2+} ion coordinated 3-fold. We note that at the site where adsorption is to take place—i.e. above the top iridium atom— there is a strong decrease in the electron density. Consequently, the repulsion with the 5σ orbital decreases. Only for very short distances the increase shown by the cloverleaf structure around the top iridium atom becomes noticeable. Furthermore, we see in Table 5 that the reduction of the steric repulsion increases with increasing coordination of the CO molecule, as might be expected.

Table 4. Adsorption energies and their decomposition (in eV) of CO on a tetrahedral Ir_4 cluster with a Mg^{2+} ion at the opposite side.

geometry	adsorption energy	steric repulsion	interaction energy
1-fold	0.98	0.18	1.16
2-fold	2.35	5.62	7.93
3-fold	2.42	4.78	7.19

Table 5. Change in the adsorption energies of CO and their decomposition (in eV) upon introduction of the Mg^{2+} ion. Shown are values with minus values without the cation.

geometry	adsorption energy	steric repulsion	interaction energy
1-fold	0.05	−1.53	−1.48
2-fold	−0.07	−1.86	−1.96
3-fold	0.03	−3.16	−3.11

The interaction energy becomes less too. This again may be explained (in part) on electrostatic grounds. Firstly, as the steric repulsion has been reduced, less can be gained by a further reduction of this repulsion. Hence, the contribution of the σ orbitals to the interaction energy is less than in the absence of the cation. Secondly, the π donation into the $2\pi^*$ orbitals of the CO molecule is decreased. As this π donation involves a movement of electrons away from the cation, it is energetically less favourable. Both the changes in the σ and the π orbitals thus yield less interaction energy in the presence than in the absence of the cation. The fact that the changes in the steric repulsion and the interaction energy almost cancel seems to be accidental.

The changes of the gross populations of the 5σ and $2\pi^*$ orbitals are in agreement with the explanations given above (see Table 6). The gross population of the $2\pi^*$ orbitals decreases

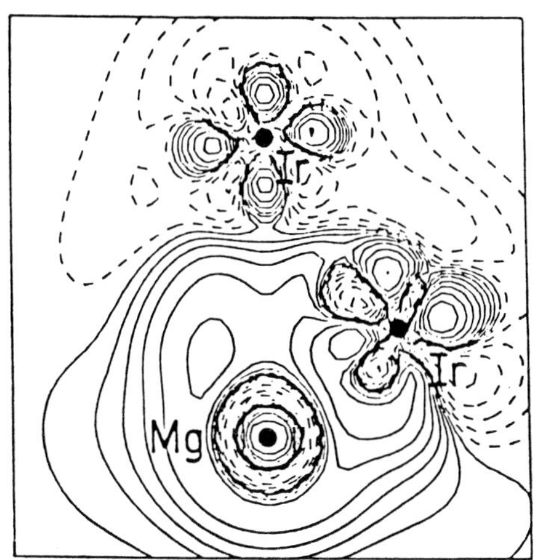

Fig. 4. Contour plot of the electron density difference $\rho(Mg^{2+}–Ir_4) - \rho(Ir_4) - \rho(Mg^{2+})$ for 3-fold adsorption of Mg^{2+}. Solid and dashed lines are to be interpreted as in Fig. 3. Subsequent contours correspond to ±0.005, ±0.011, ±0.020, ±0.035, ±0.058, ±0.096, ±0.159, ±0.262, and ±0.432 electrons per $Å^3$.

when the cation is added. The decrease is somewhat larger for the 1-fold as for the 2- and 3-fold adsorption geometries. The reason for this is probably that, as the distance between the metal cluster and the CO molecule is largest for the 1-fold adsorption geometry, the π donation entails a larger displacement of electrons away from the cation. The gross population of the 5σ orbital has increased. This agrees with the observation that it is less necessary to reduce repulsion. It is however not in agreement with the Blyholder mechanism where electrons are donated into empty metal orbitals. If this would be a favourable thing to occur without the cation, then a cation would only enhance this effect by pulling σ electrons more to the metal cluster, and thus yield a smaller gross population of the 5σ orbital. This means that σ donation is less important than reduction of steric repulsion.

Table 6. Gross populations of the CO orbitals with and without the Mg^{2+} ion. The change in population is defined as the population with minus the population without the cation.

geometry	orbital	without cation	with cation	change
1-fold	5σ	1.79	1.83	0.04
	$2\pi^*$	0.32	0.04	−0.28
2-fold	5σ	1.46	1.50	0.04
	$2\pi^*$	0.69	0.53	−0.16
3-fold	5σ	1.50	1.56	0.06
	$2\pi^*$	0.75	0.56	−0.19

Although the adsorption energies of the CO molecule do not change, other properties do change. As the occupation of the $2\pi^*$ orbitals decreases, the C–O bond becomes stronger, and hence the CO molecule will be less likely to dissociate. The change in the C–O bond strength should be observable with infrared spectroscopy; i.e. the stretch frequency should increase.

For the hydrogen molecule we have found that for all adsorption geometries the adsorption energy increases (for the 2-fold side-on geometry we have only an upper bound for adsorption energy of 0.33 eV). The changes are largest for the 3-fold adsorption geometry (see Tables 7, 8, and 9), where we have found a large change in the hydrogen molecule-metal cluster distance. For the 1-fold adsorption where the hydrogen molecule-metal cluster distance does not change very much we see that the steric repulsion decreases, just as we found for the CO molecule. For the other adsorption geometries the steric repulsion increases. We observe however at the same time a (much) smaller distance to the metal cluster. (We found for the hydrogen molecule-metal cluster distance for end-on 2.02, 1.79, and 1.39 Å, for 1-fold, 2-fold, and 3-fold adsorption, respectively, and for side-on 2.22, and 1.42 Å for 1-fold, and 3-fold adsorption, respectively.) Calculations at the same distance showed that, also for the 2- and 3-fold adsorption geometries the steric repulsion decreases. For example, for the 3-fold end-on adsorption with a metal cluster-hydrogen molecule distance of 2.48 Å, the change in steric repulsion is −0.44 eV.

Table 7. Adsorption energies and their decomposition (in eV) of H_2 on a tetrahedral Ir_4 cluster with a Mg^{2+} ion at the opposite side.

geometry		adsorption energy	steric repulsion	interaction energy
1-fold	end-on	0.38	0.34	0.71
	side-on	0.53	0.07	0.58
2-fold	end-on	0.42	0.37	0.78
	side-on	> 0.33		
3-fold	end-on	0.72	0.98	1.71
	side-on	0.98	2.25	3.23

Table 8. Change in the adsorption energies of H_2 and their decomposition (in eV) upon introduction of the Mg^{2+} ion. Shown are values with minus values without the cation.

geometry		adsorption energy	steric repulsion	interaction energy
1-fold	end-on	0.14	−0.01	0.12
	side-on	0.23	−0.40	−0.19
2-fold	end-on	0.25	0.05	0.29
	side-on			
3-fold	end-on	0.62	0.88	1.51
	side-on	0.95	2.25	3.20

The interaction energy increases also for all adsorption geometries, except for the 1-fold side-on adsorption. The situation is however really not so clear. This we see when we take the same hydrogen molecule-metal cluster distance with and without the cation. Taking the 3-fold adsorption we find for both the end- and side-on adsorption (metal cluster-hydrogen

Table 9. Gross populations of the H_2 orbitals with and without the Mg^{2+} ion. The change in population is defined as the population with minus the population without the cation.

geometry		orbital	without cation	with cation	change
1-fold	end-on	σ_g	1.86	1.83	−0.03
		σ_u	0.05	0.01	−0.04
	side-on	σ_g	1.89	1.86	−0.03
		σ_u	0.06	0.01	−0.05
2-fold	end-on	σ_g	1.88	1.82	−0.06
		σ_u	0.08	0.04	−0.04
	side-on	σ_g	1.96		
		σ_u	0.01		
3-fold	end-on	σ_g	1.94	1.73	−0.21
		σ_u	0.06	0.06	0.00
	side-on	σ_g	1.99	1.60	−0.39
		σ_u	0.00	0.19	0.19

molecule distance 2.47 and 2.54 Å, respectively), that the steric repulsion changes −0.44 and −0.08 eV, respectively. The interaction energy with the cation on the other hand is smaller for end-on (changes −0.28 eV), but larger for the side-on geometry (changes 0.29 eV). This means that the end-on geometry is dominated by the steric repulsion, whereas the side-on geometry is dominated by the interaction energy.

In this latter aspect the 3-fold side-on adsorption may very well be an exception. If we look at the gross occupations of the σ_g and the σ_u^* orbital of the hydrogen molecule, we find in general only small changes. Only for the 3-fold adsorption most changes are appreciable. For the 3-fold end-on adsorption the σ_g orbital depopulates a little, and the σ_u^* occupation does not change. For 3-fold side-on adsorption however both σ_g and σ_u^* occupation change much. This indicates a change in the molecular orbitals too. For both 3-fold adsorptions the H–H distance increases from 0.78 to 0.83 Å, whereas for the other geometries no change in H–H distance was found.

Dissociation of the H–H bond is facilitated by a decrease of the σ_g population and an increase of the σ_u^* populations. Hence, for the 3-fold side-on adsorption geometry, and to a lesser extent for the 3-fold end-on adsorption geometry, a cation enhances the dissociation probability. The effect is further enhanced by the more favourable adsorption energies for the 3-fold adsorption.

Just as the CO molecule, we see in Tables 10, 11, and 12 that the hydrogen atom does not seem to be much affected by the presence of the cation. The adsorption energy becomes slightly smaller for the 1-fold adsorption, but remains almost the same, with a slight tendency to increase, for the 2- and 3-fold adsorption. Also the distances to the metal cluster do not change.

Table 10. Adsorption energies and their decomposition (in eV) of H on a tetrahedral Ir_4 cluster with a Mg^{2+} ion at the opposite side.

geometry	adsorption energy	steric repulsion	interaction energy
1-fold	2.61	1.88	4.48
2-fold	4.11	3.09	7.17
3-fold	3.94	3.34	7.27

Table 11. Change in the adsorption energies of H and their decomposition (in eV) upon introduction of the Mg^{2+} ion. Shown are values with minus values without the cation.

geometry	adsorption energy	steric repulsion	interaction energy
1-fold	−0.26	−0.14	−0.40
2-fold	0.09	−0.85	−0.76
3-fold	0.02	−1.83	−1.79

Table 12. Gross populations of the H orbital with and without the Mg^{2+} ion. The change in population is defined as the population with minus the population without the cation.

geometry	orbital	without cation	with cation	change
1-fold	1*s*	0.93	0.75	−0.18
2-fold	1*s*	0.81	0.79	−0.02
3-fold	1*s*	0.84	0.86	0.02

5. CATALYTIC IMPLICATIONS

This theoretical study was stimulated by the need to understand the promoter effect of cations on metal surfaces, and the effects of charge-compensating cations in zeolites. We are primarily interested in the latter. We showed that the effect of cations depends on the adsorption site. However, also the position of the cation with respect to the admolecule matters. In a recent paper Wimmer *et al.*[24] looked at the effect of a K^+ ion adsorbed on a Ni(100) surface next to a CO molecule. They found an increased occupation of the $2\pi^*$ orbitals, and that the CO orbitals were stabilized with respect to the metal orbitals. We have found just the opposite. This again can be explained with a simple electrostatic picture. The cation stabilizes all orbitals, but those located closer to the cation are stabilized more. Hence, in our case, where the metal cluster is positioned between the cation and the CO molecule, the metal orbitals are stabilized most, whereas Wimmer *et al.* have found the opposite, as in their system the cation is closer to the CO molecule. Consequently, CO dissociation becomes more likely with cations closer to the CO molecule, and less likely with cations at the opposite side of the metal cluster.

The position of the cations with respect to metal particles in zeolites is in general not known. The cases we have studied correspond to a situation in which the cation is sandwiched between the wall of a pore and the metal cluster. The actual situation may be different; it may actually vary with the type of zeolite, and cation species. There exists a considerable literature on the behaviour of chemisorbed CO to metal particles in the channels of a zeolite as a function of cation.[25–31] Also reports on competitive adsorption of toluene and benzene have appeared.[32,33] Studies on zeolite X, Y as well as zeolite L have been reported. Whereas Pt particles in CeY and NaHY behave as Lewis acids,[32] they behave as Lewis bases in zeolite L.[33] Basicity increases with size of the earth alkali cation. This basic behaviour agrees with the observed lowered stretch frequency of CO adsorbed atop to Pt particles in zeolite L next to alkali cation.[30] Experimental results consistently indicate increased basic behaviour of transition metal particles occluded in zeolite L. The situation is less clear for the zeolites X and Y. Very high catalytic hydrogenation activities have been measured for metal particles next to highly charged cations,[27,34,35] indicating increasing Lewis acid character. Some CO adsorption studies indicate an increased CO frequency in such situation,[25–28] but also limited data are available that contradict this.[30] It therefore seems that the experiments on these zeolites warrant closer study. Our work agrees with increased Lewis acidity of metal particles in contact with cation of high charge, if in that situation the metal particle is located in between adsorbate and cation, as may be the case for zeolites X and Y. Experimental evidence seems to indicate that in zeolite L this situation does not exist. In zeolite L the adsorbate appears to be located between metal particle and cation, or the metal particle is close to the negatively charged zeolite wall with the cation in an inaccessible position.

For situations in which the cation is indeed at the opposite side of the metal cluster as the admolecule our results indicate that reactions in which the CO should not dissociate are favoured over dissociative ones. Furthermore, the dissociation reaction of the hydrogen molecule may become faster. When the rate of adsorption is rate limiting, the overall activation energy for dissociation with respect to the gas phase will be lowered, because of the increased heat of adsorption of the hydrogen molecule. Dissociation of H_2 can be considered a model of CH activation. Our results suggest that CH activation is enhanced on metal particles with a cation on the opposite side.

6. REFERENCES

1. I. Maxwell, Adv. Catal. 31 (1982) 1.
2. A.P.J. Jansen, and R.A. van Santen, J. Phys. Chem., in press.
3. E. Sanchez Marcos, A.P.J. Jansen, and R.A. van Santen, Chem. Phys. Lett. 167 (1990) 399.
4. P. Gallezot, Catal. Rev. - Sci. Eng. 20 (1979) 121.
5. M. Boudart, and G. Djega-Mariadassou, Kinetics of Heterogeneous Catalytic Reactions, (Princeton University Press, Princeton, 1984).
6. C. Besoukhanova, J. Guidot, D. Barthomeuf, J. Chem. Soc. Far. Trans. I 77 (1981) 1595.
7. A. de Mallman, and D. Barthomeuf, in Studies in Surface Science and Catalysis, eds. H.G. Karge, and J. Weitkamp (Elsevier, Amsterdam, 1989), Vol. 46.
8. G. Larsen, and G.L. Haller, Catal. Letter. 3 (1989) 103.
9. E.J. Baerends, D.E. Ellis, and P. Ros, Chem. Phys. 2 (1973) 41.
10. E.J. Baerends, and P. Ros, Chem. Phys. 2 (1973) 52.
11. E.J. Baerends, and P. Ros, Chem. Phys. 8 (1975) 412.
12. T. Ziegler, and A. Rauk, Theoret. Chim. Acta 46 (1977) 1.
13. G. Blyholder, J. Phys. Chem. 68 (1964) 2772.
14. M.J.S. Dewar, Bull. Soc. Chim. France 18 (1951) C71.
15. J. Chatt, and L.A. Duncanson, J. Chem. Soc. 2939 (1953).
16. P.S. Bagus, K. Hermann, and C.W. Bauschlicher, Jr., J. Chem. Phys. 81 (1984) 1966.
17. P.S. Bagus, C.J. Nelin, and C.W. Bauschlicher, Jr., J. Vac. Sci. Technol. A2 (1984) 905.
18. P. Gélin, A. Auroux, Y. Ben Taarit, and P.C. Gravelle, Appl. Catal. 46 (1989) 227. and references therein.
19. F. Solymosi, and J. Raskó, J. Catal. 62 (1980) 253.
20. K. Tanaka, K.L. Watters, and R.F. Howe, J. Catal. 75 (1982) 23.

21. F.J.C.M. Toolenaar, A.G.T.M. Bastein, and V. Ponec, J. Catal. 82 (1983) 35.
22. D. Post, thesis (Free University of Amsterdam, Amsterdam, 1981).
23. E.J. Baerends, private communications.
24. E. Wimmer, C.L. Fu, and A.J. Freeman, Phys. Rev. Lett. 55 (1985) 2618.
25. C. Naccache, M. Primet, and M.V. Mathieu, Adv. Chem. Ser. 121 (1973) 266.
26. P. Gallezot, J. Datka, J. Massardier, M. Primet, and B. Imelik, in Proceedings of the 6th International Congress on Catalysis, eds. G.C. Bond, P.B. Wells, and F.C. Tompkins (Letchworth, London, 1977), Vol. 2.
27. F. Figueras, R. Gomez, M. Primet, Adv. Chem. Ser. 121 (1973) 480.
28. G.D. Chukin, M.V. Landau, V.Ya. Kruglikov, D.A. Agievskii, B.V. Smirnov, A.L. Belozerov, V.D. Asrieva, N.V. Goncharova, E.D. Radchenko, O.D. Konovalcherov, A.V. Agafonoy, in Proceedings of the 6th International Congress on Catalysis, eds. G.C. Bond, P.B. Wells, and F.C. Tompkins (Letchworth, London, 1977), Vol. 1.
29. C. Besoukhanova, J. Guidot, D. Barthomeuf, J. Chem. Soc. Far. Trans. I 77 (1981) 1595.
30. A. de Mallman, and D. Barthomeuf, in Studies in Surface Science and Catalysis, eds. H.G. Karge, and J. Weitkamp (Elsevier, Amsterdam, 1989), Vol. 46.
31. L.L. Sheu, H. Knözinger, and W.M.H. Sachtler, Catal. Lett. 2 (1989) 129.
32. T.M. Tri, J. Massardier, P. Gallezot, and B. Imelik, in Studies in Surface Science and Catalysis, eds. B. Imelik, C. Naccache, G. Coudurier, H. Praliaud, P. Meriaudeau, G. Gallezot, G.A. Martin, and J.C. Vedrine (Elsevier, Amsterdam, 1982), Vol. 11.
33. G. Larsen, and G.L. Haller, Catal. Letter. 3 (1989) 103.
34. R.A. Dalla Betta, and M. Boudart, in Proceedings of the 5th International Congress on Catalysis, ed. J.W. Hightower (North-Holland, Amsterdam, 1973), Vol. 2.
35. C. Naccache, N. Kaufherr, M. Dufaux, J. Bandiera, and B. Imelik, ACS Symp. Ser. 40 (1977) 538.

R.K. Grasselli and A.W. Sleight (Editors), *Structure-Activity and Selectivity Relationships in Heterogeneous Catalysis*
© 1991 Elsevier Science Publishers B.V., Amsterdam

Structure-Reactivity Relationships in Methanol to Olefin Conversion on Various Microporous Crystalline Catalysts

T. INUI

Department of Hydrocarbon Chemistry, Faculty of Engineering, Kyoto University, Sakyo-ku, Kyoto 606, Japan

ABSTRACT

Characteristics of various zeolitic catalysts for light olefin synthesis from methanol were discussed. Narrow pore zeolites such as chabazite and ZSM-34 are effective for this purpose, however, have short catalyst life due to rapid coke deposition. Middle pore zeolite ZSM-5 has too strong acidity to produce light olefins with a high selectivity. Modified H-ZSM-5 with basic materials need high reaction temperatures and afford propylene as the major product by obeying β-scission. A pentasil type H-Fe-silicate yielded ethylene and propylene almost exclusively. Nickel containing SAPO-34 produced ethylene almost quantitatively, irrespective of its chabazite structure, indicating that when the acidity is controlled appropriately to do not produce aromatics, the selective conversion of methanol to light olefins would be achieved.

1. INTRODUCTION

After the Oil crisis investigation on olefin synthesis from alternative routes has become active. In spite of tremendous efforts, in the direct conversion of syngas to olefins by using Fischer-Tropsch type of mixed metal-oxide catalysts, the serious problem caused by Schulz-Flory probability law is still not yet overcome. In the most recently developing new route, oxidative coupling of methane, researches of fundamental stage are still going on. Thus, so far, olefin synthesis from syngas or the mixture of carbon dioxide and hydrogen *via* methanol using shape-selective microporous crystalline catalysts would be considered the most promising route for the highly selective conversion.

In the last decade, the studies on methanol to olefin conversion on zeolitic catalysts have passed several transition phases. In this paper, these phases will be followed from the viewpoint of the relationship between the structure of zeolitic catalysts and their catalytic performance in methanol conversion.

2. NARROW PORE ZEOLITE CATALYSTS

In the beginning stage of the research, narrow pore zeolites having 8-oxygen member rings like as chabazite were chosen as the catalyst(1,2). One can easily think of and expect a molecular sieving effect for light olefins through such narrow size pores. However, selectivity to light olefins on these catalysts was insufficient, and deactivation by coke formation was very fast.

Offretite-Erionite intergrowthed zeolite ZSM-34 was announced in patent Literatures by Mobil(3,4) and they confirmed that this narrow pore zeolite exhibited a very high selectivity to ethylene, i.e. about 50% in the begging of the reaction run, but this zeolite needed extraordinary long crystallization time up to 196 days for its synthesis, and the deactivation by coke formation was very fast.

In order to improve the crystallization method, the author et al.(5,6) analyzed what happened in such long crystallization period, and found out that the mixed gel converts to non-crystalline spherical particles within three days. These spherical particles have an activity to convert methanol into dimethyl ether exclusively. Then these precursor particles change into zeolitic crystallites while maintaining the spherical shape. The activity of hydrocarbon synthesis increased with an increase of the crystallization time. Within one month crystallinity reached 100%, and then gradually increased a complexity in crystalline structure. Accompanied with this, ethylene selectivity increased but simultaneously light paraffinic hydrocarbons also increased. This indicates that offretite-erionite intergrowth advanced and acidity also strengthened.

When the gel mixture is directly heated up, sodalite crystal was mixed and ethylene selectivity was low, but the activity was rather higher than that for 25 days crystallization at standard crystallization temperature. On the other hand, the mixed gel for precursor making was maintained at 100°C for 3 days, and then the crystallization temperature was raised up, better activity and better olefin selectivity were obtained. Total hydrocarbons formed until deactivation, ΣHC, begins increased with an increase of the activity, indicating that number of acidic sites, which responsible for hydrocarbon synthesis, increased.

We have tried many factors other than crystallization temperature; kind of organic templates, composition of the gel, and addition of seed crystals. The size of crystals formed decreased sensitively at the stage of mixed gel preparation(6) and uniformity of the crystals increased with an increase of the addition of small amount fine seed crystal particles(6).

It was noteworthy that when some kinds of catalyst metal were previously dispersed on the seed materials, and then these metal-loaded seed materials were used as the seed for crystallization(8). The resulting crystals exhibited a longer catalyst life until deactivation by coke. The dispersed metal played a role of coke combustion during the methanol conversion reaction. However, essentially the coke formation was inevitable in these kinds of narrow pore zeolite.

3. MEDIUM PORE ZEOLITE CATALYSTS

Naturally, then medium pore or large pore zeolites were considered as alternative candidates for olefin synthesis catalysts, like as, 10-member oxygen ring ZSM-5, and 12-member oxygen ring Mordenite or Faujasite-type zeolite like as Y.

However, many researchers, for instance, Dajaive et al.(9) demonstrated that only ZSM-5 zeolite can continue its activity in methanol conversion during the significant time on stream. We have also observed the large difference in catalyst life in n-hexadecane conversion on these zeolite catalysts(10). Fresh catalysts of protonated M and Y exhibited enough activity to convert into other hydrocarbons, but very rapid deactivation occured, whereas pentasil-type zeolites maintained their initial activity without any deactivation at least during the test up to 10 hours on stream.

The products of initial stage were compared for each catalyst, and it was found that the rapidly deactivated catalysts yielded a little aromatics, which are precursor of coke. It seems to be contradictory. However, this question was solved when we saw the result on the amount of coke accumulated during the conversion reaction. Inside the pores of H-M and H-Y, considerable amounts of coke were deposited. To the contrary, on the pentasil-type zeolites, the deposited coke was very little although much more aromatics formed.

Evidently, these results suggest that not only pore diameter of zeolite but also other characteristics of pore structures, like as the dimension of pore connection and, especially, absence of supercage, in which space fused ring aromatics can be grown, is important for longer catalyst life.

Therefore, many researches studied on ZSM-5 type zeolite; however, a typical H-ZSM-5 itself has strong acid sites, and a considerable part of methanol can be converted to aromatics, and selectivity to intermediate olefins are low. In order to increase olefin selectivity the strong acid sites were modified by basic materials such as Ca, Mg, Zn, Sr, B and P(Phosphite)(11-18), or Al content in ZSM-5 was decreased(19). As the selectivity to olefin increases but the activity decreases, the reaction temperature was raised up to 600°C from around 400°C. Unfortunately, on these catalysts the major olefin in the products was occupied with propylene by the occurence of β-scission at these higher temperatures, and the most valuable ethylene was produced fairly little.

4. IRON-SILICATE CATALYST HAVING PENTASIL STRUCTURE

In order to change the catalytic property of H-ZSM-5, we tried isomorphs substitution of Al in ZSM-5 with other transition metals from the stage of gel mixture before crystallization(20). First, we modified crystallization method. In most of conventional synthesis method, for example, the temperature is kept at a constant level for long time. In stead of this conventional method, we tried a rapid crystallization method(21). Temperature was raised to keep constant of the crystallization rate. Usually, we achieved crystallization of ZSM-5 within only 4-hour hydrothermal treatment. Crystals synthesized by the rapid crystallization method were small and uniform and the activity of methanol conversion was much more higher. The temperature to convert methanol 100% decreased 100°C compared with conventional one. The product distribution shows that aromatics increased and olefins decreased indicating that the acid strength increased and a sequence of consecutive reactions of methanol progressed more deeply.

The other important aspect of the rapid crystallization method is to be able to incorporate transition metals other than Al. First, we carefully tried a small amount of metal incorporation instead of Al, and found even only Si/metal atomic ratio 3200, the crystal habit was evidently different from each other(20). It means that other metal ingredients act like as seed materials and accelerate the nucleation of crystals.

The results of methanol conversion on these metallosilicates involved various kinds of metals with Si/Metal atomic ratio 3200 showed distinct difference from that of Al-silicate, i.e. ZSM-5. Ga-silicate, for example, yielded the maximum gasoline fraction, and to the contrary, Fe-silicate yielded the maximum level of olefin fraction. The order of the magnitude for olefins and gasolines fraction was roughly inverse relationship(22). This fact must reflected the acid strength of metallosilicates.

Consequently, we chose Fe-silicate for the sake of selective olefin synthesis. The concentration of iron was varied and confirmed that iron can be incorporated into the framework of silicate up to 10wt% as Fe_2O_3 by using various kinds of measurement like as XRD, solid NMR, UV, IR, TG, SEM, BET surface area, AA, UV-X ray(22,23). Distribution of iron in the Fe-silicate crystallites was measured by EPMA for crude crystals and after pulverized to 0.2 μm by means of agate mill. In dilute concentrations of Fe, Fe distributes much more around core parts of the crystallites than the outer surface zone of the crystals. This means that the outer surface does not have any acidity and catalytic activity caused by iron. It would contribute to decrease non-shape selective reaction on the outer surface of the crystallites and improve the product selectivity.

Since Fe ingredient works at the nucleation stage, and then we studied more precisely the condition of Fe gradient, i.e. effects of Fe-salt kinds and concentration of the Fe solution in the gel mixture preparation. We found that the optimum condition is exist to exert the high olefin selectivity, in which the effect of LV is very large, and can be obtained highly selective synthesis of C_2= and C_3= at a higher LV condition at proper reaction temperature around 300°C. This means that C-C chain growth stops at C_3 and the reaction does not involve formation of the aromatics. This gives a sharp contrast to the high-temperature active catalysts like as a modified H-ZSM-5 with alkaline earth metal oxides. Indeed, since Fe-silicate catalyst does not yield aromatics, coke formation, and consequently, deactivation does not occur. However, the catalytic activity of Fe-silicates in methanol conversion is still not large, and ethylene selectivity is insufficient from the viewpoint of quantitative conversion of methanol to ethylene.

5. METALLOSILICOALUMINOPHOSPHATE CATALYSTS

Silicoaluminophosphates have a very weak acidity compared

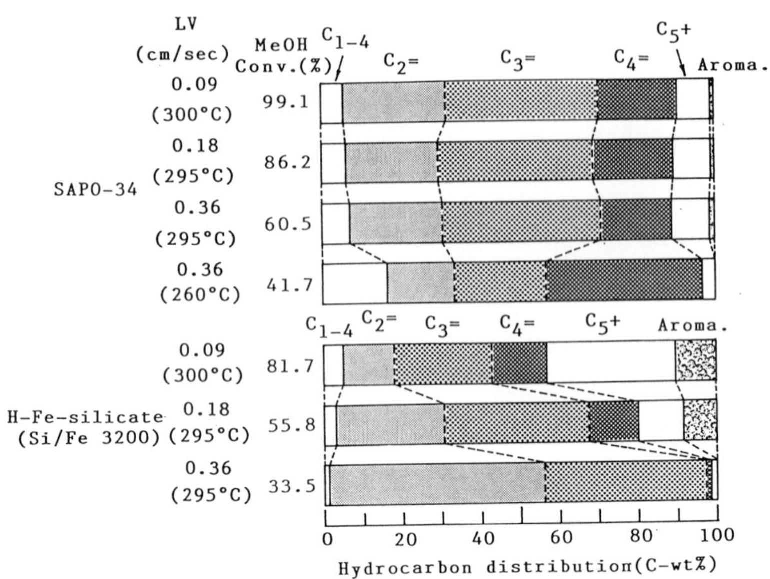

Fig.1 Comparison of the effect of linear velocity on the product distribution for SAPO-34 and H-Fe-silicate.

with zeolites, and would have a potential of methanol to olefin conversion. Especially, SAPO-17 and 34 were noted, because pore structures of these SAPOs are similar to erionite and chabazite structures, respectively, and the acidities of these SAPOs are much weaker than those of corresponding zeolites(24).

SEM photographs of these materials showed that SAPO-34 corresponded to typical crystal habits of chabazite, however, SAPO-17 was somewhat different from the crystal habit of erionite but similar to that of mordenite. The XRD patterns of SAPO-34 synthesized in this study coincided with SAPO-34 described in patent literature(25), but SAPO-17 synthesized here contained SAPO-5 structure. The NH_3-TPD profiles for each SAPOs prepared by slow crystallization and rapid crystallization of this study coincide with each other, respectively. SAPO-34 has much higher acidity than SAPO-17. Notwithstanding of acid strength, product distribution of methanol conversion seems to be strange contradiction. SAPO-17, which has much weaker acidity, can convert methanol more deeply than SAPO-34. This must be attributed to their different pore structures. SAPO-17 contains SAPO-5 structure that is similar to mordenite structure. This permits formation of fused-ring aromatics, and furthermore, it contains one-dimensional pore structure and therefore, effective diffusivity is very low. The effect of linear velocity on the product selectivities was compared between SAPO-34 and Fe-silicate, and a striking contrast between them was found as shown in Fig.1. The product distribution on Fe-silicate was affected by linear velocity of the reaction gas sensitively, and the selectivity to light olefins increased distinctly with an increase of linear velocity. On the other hand, that on SAPO-34 accepted almost no effect. The reason would be attributed to the big difference in the effective diffusivity measured by HETP method; that of Fe-silicate is about 3 times compared with that of SAPO-34(24). However, the catalytic activity of Fe-silicate was lower than that of SAPO-34.

In order to modify the product selectivity, and then we tried to incorporation of metals having larger ionic radius than Al into the framework of SAPO-34 expecting that expansion of pore size and decrease in acidity similar to the case of Fe-incorporation(22). Ni has a larger ionic radius and would not be easily incorporated into the framework, however, as shown in Fig.2, XRD patterns and BET surface areas of Ni-SAPO-34s having Si/Ni ratios 100 and 40 were similar to that of SAPO-34, indicating that Ni could be incorporated. NH_3-TPD profiles shown in Fig.3 indicates that the acidity is weakened with an increase of Ni incorporation.

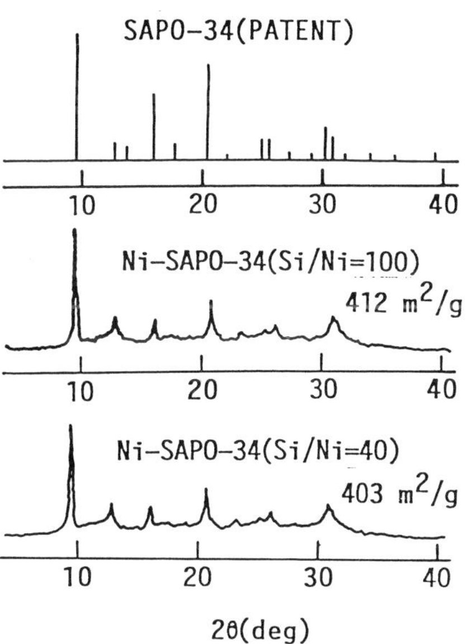

Fig.2 XRD patterns and BET surface area of Ni-SAPO-34 catalysts.

Performance of Ni-SAPO-34s were compared with that of SAPO-34. The results are shown in Fig.4. The ethylene selectivity increased with an increase of Ni content in SAPO-34, which consistents with the decrease in the amounts of strong acid sites, and an extraordinary high selectivity to ethylene from methanol was achieved by Ni-SAPO-34 having Si/Ni ratio 40. The temperature dependence of the product distribution is shown in Fig.5. The maximum selectivity to ethylene , almost 90%, could

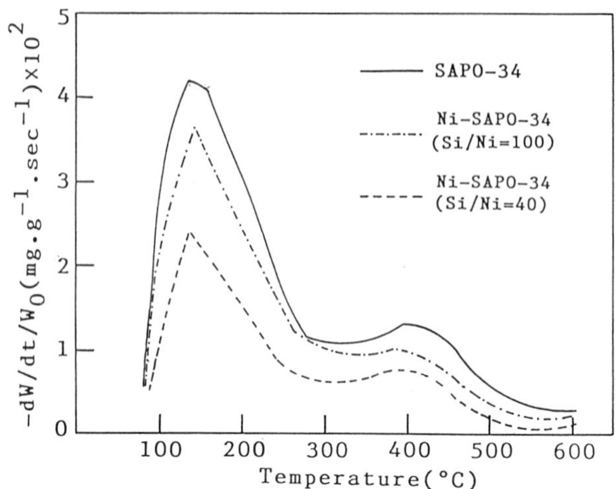

Fig.3 TPD profile of desorped NH_3 from the SAPO-34 and Ni-SAPO-34 catalysts.

be obtained at around 450°C, and above that temperature formation of CO, H_2, and CH_4 increased. CO and H_2, and CH_4 were attributed to the products caused by methanol decomposition and their consecutive reaction, respectively on the catalytic site of Ni in the Ni-SAPO-34. The catalytic performance of Ni-SAPO-34 having Si/Ni ratio 40 at 450°C shown in Fig.5 was maintained as long as 13 h as shown in Fig.6. In the beginning, the selectivities to propylene and others decreased gradually, however ethylene selectivity was consistently maintained at 90% without decrease of the activity. The initial decrease in selectivity to C_3+ hydrocarbons means that the catalytic sites which can form higher carbon number hydrocarbons were covered by some retardative

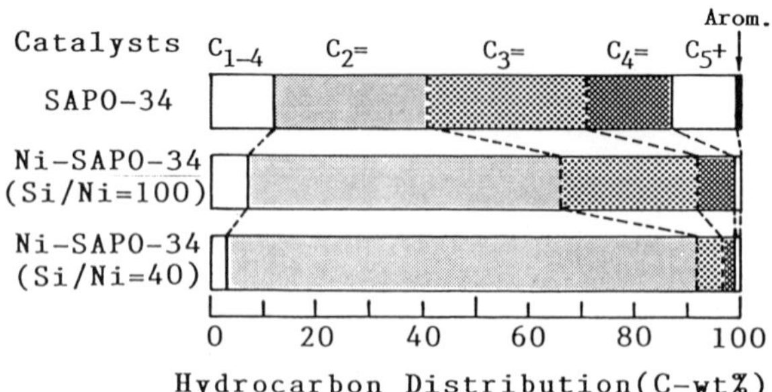

Fig.4 Hydrocarbon distribution of methanol conversion on different Si/Ni ratios of Ni-SAPO-34.
Feed: 20% MeOH and 80% N_2
Reaction condition: 450°C, GHSV 2000 h^{-1}

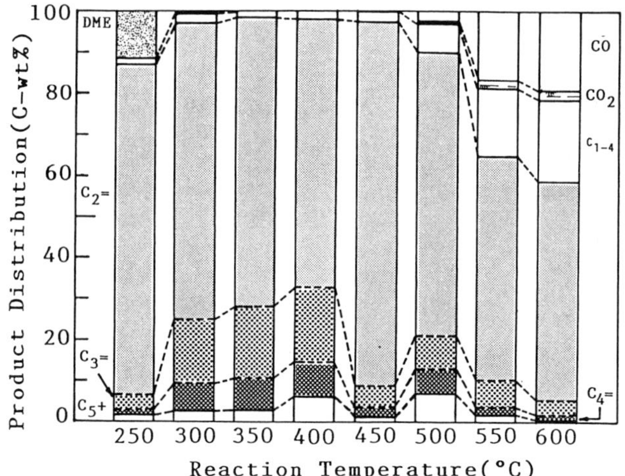

Fig.5 Temperature dependence of the product distribution on the Ni high-content Ni-SAPO-34(Si/Ni=40).

adsorbates like as precursor of coke, and only the active sites which can form ethylene remained. Since any consecutive reaction could not occur, no deactivation caused by coke accumulation was realized.

In conclusion, an extraordinary high selectivity to ethylene, as high as 90%, was achieved without coke formation in total methanol conversion condition on the catalyst of Ni-SAPO-34 having Si/Ni ratio 40. The catalyst has a similar pore structures to chabazite, however it has a considerably weaker

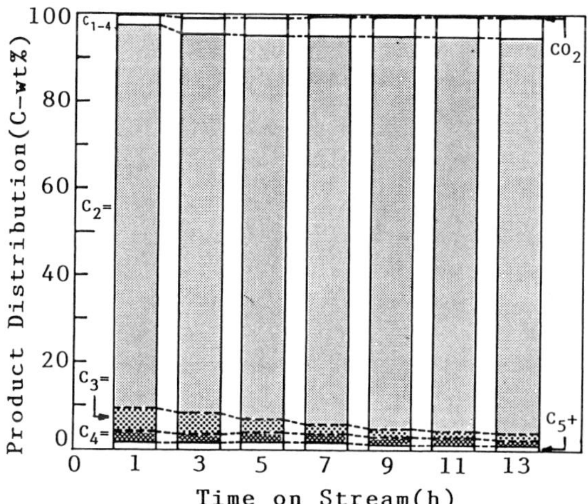

Fig.6 Prolonged test of methanol conversion for Ni-SAPO-34(Si/Ni=40).

acidity than chabazite and even more SAPO-34 having the same Si content by incorporation of Ni into the framework. It is noteworthy that even though the microporous crystalline catalyst having large cavity, which allows coke formation, when the acidity of the catalyst was controlled precisely, the carbon-carbon formation reaction can be controlled precisely. This fact would extend to the design of zeolitic catalysts for wider selective hydrocarbon conversion.

REFERENCES

1. C.D. Chang, and A.J. Silvestri, J. Catal., **47**, 249(1977).
2. R.G. Anthony, and B.B. Singh, Hydrocarbon Process., Mar., 85 (1981).
3. M.K. Rubin, E.J. Rosinski, and J.C. Plank, U.S. Pat. 4,086,186(1978).
4. Mobil Oil Corp., Japan Kokai 53-58499(1978).
5. T. Inui, and Y. Takegami, Hydrocarbon Process., Nov., 117 (1982).
6. T. Inui, T. Ishihara, N. Morinaga, G. Takeuchi, H. Matsuda, and Y. Takegami, Ind. Eng. Chem. Prod. Res. Dev., **22**, 26 (1983).
7. T. Inui, N. Morinaga, and Y. Takegami, Appl. Catal., **8**, 187(1983).
8. T. Inui, G. Takeuchi, and Y. Takegami, Appl. Catal., **4**, 211 (1982).
9. P. Dejaifve, A. Auroux, P.C. Gravelle, J.C. Vedrine, Z. Gabelica, and E.G. Derouane, J. Catal., **70**, 123(1981).
10. T. Inui, J. Jpn. Petrol. Inst., **33**, 198(1990).
11. W.W. Kaeding, and S.A. Butter, J. Catal., **61**, 155(1980).
12. S.A. Butter, U.S. Pat. 3,979,472(1976).
13. W.W. Kaeding, U.S. Pat. 4,049,573(1977).
14. R.J. Mc Intosh, and D. Seddon, Appl. Catal., 6, 307(1983).
15. J.C. Vedrine, A. Auroux, P. Dejaifve, V. Ducarme, H. Hoser, and S. Zhou, J. Catal., **73**, 147(1982).
16. C. Chen, J. Liang, Q. Wang, G. Cai, S. Zhao, and M. Ying, Proc. 7th Intern. Zeolite Confer., Tokyo, Japan, 909(1986).
17. H. Okado, H. Shoji, K. Kawamura, Y. Kohtoku, Y. Yamazaki, T. Sano, and H. Takaya, Nippon Kagaku Kaishi., 25(1987).
18. W. Holderich, H. Eichhorn, R. Lehnert, L. Marasi, W. Mross, R. Reinke, W. Ruppel, and H. Schlimper, Proc. 6th Intern. Zeolite Confer., Reno, U.S.A., 545(1983).
19. E. Kikuchi, R. Hamana, S. Hatanaka, Y. Morita, J. Jpn. Petrol. Inst., **24**, 275(1981).
20. T. Inui, O. Yamase, K. Fukuda, A. Itoh, J. Tarumoto, N. Morinaga, T. Hagiwara, Y. Takegami, Proc. 8th Intern. Congr. Catal., Berlin, 1984, vol III, p.569.
21. T. Inui, ACS Symposium Series 398, Zeolite Synthesis, M.L. Occelli and H.E. Roleson eds., Amer. Chem. Soc., 479(1989).
22. T. Inui, H. Matsuda, O. Yamase, H. Nagata, K. Fukuda, T. Ukawa, and A. Miyamoto, J. Catal., **98**, 491(1986).
23. T. Inui and M. Inoue, Hyomen, in press.
24. T. Inui, H. Matsuda, H. Okaniwa, A. Miyamoto, Appl. Catal., **58**, 155(1990).
25. Union Carbide Corporation, Japan Kokai, 59-35018(1984).

R.K. Grasselli and A.W. Sleight (Editors), *Structure-Activity and Selectivity Relationships in Heterogeneous Catalysis*
1991 Elsevier Science Publishers B.V., Amsterdam

TITANIUM SILICALITE: A NEW SELECTIVE OXIDATION CATALYST

Bruno NOTARI

ENI- Ricerca e Sviluppo

20097-San Donato Milanese,MILANO,ITALY

1. Introduction

The activity of titanium based catalysts for the oxidation of organic compounds is well known.

H.Wulff et al in 1971 (1) patented for Shell Oil a process for the selective epoxidation of propylene with hydroperoxides like ethylbenzene hydroperoxide (EBH) or tertiary-butyl hydroperoxide (TBH) with the use of a catalyst made of TiO_2 deposited on high surface area SiO_2.A Shell Oil plant for the production of 130.000 Tons/y of propylene oxide at Moerdijk,Holland,is based on this technology.

Sheldon et al.(2) have shown that in the epoxidation of olefins with TBH,compounds like titaniumacetylacetonate or tetra-n-butyltitanate containing Ti^{IV} produce epoxides with extremely high selectivities (98%),even though the rate of reaction is generally lower with respect to Mo(VI) or V(V) catalysts.

Hydroperoxides have been considered better oxidation agents with respect to H_2O_2 in view of their excellent thermal and chemical stability,the high selectivities towards desired products and their solubility in organic solvents (3).However the industrial interest in the use of H_2O_2 for selective oxidations remains high in view of the advantages offered by this oxidant namely low cost per oxygen atom and absence of by-products.The safe use of H_2O_2 in industrial plants is possible only if H_2O_2decomposition catalysts like Fe salts are rigorously excluded.Every single part of the plant must be designed in order to comply with this requirement.

Shirmann et al. in 1977 (4) patented homogeneous catalysts containing B_2O_3,As_2O_3 and MoO_3 which,under particular anhydrous conditions,could perform epoxidations of olefins with H_2O_2 in organic solvents with high selectivities.

A research program has been carried out at ENI directed towards the synthesis of zeolite-like materials:a new microporous material made of SiO_2 and TiO_2 has been obtained (5,6,7) which turned out to be a very efficient heterogeneous catalyst for selective oxidations with H_2O_2. Its structure closely parallels that of silicalite-1: it has been described as a silicalite in which isomorphous substitution of Si^{IV} with Ti^{IV} has taken place,and therefore named Titanium Silicalite-1 or TS-1. The changes in unit cell dimensions brought about by Ti in the solid agree well with the values that can be calculated by the difference between Ti-O and Si-O bond length.The presence in the IR spectrum of an absorption band at 960 cm^{-1} which is absent in silicalite has been ascribed to the presence of Ti in the solid. The ^{29}Si MAS NMR spectrum gives a signal at -115 ppm absent in silicalite,and therefore considered characteristic of TS-1. Finally,the distribution of Ti along the crystal has been shown to be perfectly homogeneous (8).

The catalytic properties of TS-1 are of great scientific and technological interest:using H_2O_2 it is possible to perform the selective oxidation of olefins and diolefins to epoxides (9,10,11),the hydroxylation of aromatic compounds (12),the selective oxidation of primary alcohols to aldehydes and secondary alcohols to ketones (13),the ammoximation of ketones (14,15). Heterogeneous catalytic systems offer technological advantages in industrial applications with respect to homogeneous systems:simple separation and recovery of the catalyst from the reaction mixture, its recycle and eventual regeneration once deactivated, easier recovery of reaction products.

In Fig.1 conversion and selectivity values obtained in different reactions are given.

Fig.1. TS-1 CATALYZED OXIDATIONS

REACTION	H_2O_2 conversion	Selectivity	Ref.
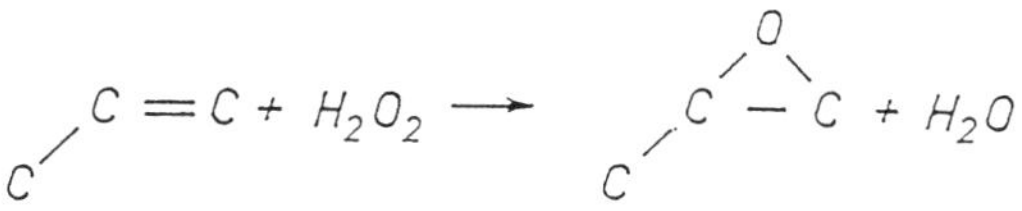 $C\text{-}C{=}C + H_2O_2 \rightarrow C\text{-}C(O)C + H_2O$ (epoxide)	99	97,8	(9)
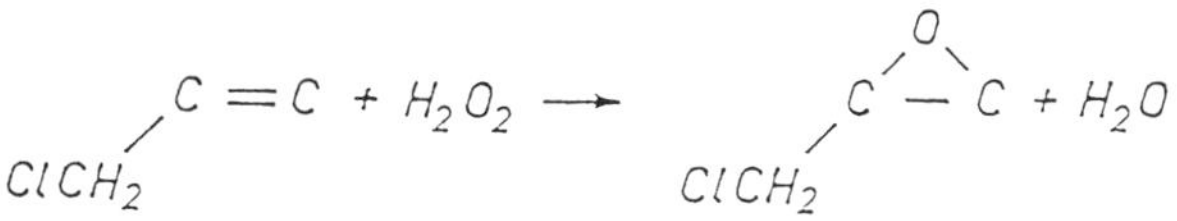 $ClCH_2\text{-}C{=}C + H_2O_2 \rightarrow ClCH_2\text{-}C(O)C + H_2O$ (epoxide)	98	93	(10)
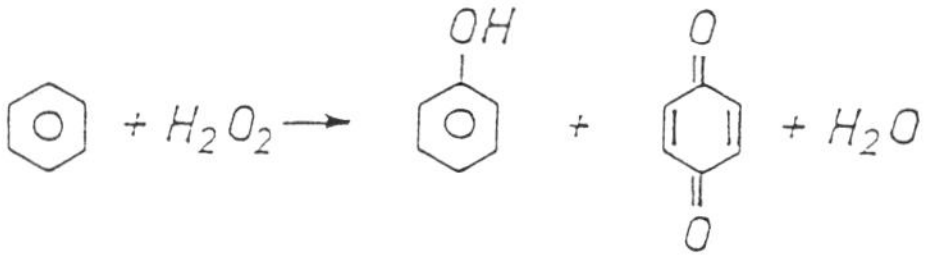 benzene $+ H_2O_2 \rightarrow$ phenol (OH) + benzoquinone (O, O) $+ H_2O$	100	76 ÷ 24	(12b)
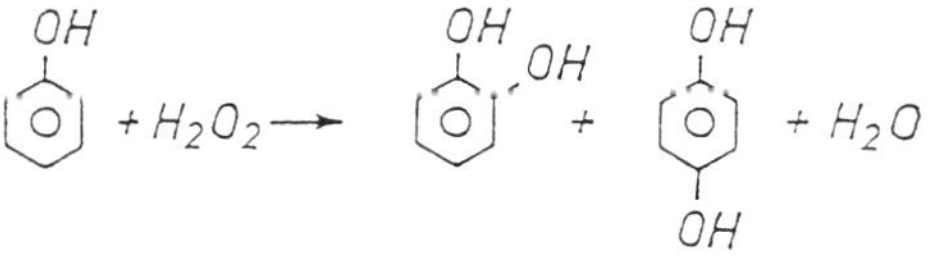 phenol (OH) $+ H_2O_2 \rightarrow$ catechol (OH, OH) + hydroquinone (OH, OH) $+ H_2O$	98	70 90	(12a)
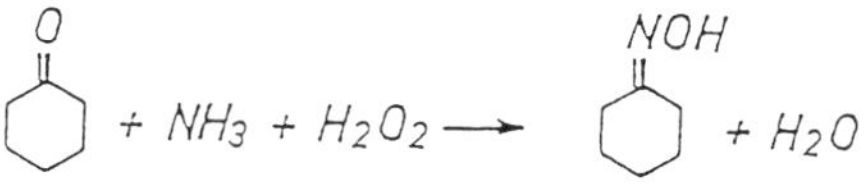 cyclohexanone (O) $+ NH_3 + H_2O_2 \rightarrow$ cyclohexanone oxime (NOH) $+ H_2O$	99,9	93,2 98,2	(14)

Such high selectivities at high H_2O_2 conversions can hardly be obtained with other catalysts; silicalite itself is rather unreactive, while on other Ti^{IV} containing catalysts H_2O_2 either does not react or, under more severe conditions, is decomposed into H_2O and O_2.

In the hydroxylation of benzene in anhydrous solvent the selectivity

to phenol is reduced because of the consecutive reaction of phenol to p-benzoquinone,but clearly a more careful choice of operating conditions could easily increase the phenol yield.
The ammoximation reaction of cyclohexanone to cyclohexanone oxime,as well as other reactions,are being carefully studied and possibilities of industrial applications evaluated.

The production of hydroquinone and catechol by TS-1 catalyzed hydroxylation of phenol with H_2O_2 appeared competitive with respect to existing industrial processes. A new industrial process has been developed based on TS-1 and a plant for the production of 10.000 tons/y of diphenols has been built in Ravenna,ITALY (7). It operates since 1986 with excellent results.A plant for the industrial production of TS-1 has also been built to provide the diphenols plant with the required amount of catalyst.

2.Synthesis of TS-1

The study of the catalytic properties of any material requires that the product is obtained always with the same chemical composition,structure and catalytic activity,and this has been a serious problem in catalysis. But when the catalyst must be used in an industrial plant,this problem becomes vital. The whole plant is designed on the assumption that the desired reaction takes place with the rate and selectivity defined in the project.Much care has therefore been given to the synthesis of this new catalyst,taking into account all variables that could influence the final result.

The major variables are:

- -reagents used
- -purity,particularly effect of alkalies
- -crystallite dimensions
- -non-framework TiO_2 effects
- -crystallite agglomeration

2.1 Reagents

Reagents to be employed in the synthesis must be selected between a very large number of possible alternatives.With the use of tetraethylsilicate (TES) as the source of SiO_2,tetraethyltitanate (TET) as the source of TiO_2 and tetrapropylammonium hydroxide (TPAOH) as the base,indicated in the first patent (5),a high degree of reliability could be obtained,and these reagents have therefore been applied also for the industrial production in spite of their rather high prices.It soon appeared that one of the key features was the purity of reagents and in particular the effect of even minute traces of alkalies:this required the development of a process for the production of high purity TPAOH since commercial products then available proved unsatisfactory. The process (17) is very efficient and has been developed for the industrial production.

When TES and TET are brought in contact,mixed oligomers are formed:but when acqueous TPAOH is added,under certain conditions a precipitate can form. It has been reported (18) that hydrothermal treatment of clear solutions produces by crystallization orthorombic TS-1,while hydrothermal treatment of mixtures containing a precipitate produces monoclinic silicalite. A possible explanation is that upon reaction with TPAOH,hydrolysis products containing Ti are formed: these products do not redissolve during the subsequent operations and therefore the Ti is not available for crystals formation. As a consequence of this segregation the Ti containing compounds undergo independent transformations and appear in the final calcined product as Ti oxides. It is therefore of the utmost importance to prevent the formation of precipitates when the TES,TET and TPAOH mixture is prepared. A procedure which appears satisfactory involves mixing TES and TET and cooling to 273°K before the addition of TPAOH. Alternatively the mixture of TES and TET is added with vigorous stirring to an acqueous concentrated solution of TPAOH (15%). Both procedures have proved adequate.

2.2 Effect of alkalies

Synthesis of TS-1 in the early experiments gave erratic results: the purity of the TPAOH base used in the different experiments was suspected to be responsible and it was hypothesized that alkalies could have an influence on the crystallization process. To clarify this point experiments were carried out with pure TPAOH and the same base to which controlled amounts of alkalies were added as indicated in columns 2 and 3 of Table 1. X ray diffraction and catalytic activity in the hydroxylation of phenol were used to measure the properties of the products obtained.

Table 1: Effect of alkalies

SiO_2/TiO_2	Na^+,ppm	K^+,ppm	Rx	H_2O_2 yield
50	0	0	orthorombic	79,5
50	0	1060	orthorombic	55.0
50	0	3530	orthorombic	23.0
50	0	7060	monoclinic	0.0
50	1765	0	orthorombic	42.0
50	3529	0	orthorombic	22.0

The catalyst obtained with pure TPAOH has the orthorombic structure and gives a high yield of H_2O_2. When Na^+ or K^+ are added the yield decreases and the magnitude of the effect is a function of the amount of added alkalies. Changes in the crystal structure follow a different trend: at low alkalies content the orthorombic structure is maintained; at high alkalies values it suddenly changes to monoclinic like silicalite, but the yield has dropped to zero.

J.El Hage-Al Asswad, J.B.Nagy, Z.Gabelica and E.G.Derouane have independently reached similar conclusions (19).

2.3 Crystallite dimensions

Since the early catalytic experiments it appeared that the results were also influenced by the crystallite dimensions,with the best performances in the $0.2\div0.3\mu$ range. Larger dimensions produced lower reaction rates and lower selectivities. With the reagents and method indicated,it was sufficient to regulate hydrothermal treatment temperature and time,433°K and 3 hours,to obtain the desired size.

The separation of the crystals from mother liquors containing hydrolysis products of TES and TET must be carefully conducted:repeated washings are necessary to remove non-framework TiO_2.

2.4 Agglomeration

In order to be used in an industrial plant the catalyst must be shaped in particles of at least $20\div30\mu$ of high mechanical resistance.Only when these requirements are satisfied it can be successfully used in the plant and survive the very severe regeneration procedures which must be periodically carried out to remove carbonaceous deposits and restore catalytic activity. The use of Ludox silica,silicates or other bonding agents has been unsuccessful.

The solution to the problem has been found (20) with a procedure that brings about the formation of a thin layer of silica coating every single crystal and connecting all crystals of a particle together: this is obtained by dispersing the crystals into a TPA-silicate solution, transforming this suspension into particles of the desired dimensions through spray-drying and finally decomposing the organic silicate.

The catalytic properties of the material so obtained are not significantly different from those of the 0.2μ crystallites,while the silica layer improves the mechanical properties of the particles making them satisfactory for industrial use.

2.5 Kraushaar-van Hooff method

Recently Kraushaar and van Hooff (21) have described a new method for the production of TS-1 based on the reaction of a Ti^{IV} compound, typically $TiCl_4$, with a defective silicalite or a ZSM-5 from which Al^{III} has been removed by HCl treatment. The Ti^{IV} compound is contacted with the solid in the gas phase at 400-500 °C in a stream of nitrogen.

The formation of TS-1 has been demonstrated by the changes occurring in the X-Ray diffraction pattern, the IR spectra and the ^{29}Si MAS NMR spectra, all of which produce the patterns characteristic of classical TS-1. Also the catalytic properties are identical with those of TS-1, as shown by the results of the hydroxylation of phenol with H_2O_2. However, even small amounts of non-framework TiO_2 dramatically change the catalytic performances: most of the H_2O_2 is decomposed to H_2O and O_2, the yield of diphenols drops to almost zero and tars are formed. The risk of non-framework TiO_2 formation is definitely high in this method as a consequence of the hydrolysis of the Ti^{IV} compound.

The authors suggest that the new method could be of some value for titanium containing zeolites with structures different from silicalite, for instance large pore zeolites which could be useful in the oxidation of large molecules which cannot be oxidized with TS-1.

3. Structure and catalytic activity of TS-1

The catalytic activity of TS-1 must no doubt be ascribed to the presence of Ti^{IV} : silicalite under the same experimental conditions is totally inactive. In order to explain the peculiar performances of TS-1 it has been proposed (7) that these Ti^{IV} are isolated from each other by long sequences of -O-Si-O-Si-O-.

Different hypothesis have been proposed concerning the coordination of Ti^{IV} in the solid. Because of the analogy with the closely related TiO_2/SiO_2 catalyst, the possibility that Ti^{IV} are present as titanyl groups >Ti=O or the corresponding hydrated form with contiguous Si-OH groups has been considered:

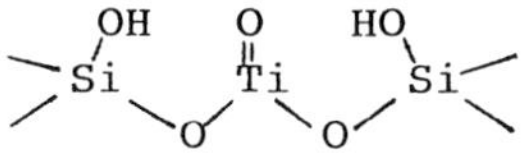

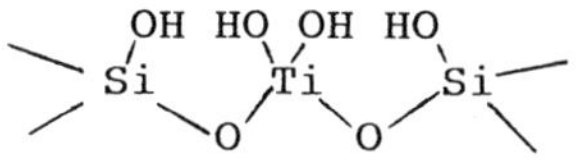

```
    OH     O     HO                 OH  HO  OH  HO
  \ /      ||      \ /            \ /     \ /      \ /
 —Si      Ti      Si—            —Si      Ti       Si—
  /  \   /  \   /  \              /  \   /  \    /  \
      O       O                       O       O
```

Another possibility is that Ti^{IV} are present in tetrahedral coordination of oxygens like Si^{IV}:

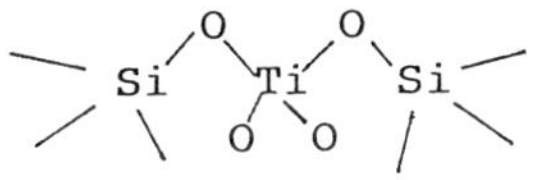

```
          O       O
        /   \   /   \
  — Si       Ti       Si —
  /   \     /  \     /   \
           O    O
```

The 960 cm^{-1} absorption band is in favour of the >Ti=O group, since it comes very close to the stretching frequency of the >Ti=O group (975 cm^{-1}): however Boccuti, Rao, Zecchina, Leofanti and Petrini (22) pointed out recently that this absorption is better explained as the Si-O stretching modified by the presence of Ti. The same authors from the examination of the UV-Vis spectra pointed out that the >Ti=O group should have an electronic transition at 25.000 - 35.000 cm^{-1} which is absent in TS-1, while the electronic transition at 48.000 cm^{-1} which is present must be assigned to Ti^{IV} tetrahedrally coordinated by -OH and -O-Si groups. Upon heating at temperatures above 373°K a gradual loss of water is observed.

On the basis of these observations they propose structures of the type:

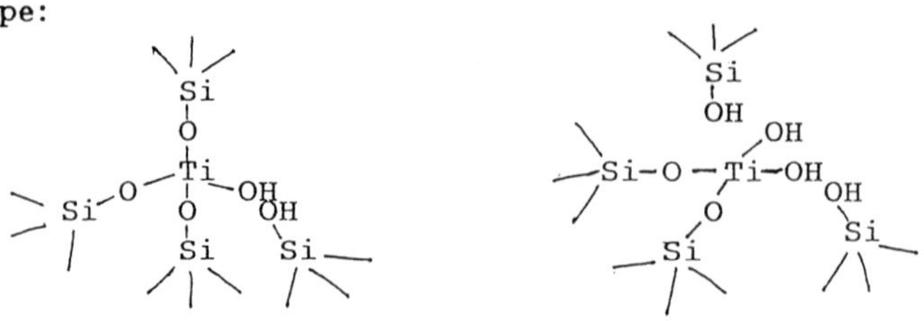

[Structures: Ti bonded through O to Si groups, with one or two Ti-O-Si bonds hydrated to Ti-OH and Si-OH]

in which one or two Ti-O-Si bonds of the crystalline structure are hydrated, forming surface titanols and silanols groups which can reversibly dehydrate:

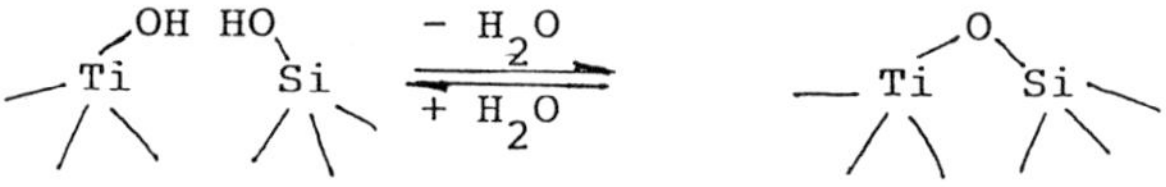

$$\text{Ti-OH} + \text{HO-Si} \underset{+H_2O}{\overset{-H_2O}{\rightleftharpoons}} \text{Ti-O-Si}$$

It should be noted that the doubly hydrated form is very similar to the hydrated titanyl form: distinction between the two could therefore be only apparent.

By analogy with the reaction of soluble Ti^{IV} compounds with H_2O_2 (22), the mechanism by which TS-1 acts as an oxidation catalyst with H_2O_2 could consist in the interaction of Ti^{IV} of the solid with H_2O_2 to form a surface peroxotitanate (7). In a second stage the surface peroxotitanate can perform the oxidation of the oxidizable organic products: if these are indicated by Red, we have:

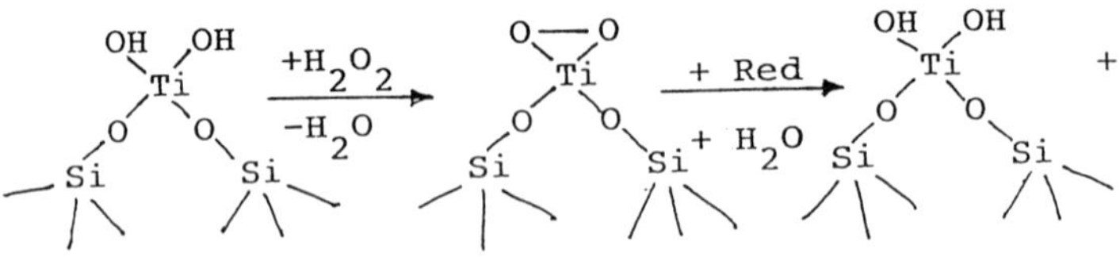

$$\text{(SiO)}_2\text{Ti(OH)}_2 \underset{-H_2O}{\xrightarrow{+H_2O_2}} \text{(SiO)}_2\text{Ti(O}_2\text{)} \xrightarrow[+H_2O]{+ \text{Red}} \text{(SiO)}_2\text{Ti(OH)}_2 + \text{Red}\cdot\text{O}$$

According to this proposal, the high selectivity of TS-1 should be ascribed to the fact that H_2O_2 can be decomposed into H_2O and O_2 only when two or more Ti^{IV} are in near-neighbour positions,a very unlikely possibility in TS-1. This results in a low decomposition rate of H_2O_2 which favours the transfer of peroxidic oxygen to the organic compounds.

The problem of the role of acidity in the oxidation reaction has been examined. To this end silicalites containing both Ti^{IV} and Al^{III} ,or Fe^{III} or Ga^{III} have been synthesized (24,25,26) and used in the epoxidation of propylene.It is well known that trivalent elements introduced in the framework impart definite acidic character to the material. The results obtained under very similar experimental conditions are given in Table 2.

Table 2. Epoxidation of propylene

Catalyst	T,°K	$C{-}\overbrace{C{-}C}^{O}$	$C{-}C(OH){-}C(OCH_3)$	$C{-}C(CH_3O){-}C(OH)$	$C{-}C(OH){-}C(OH)$
TS-1	313	97.7	1	0.2	1.1
Ti-Fe-Si	313	80	11	5.5	3
Ti-Ga-Si	293	6.5	56.1	37.3	

The effect of the acidity created by the trivalent elements is evident: a substantial amount of the initially formed epoxide undergoes the typical acid catalyzed addition of water or methanol to the epoxide ring.This reaction is present only to a very limited extent when TS-1 is used, and this could be considered an indication of a very weak acidity of this material. But the fact that epoxidation selectivity can be increased by treatment of TS-1 with modifying agents like $Cl{-}Si{-}(CH_3)_3$

or CH_3COONa (16) can be regarded as evidence that this weak acidity must be attributed to surface silanol groups which are transformed by the modifying agents into inactive $Si-O-(CH_3)_3$ or Si-ONa groups, while the catalytic activity due to Ti is not affected.

Also in gas phase reactions TS-1 does not show activity for typical acid catalyzed reactions like methanol transformation into hydrocarbons or olefin isomerization.

Assuming that Ti^{IV} is distributed statistically in all tetrahedral positions,it can be easily seen that even for crystallite sizes of 0.2μ the great majority of Ti^{IV} is located inside the pore structure. Assuming that every Ti^{IV} is a catalytic centre with equal activity,diffusion limitations for molecules of different sizes should be observed.

This is in fact the case. It has been shown (27)that the rate of oxidation of primary alcohols decreases regularly as the chain length increases,while for iso-butyl alcohol a sudden drop in the rate is observed.Also the reactivity order of olefins on TS-1 is different from the order observed with homogeneous electrophilic catalysts,while as already indicated very bulky molecules are unreactive when TS-1 is used as the catalyst.All these facts can only be interpreted as due to diffusion limitations of the bulkier molecules, which means that the catalytic sites are located inside the pore structure of the solid.

4. Conclusions

A new microporous solid material has been obtained made of TiO_2 and SiO_2 (TS-1) which has a silicalite-1 structure modified by isomorphous substitution of Si^{IV} with Ti^{IV}. Its synthesis takes place in the presence of tetraalkylammonium bases under carefully controlled conditions.

TS-1 has unique properties as heterogeneous oxidation catalyst for the oxidation of organic compounds with H_2O_2:very high selectivities are obtained and this parallels the behaviour of Ti^{IV}based homogeneous catalysts.

It is proposed that the oxidation reactions proceed through the formation of a surface peroxotitanate by interaction of framework Ti^{IV} with H_2O_2, and the subsequent transfer of the oxygen from the peroxotitanate to the oxidizable organic products. The difference with respect to other Ti^{IV} containing catalysts is attributed to the fact that in TS-1 all Ti^{IV} are isolated from each other, with the consequence that the rate of H_2O_2 decomposition is reduced thus favouring the selective oxidation of the organic products.

The production of diphenols from phenol and H_2O_2 on TS-1 has proved competitive with other industrial processes and a plant has been built which operates since 1986 with excellent results.

The discovery of TS-1 and its peculiar catalytic properties constitutes a significant contribution to the knowledge of silica-based zeolite-like materials containing elements different from Al^{III} and opens new technological possibilities for oxidation processes with H_2O_2.

References

1) H.Wulff et al,USP 3,642,833; 3,923;843; 4,021,454; 4,367,342; Brit.Pat. 1,249,079.
2) a)R.A.Sheldon and J.A.van Doorn,J.Cat. 31 (1973) 427
b)R.A.Sheldon,J.A.van Doorn,W.A.Shram and A.J.De Jong,ib.31(1973) 438
3) R.A.Sheldon in "The Chemistry of Functional Groups,Peroxides", Ed.S.Patai 1983 J.Wiley p.163
4) J.P.Shirmann et al. Ger.Pat.2.752.626;2.803.757;2.803.791.
5) M.Taramasso,G.Perego and B.Notari,U.S.P.4,410,501
6) M.Taramasso,G.Manara,V.Fattore and B.Notari,U.S.P.4,666,692
7) B.Notari,Stud.Surf.Sci.Catal. 37,413 (1987)
8) G.Perego,G.Bellussi,C.Corno,M.Taramasso,A.Esposito in Y.Murakami, A.Iijima,J.W.Ward (Eds) Proc.Seventh Int.Conf.on Zeolites,Tokyo 1986,Tonk Kodanska p.129
9) C.Neri,A.Esposito,B.Anfossi and F.Buonomo,Eur.Pat.100.119
10) C.Neri,B.Anfossi and F.Buonomo,Eur.Pat. 100.118
11) F.Maspero and U.Romano,Eur.Pat. 190.609
12) a) A.Esposito,M.Taramasso,C.Neri and F.Buonomo,Brit.Pat.2.116.974
b) A.Thangaray,R.Kumar and P.Ratnasamy,App.Cat. 57(1990)L1.
13) A.Esposito,C.Neri and F.Buonomo,U.S.P. 4,480,135
14) P.Roffia,M.Padovan,E.Moretti and G.De Alberti,Eur.Pat.208.311
15) P.Roffia,M.Mantegazza,A.Cesana,M.Padovan and G.Leofanti,XVI Italian National Chemistry Congress,Oct.1988,p.259

16) M.G.Clerici and U.Romano,Eur.Pat.230.949
17) F.Buonomo,G.Bellussi and B.Notari U.S.P.4,578,161
18) B.Kraushaar-Czarnetzki and J.H.C.van Hooff,Cat.Lett. 2(1989)43
19) J.El Hage-Al Asswad,J.B.Nagy,Z.Gabelica and E.G.Derouane,8th Int. Zeol.Conf.July 1989
20) G.Bellussi,M.Clerici,F.Buonomo,U.Romano,A.Esposito and B.Notari, Eur.Pat.200.260
21) B.Kraushaar and J.H.C.van Hooff,Cat.Lett. 1(1988) 81
22) M.R.Boccuti,K.M.Rao,A.Zecchina,G.Leofanti and G.Petrini,Stud.Surf. Sci.Catal. 48,(1989) 133
23) a) O.Bortolini,F.Di Furia and G.Modena,J.Mol.Cat. 16(1982) 69
b) G.Amato,A.Arcoria,F.P.Ballistreri,G.A.Tomaselli,O.Bortolini, V.Conte,F.Di Furia,G.Modena and G.Valle,J.Mol.Cat. 37(1986) 165
24) G.Bellussi,A.Giusti,A.Esposito and F.Buonomo,Eur.Pat.A.266.257
25) G.Bellussi,M.G.Clerici,A.Giusti and F.Buonomo,Eur.Pat.A.266.258
26) G.Bellussi,M.G.Clerici,A.Carati and A.Esposito,Eur.Pat.A.266.825
27) U.Romano,A.Esposito,F.Maspero,C.Neri,M.G.Clerici in "New Development in Selective Oxidation,Paper B-1,Rimini 1989

R.K. Grasselli and A.W. Sleight (Editors), *Structure-Activity and Selectivity Relationships in Heterogeneous Catalysis*
© 1991 Elsevier Science Publishers B.V., Amsterdam

POTENTIAL OF ZEOLITES AS CATALYSTS IN ORGANIC SYNTHESIS

WOLFGANG F. HOELDERICH

BASF AKTIENGESELLSCHAFT, Ammonia Laboratory, D - 6700 Ludwigshafen, FRG

ABSTRACT

Zeolite catalysts offer a broad range of possibilities for carrying out reactions with high selectivity in the synthesis of organic intermediates and fine chemicals, i. e. compounds possessing functional groups. The potential afforded by zeolite catalysts is illustrated by examples taken from both academic publications and the patent-literature. An extremely important aspect is the fact that zeolites contribute to the development of processes which are environmentally more friendly.

1. INTRODUCTION

Zeolites have a wide range of applications. At present, they are employed as phosphate substitutes in detergents, as absorbents for the separation and purification of substances and as catalysts (1). In the future, new uses for zeolites will be in the fields of semiconductor (2) and special sensor technology (2, 3), as membranes (2, 4 - 6) and optical storge devices (2, 7) and as components of plastics (8 - 11).

Zeolite catalysts find major technical application in refining and petrochemistry (12 - 23). FCC catalysts based upon Y-Zeolites are the most important, followed by the bifunctional Y-Zeolite catalysts for hydrocracking (23).

Over the past ten years, remarkable and rapid progress has been made in the use of zeolites in the organic synthesis of intermediates and fine chemicals (24 - 33). Pentasil zeolites in particular are very successfully used here. This new area represents a second promising development in zeolite catalysis in addition to refinery technology and petrochemistry, which continues to be of great economic importance. In this review article, the significance and the potential of the zeolite catalysts in this second field of use will be discussed in more detail. Zeolite catalysts offer the following possibilities:

- Improvement of existing processes by simple exchange of conventional catalysts.
- Introduction of commercially interesting reactions into industry, these reactions being ones which had previously been unsuccessful owing to catalyst problems, such as insufficient activity, selectivity and catalyst life.
- Changeover from homogeneous to heterogeneous catalysis, particularly when this is required because of environ mental problems or technical problems, such as separation of the catalyst.
- Heterogenization - immobilization - of homogeneous catalysts to avoid separation problems.
- Combination of several individual reactions into one synthesis step, ie. shortening the synthesis route by using multifunctional catalysis with zeolites.
- Opening up previously unknown synthesis routes, this being probably the most important economic aim.
- Time-saving catalyst development by computer graphics or computer-aided catalyst design.
- Contribution to environmental protection and to energy saving.

2. RESULTS AND DISCUSSION

2.1 Improvement of existing processes with zeolite catalysts

The example in which acidity and shape selectivity of the zeolite catalyst play a role comes from the area of elimination reactions, of which the dehydration reactions are the most widely encountered ones.

$$HOOC\text{-}(CH_2)_4\text{-}COOH \xrightarrow[-\ 4\ H_2O]{+\ 2\ NH_3} NC\text{-}(CH_2)_4\text{-}CN \qquad [1]$$

Let us consider the preparation of adipodinitrile from adipic acid and ammonia, which is carried out industrially on a large scale in the gas phase by the fluidized-bed (BASF) and fixed-bed processes (ICI) [equation 1]. Over the conventional catalysts, selectivity-reduced cyclization to cyclopentanone and cyanocyclopetanoneimine occurs (Table 1).

TABLE 1
Adipodinitrile from adipic acid

Catalyst	Na/P-B zeolite	Na/P-SiO_2
Adipodinitrile	94.0 % (mol/mol)	83.4 % (mol/mol)
Cyanovaleric acid	0.7 % (")	3.0 % (")
Cyanovaleramide	2.8 % (")	0.6 % (")
> Desired products	97.5 % (")	87.0 % (")
Cyanocyclopenta-noneimine	0.5 % (")	2.4 % (")
Cyclopentanone	0.9 % (")	4.3 % (")

Conditions: fluidized bed, 400 °C

Secondary reactions of this type are suppressed if a phosphorus-modified pentasil zeolite is used, as shown by comparison with the conventional silica-based catalyst. The amount of cyclic compounds is reduced from 6.7 mol % to 1.4 mol %, and the yield of desired product is about 10 % higher (30). This is evidently due to the transition state shape selectivity of the pentasil zeolite preventing cyclization.

2.2 Introduction of commercially interesting reactions with zeolite catalysts

Zeolites can catalyze both double bond isomerization and skeletal isomerization (28, 30).

$$R^1R^2R^3C\text{-}CHO \longrightarrow R^1R^2HC\text{-}\overset{O}{\overset{\|}{C}}\text{-}R^3 \qquad [2]$$

An example of skeletal isomerization is the aldehyde/ketone rearrangement over zeolites [equation 2], in which the effects of isomorphous substitution and of shape selectivity on the course of the reaction and the advantages compared with conventional catalysts are clearly evident (30).

The preparation of ketones from aldehydes is desirable since the latter are readily available, for example via the oxo synthesis. Isomerizations of this type, for example over catalysts of mixed oxides containing tin, molybdenum and copper, are known. The disadvantages here are that only low selectivities are achieved at satisfactory conversions, and the best results with regard to selectivity and catalyst life can be obtained only with the addition of steam. Hence, in the industrial production of asymmetrically substituted ketones, it was necessary as a rule to rely on the condensation of different organic acids with decarboxylation. In this process, the inevitable production of symmetrically substituted ketones and of carbon dioxide is a disadvantage. Aldol condensation with subsequent hydrogenation is another possibility but requires two reaction stages.

TABLE 2
Aldehyde/ketone rearrangement

Educt	Zeolite	Conditions	Product	Conv.	Selec.
2-phenyl-propanal	Boron[a)]	400 °C 0.8 h-1	1-phenyl-propan-2-one	63 %	97 %
2-phenyl-propanal	Iron[a)]	400 °C 2 h-1	1-phenyl-propan-2-one	98 %	95 %
2-phenyl-propanal	Iron[b)]	400 °C 2 h-1	1-phenyl-propan-2-one	100 %	87 %
2-phenyl-2-methyl-propanal	Boron[a)]	400 °C 2 h-1	4-phenyl-butan-2-one	59 %	85%

a) Pure, without binder
b) With boehmite as a binder, in a weight ratio of 60 : 40

By using zeolite catalysts - in particular of the pentasil type, it is possible to obtain high yields in this isomerization (Table 2). An advantage is that there is no need to add steam.

In the conversion of 2-phenylpropanal to phenylacetone, 97 % selectivity and a conversion of 63 % are achieved over B pentasil zeolite. The Fe pentasil zeolite is more active; in spite of higher space velocity, the conversion is increased to 95 % while retaining a selectivity of 95%. Zeolites molded with Al_2O_3-containing binders are less suitable than the pure zeolite catalysts for this purpose. Al pentasil zeolites, such as ZSM-5, are also unsuitable; they have only moderate selectivity at satisfactory conversions.

In the competition of the various groups, exclusively the migration of the small methyl group is observed. This is promoted by the restricted transition state shape selectivity (Table 2).

2.3 Changeover from homogeneous to heterogeneous catalysis

From the point of view of the acidity, zeolite catalysts can replace the following catalysts in electrophilic and nucleophilic substitution reactions:

- homogeneous Lewis acid catalysts, such as $AlCl_3$ and $FeCl_3$
- conventional acidic catalysts, such as Al_2O_3 and SiO_2
- mineral acids and organic acids.

Thus, zeolites can be used to catalyze Friedel-Crafts alkylation reactions and acylation reactions. Here, there are a very great number of reactions from a wide range of industrial and university laboratories; the most well known one is the Mobil-Badger process.

$$\text{C}_4\text{H}_3\text{X (5-ring)} + (CH_3CO)_2O \longrightarrow \text{C}_4\text{H}_3\text{X}\text{-}\overset{O}{\overset{\|}{C}}\text{-}CH_3 + CH_3COOH \qquad [3]$$

X = S, O or NH

To date, virtually only processes involving homogeneous catalysis by Lewis acids have been known for the acylation of heteroaromatics, such as thiophene, furan and pyrrole, in accordance with equation 3. Recently, BASF has found (25) that this acylation can be carried out with very high selectivity in the gas phase over zeolite catalysts. The reaction of thiophene with acetic anhydride at 250 °C over a boron zeolite of the pentasil type leads to 2-acetylthiophene with 99 % selectivity at a conversion of 24 %. Over a somewhat more active Ce-doped boron zeolite (200 °C, WHSV = 2.2 h^{-1}), 2-acetylfuran is formed with 99 % selectivity at a conversion of 23 %. On the other hand, in the case of pyrrole, which tends to undergo polymerization, the acidity of the catalyst as well as the temperature must be reduced in order to achieve high selectivity. A boron zeolite doped with 0.2 % by weight of Cs gives 2-acetylpyrrole with 98 % selectivity and 41 % conversion at 150 °C. Although the reaction type is the same, the catalyst is matched to the particular substrate by doping; the acidity is controlled. In these zeolite-catalyzed reactions, as in the homogeneously catalyzed reactions, the acylation takes place virtually exclusively in the 2-position of the heteroaromatic. In these acylations of the heteroaromatics, the acidity of the zeolites is important; the shape selectivity plays a minor role, if any at all.

Both in the alkylation and in the acylation of aromatics and heteroaromatics in the presence of zeolite catalysts at elevated temperatures, the use of Lewis acids is avoided; this has advantages with regard to the recovery of the heat of reaction (the reaction takes place at high temperatures); furthermore, there are no corrosion, separation and waste problems, and the catalysts can be readily regenerated. This makes a contribution to energy saving and environmental protection.

2.4 Heterogenization/Immobilization of homogeneous catalysts

The example for the immobilization of homogeneous catalysts with the aid of zeolites comes from the area of zeolite-catalyzed oxidation reactions, which have attracted wide attention over the past few years. In the oxidation catalysts, the zeolites serve as carriers for active components, such as Pd, Cu, Ag, V, P, Ni and Mo, in order to perform oxidation and ammonoxidation reactions in the presence of elemental oxygen (e. g. 25). The use of zeolites directly as a catalyst for the oxidation reactions with H_2O_2 as an oxygen donor (e.g. 25 and references cited there) has been the subject of considerable research in the past 6 years.

A more recent focus of attention has been the oxidation reactions with O_2 in the presence of so-called enzyme mimics (25, 34 - 38). The most recent discoveries in connection with the selective partial oxidation of nonactivated alkanes over metal phthalocyanine complexes (MPc) enclosed in faujasite super-cages are very promising and many potential applications for these inorganic simulations of enzymes, for example of the natural monoxygenase enzyme cytochrome P 450. MPc complexes are synthesized in the zeolite framework by subjecting the zeolite to metal ion exchange and then treating it with molten dicyanobenzene. These "ship-in-a-bottle" complexes cannot leave the zeolite without destroying the framework. Such zeolite catalysts, whose super-cages serve as a sort of reaction flask with molecular dimensions, continue to possess shape selectivity, reactant selectivity, regioselectivity and stereoselectivity.

The expected substrate selectivity of FePc-NaY zeolites is demonstrated in a competing oxidation reaction of cyclohexanes and cyclododecanes in CH_2Cl_2 solution with phenyl hypoiodide and air as an oxidizing agent at room temperature to give the corresponding alcohols (34, 35). The oxidation rate of the sterically less bulky cyclohexane is about twice as high as that of the cyclododecane (62 : 38). Additional reduction of the pore diameter by replacing Na^+ with Rb^+ increases the selectivity for the smaller substrate to a ratio of 90 : 10. On the other hand, the same oxidation rates are obtained for both reactants over homogeneous FePc.

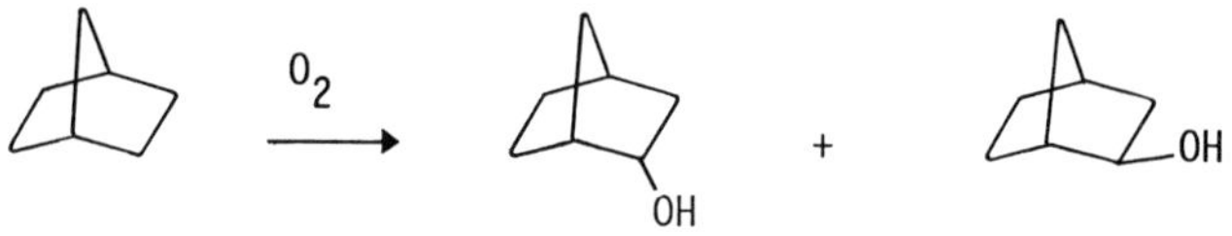

O_2 [4]

"Ship-in-a-bottle" complexes also have stereoselectivity. In the oxidation of norbornane according to equation 4, the ratio of exo-norbeneol to endo-norbeneol is about 5 over the FePc-NaY zeolite but about 9 over FePc. This preferred oxidation over one of the two diastereotopic hydrogens is a consequence of the relative orientation of the substrate with respect to the catalyst. The inorganic enzyme analogs, ie. "ship-in-a-bottle" complexes, thus make it possible to guide organic reactions, such as partial oxidation, in directions which are less preferred in the case of homogeneous catalysts. These zeolite catalysts constitute a further step in "enzyme modeling". These "ship-in-a-bottle" complexes are also interesting from the point of view of heterogenization of homogeneous catalysts.

2.5 Combining several individual reactions into one synthesis step

The relationship between catalysis and cost-efficiency in chemical processes is based on the selectivity of the catalyst, simple elimination of byproducts, safety considerations and the small number of synthesis stages.

Multifunctional catalysis, in which reactions consisting of several reaction steps are carried out by a shorter synthesis route, is becoming increasingly important in organic synthesis. Zeolite catalysts, too, help to combine several catalytic steps and tailor them optimally to one another (25, 28).

2.5.1 Dehydroisomerization

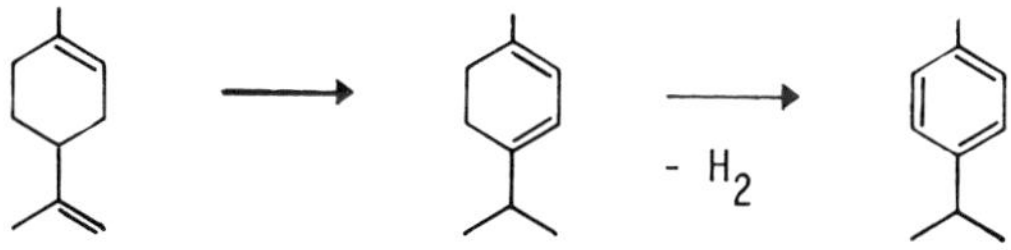

$- H_2$ [5]

In addition to their isomerization properties, zeolites are capable of catalyzing dehydrogenation reactions, in particular in the presence of oxygen, such as the dehydrogenation of ethylbenzene to styrene (38). It is possible to combine these properties in order to rearrange double bonds in an aromatic system (28). Examples are the transformations of vinylcyclohexene to ethylbenzene and -limonene to p-cymene [equation 5]. In the latter case, over a boron zeolite at 200 °C, p-cymene is obtained with 21 % selectivity and 100 % conversion. A zeolite doped with 1.5 % of Pd and 3.5 % of Ce results in an increase in selectivity to 87 %. This reaction takes place in the absence of hydrogen acceptors, such as oxygen. By using the multifunctional zeolite, it is possible not only to carry out isomerization but also to effect dehydrogenation in a so-called dehydroisomerization step.

2.5.2 Dehydration and Wagner-Meerwein rearrangement

$$CH_2(OH)-CH(OH)-CH_2-CH_2-CH_2(OH) \longrightarrow \text{2-(hydroxymethyl)tetrahydrofuran } (CH_2OH) \longrightarrow \text{2,3-dihydropyran} \qquad [6]$$

It is known (39) that alkyl-substituted 1,2,5-pentanetriols can be converted to 2-hydroxymethyltetrahydrofurans in a yield of 95 % in the presence of p-toluenesulfonic acid in accordance with equation 6, the resulting furans being subjected to a rearrangement reaction in a second reaction step at about 320 °C over alumina in the gas phase to give 2,3-dihydropyrans. The disadvantage of this procedure is that it is carried out in two stages, and the yield in the rearrangement step is only 50 %. The dehydration of, for example, 1,2,5-pentanetriol over zeolite catalysts leads preferentially directly to 2,3-dihydropyran. At 350 °C and a WHSV of 2.2 h^{-1}, a conversion of 73 % and a selectivity of 70 % are achieved over the boron zeolite (H-form). The intermediate 2-hydroxymethyltetrahydrofuran is also obtained as a byproduct. By doping this catalyst with, for example, 3.1 % by weight of W, it is possible to increase the activity and selectivity under identical reaction conditions. 2,3-dihydropyran is formed with 85% selectivity at 100 % conversion (28). It is evident that, when a bifunctional zeolite catalyst is used, dehydration and subsequent Wagner-Meerwein rearrangement take place in one step.

2.6 Discovering unknown synthesis routes

Discovering unknown synthesis routes is extremely exciting for the preparative chemist and success brings him considerable satisfaction. Zeolite catalysts make it possible in specific cases.

2.6.1 Benzamine rearrangement

The synthesis of picolines by rearrangement of aminated aromatics is a new, interesting reaction using zeolite catalysts. Aniline can be converted into -picoline in the presence of NH_3 (NH_3/aniline = 1.5 molar) at 510 °C, 2,860 KPa and a WHSV of 1.1 h^{-1} over H-ZSM 5. A conversion of 13 % and a selectivity of 52 % are obtained for α-picoline (40). The presence of NH_3 is necessary in order to obtain a high picoline content, since the main product in the absence of NH_3 is diphenylamine.

[7]

The benzamine rearrangement over zeolite catalysts is of particular interest in the production of aminopyridines from 1,3-diaminobenzenes [equation 7], in which the nitrogen of one amino group migrates into an aromatic system (41). A mixture of 1,3-diaminobenzene and NH_3 in a molar ratio of 1 : 60 reacts at 350 °C and 190 bar over H-ZSM 5 to give 2- and 4- aminopyridines with a selectivity of 83% and a conversion of 43 %. Comparison with silica-alumina or Al_2O_3 under the same reaction conditions (16 - 29 % conversion, 57 - 89 % selectivity) demonstrates the excellent properties of the zeolites over other acidic catalysts without a zeolite structure. This is a valuable new route to aminopyridines, which have been obtainable to date only by reacting sodium amide with pyridine in a complicated Tschitschibabin reaction. The reaction mechanism is not yet clear, but all suggested possibilities include the addition of protons at the aromatic nucleus to form carbonium ions (42). At any rate, it is surprising that an aromatic ring is cleaved at elevated temperature and pressure in the presence of NH_3 and an acidic zeolite heterogeneous catalyst.

2.6.2 Acylation of imidazole

Direct C-acylation of imidazole and pyrazoles in Friedel-Crafts reactions were previously unknown. It was previously necessary to rely on other more expensive methods of preparation (25). This direct acylation in the gas phase has been made possible by zeolite catalysts (25). For example, if a mixture of 2-methylimidazole and acetic acid or acetic anhydride is reacted at 400 °C over a pentasil zeolite, the result is a conversion of 63 % and a selectivity of 85 % for 2-methyl-4-acetylimidazole.

Other examples of the discovery of new synthesis routes with zeolite catalysts, including the phosphorylation of olefins (43), have also become known. The future is likely to bring exciting new results.

2.7 Time-saving catalyst development with computer graphics

The use of computer databases and computer-controlled plants facilitates daily laboratory work. The recently set up expert systems for catalysts (44) help to reduce the number of experiments by rapid preselection, and accelerate catalyst development. In the structure determination of zeolites or non-zeolite molecular sieves (APOs, SAPOs), computers provide a valuable and fast service in that zeolite models are designed graphically on the screen, and the corresponding X-ray diffraction spectra are calculated and are compared with the experimental powder diffraction patterns (Rietvelt method, DLS modelling).

However, the computer can also act as a direct aid in the development of zeolite catalysts. An example from the area of acidic/basic zeolite catalysis is given below.

The effect of the acidity and basicity of a zeolite catalyst on the course of the reaction can be clearly demonstrated in the reaction of toluene with methanol or an olefin. Acidic zeolites catalyze the alkylation at the aromatic nucleus, and xylene, mainly p-xylene, is obtained owing to the shape selectivity. On the other hand, in the presence of basic zeolites, side-chain alkylation to ethylbenzene and styrene (26, 45) occurs, as in the case of other basic heterogeneous catalysts. Recent investigations have shown that the zeolite catalysts which possess both acidic and basic centers are most suitable for these side-chain alkylation reactions. The acidic center is intended to stabilize the adsorption of the aromatic nucleus at the zeolite. The basic center is intended to abstract the benzylic proton and to facilitate the attack by formaldehyde, which is also formed at the basic center by dehydrogenation of methanol. Initially, styrene is obtained from formaldehyde and the activated toluene. The styrene reacts with H_2 (from the methanol) to form ethylbenzene.

This example of side-chain alkylation of toluene with methanol serves not only to demonstrate the interaction between basic and acidic centers in a zeolite but also to illustrate another type of zeolite catalysis or of catalysis in general. The key phrases "computer graphics" or "computer-aided catalyst design" are used to describe this novel catalyst research. With the aid of computer graphics, it is possible to simulate zeolite structures and produce images of these structures. Computer graphics can also be used to produce pictures of molecules such as toluene in the zeolite pores or cages.

A. Miyamoto et al. of Kyoto University used the computer to show (45) that, in an RbLi exchange NaX zeolite, the distance from the strongly basic center Rb to the weakly acidic center at Li optimally matches the molecular dimensions of the toluene. Fixing the toluene and abstracting the benzylic proton are optimally tailored to one another in the X zeolite. In the Y zeolite, ZSM-5 and mordenite, on the other hand, this good match is absent. The computer prediction is in agreement with the experimental findings.

The computer image allows us to model complicated processes and visualize them in graphic form, to extend our range of ideas. Finally, "computer-aided catalyst design" facilitates the choice of suitable catalysts and reduces the number of experiments. Because of the well defined structure and the available structural data, this type of catalyst development is more promising for zeolite catalysts than for conventional catalysts.

2.8 Contribution to environmental protection and to energy saving

Examples here are Friedel-Crafts alkylations and acylations, in which the zeolite catalyst replaces the homogeneous Lewis acid. Zeolite catalysts can also be used in place of mineral acids and organic acids. However, these catalyst substitutions also involve changing the process from homogeneous to heterogeneous catalysis.

The advantages of a zeolite-catalyzed heterogeneous process over the homogeneous method will be demonstrated using the production of -caprolactam - one of the most important fiber intermediates - as an example. The traditional synthesis route involves oximation of cyclohexanone with hydroxylamine sulfate followed by Beckmann rearrangement of the oxime in concentrated sulfuric acid. In the oximation and rearrangement stage, about 2 - 4 t of ammonium sulfate per t of ε-caprolactam are inevitably obtained as a byproduct (32). Other problems associated with this synthesis route are connected with the handling of fuming sulfuric acid, the highly carcinogenic hydroxylamine and the corrosion of the materials. In order to avoid these problems and the formation of ammonium sulfate, only a limited amount of which can be used in the fertilizer industry, attempts have been made for many years to change over from a homogeneous to a heterogeneous catalytic process. Two solutions in this context are described in the literature.

2.8.1 Oxidation reaction with H_2O_2 and NH_3

The discovery of the weakly acidic titanium zeolite TS-1 has led to considerable progress in the area of oxidation reactions with H_2O_2 as an oxygen donor (25) over the past 6 years. Very recently, the use of these titanium zeolites for oximation with ammonia has also been reported (25, 32, 46, 47).

$$\text{cyclohexanone} \xrightarrow[- H_2O_2]{+ NH_3} \text{cyclohexanone oxime (=NOH)} + \text{peroxydicyclohexylamine (O–O, N)} \quad [8]$$

A remarkable reaction is the reaction of cyclohexanone with ammonia and H_2O_2 in the liquid phase to give cyclohexanoneoxime in accordance with equation 8 (46, 47), roughly corresponding to the oximation stage in the production of ε-caprolactam. In an autoclave experiment at 60 °C and 700 mmHg gage pressure, 95 % of the cyclohexanone are converted to the oxime with 80% selectivity and to peroxydicyclohexylamine with 15 % selectivity. This route avoids the coproduction of ammonium sulfate and carcinogenic hydroxylamine and is environmentally friendly and therefore of industrial interest.

2.8.2 Beckmann rearrangement

Over the past twenty five years, considerable efforts have been made to catalyze the Beckmann rearrangement using zeolites (32). The experiments to date have been unsuccessful owing to problems with the catalyst life. The recently published experiments by Sumitomo and UCC are of interest. H. Sato et al. were able to show (48 - 50) that both the catalytic activity and the selectivity for lactam formation increase with increasing Si/Al ratio in H-ZSM 5 catalysts. The catalyst life also increases with increasing Si/Al ratio.

TABLE 3
Comparison of silanated and non-silanated H-ZSM 5 in the conversion of cyclohexanoneoxime to ε-caprolactam[a)]

Catalyst	Life [h]	Conv. [%]	Selec. [%]
Silanated[b)]	3.3	100	95.0
	31.0	98.2	95.0
Non-silanated	3.3	100	79.7
	27.0	95.8	89.4

a) Reaction conditions: 8 % by weight of oxime solution in benzene,

350 °C, WHSV = 11.7 h^{-1}, 1 atm, CO_2 as carrier gas, oxime : CO_2 : benzene = 1 : 5.6 : 18.3 moles

b) H-ZSM 5 with Si/Al = 1 : 600, treated with trimethylchlorosilane at 350 °C for 4 h

A further improvement is obtained if the acidity of the external surface of the H-ZSM 5 is reduced by treatment with organometallic compounds, such as trimethylchlorosilane. The comparison in Table 3 between silanated and non-silanated H-ZSM 5 shows the advantageous effects of this silanation treatment on the catalyst life and the selectivity for ε-caprolactam.

In line with the idea of reducing the acidity of the zeolites in order to achieve high selectivity and a long catalyst life, the weakly acidic non-zeolite molecular sieves, for example the medium-pore SAPO-11 or SAPO-41, were used for the Beckmann rearrangement (51). Over SAPO-11, a 5 % strength solution of cyclohexanoneoxime in acetonitrile reacts at 350 °C, under atmospheric pressure and at a WHSV of 10.8 h^{-1} to give ε-caprolactam with 95% selectivity and a conversion of 98 %.

Although progress has been made with zeolite and non-zeolite molecular sieves, these alternatives cannot compete with the current, homogeneously catalyzed industrial process; the catalyst lives are still much too short.

3. CONCLUSION

In view of the numerous reactions which have been published, one could draw the conclusion that zeolites - either as catalysts or as supports for active components - are suitable for catalysing all chemical reactions in more or less high yield. The main application is still in the field of acid catalysed reactions, although recently the advent of basic zeolites has opened up the possibility of base catalysed reaction paths (38).

There are over 30 new refinery and chemical processes based upon zeolite catalysts, which are either already in commercial operation or are in the development stage. Many details regarding the technical use of zeolite catalyts in the areas of refining and petrochemistry are known from descriptions of the operating conditions and economics of the processes. On the other hand, published data regarding the technical employment of zeolite catalysts in the synthesis of organic intermediates and fine chemicals are sparse; the field is relatively new and has developed only in the last ten years. One exception, however, is the oxidation of phenol to hydroquinone with H_2O_2 or TS-1; in this case technical information relating to a 10 000 t/a. plant has appeared (e.g. refs. in 37). There have also been reports from Japan that the production of methylamines from methanol and ammonia and of cyclohexanol from cyclohexene and water are carried out on a technical scale using zeolite catalysts.

The desire to raise product yield and to lower process costs were always the driving force for catalyst development; this is also true of zeolite catalysts. In recent times the need to protect our environment has induced the chemical industry to develop new, highly selective catalysts, which yield purer products and avoid side reactions leading to the formation of undesirable and sometimes toxic by-products. The Freedonia Group in Cleveland, USA, sees catalysts for environmental protection as the most rapidly growing market segment, with estimated sales in 1992 of approx. 985 Million Dollars out of a total catalyst market of approx. 2300 Million Dollars (22). Zeolite catalysts will have an important share; they help - as various examples have shown - to protect our environment and also to save energy. They stand for "clean" chemistry.

REFERENCES

1 L. Puppe, Chem. unserer Zeit 20 (1986) 117
2 G. A. Ozin, A. Kuperman and A. Stein, Angew. Chem. 101 (1989) 373
3 T. L. Pettit and M. A. Fox, J. Phys. Chem. 90 (1986) 1353
4 H. J. L. Te Henepe, D. Bargeman, M. H. V. Mulder and C. A. Smolders, Stud. Surf. Sci. Catal. 39 (1988) 411
5 D. T. Hayhurst, P. J. Melling, W. J. Kim and W. Bibbey, A. C. S. Symp. Ser. 398 (1989) 233
6 J. C. Jansen, C. W. R. Engelen and H. van Bekkum, A. C. S. Symp. Ser. 398 (1989) 257
7 S. D. Cox, T. E. Gier, G. D. Stucky and J. Bierlein, J. Amer. Chem. Soc. 110 (1988) 2986
8 J. Shiichi and Y. Yamamoto, JP 61.145.241 (02.07.1986), Adeka Argus Chem. Co.
9 T. Imahama and Y. Tanaka, JP 60.192.742 (01.10.1985), Toyo Soda MfG. Co
10 H. Zenji, H. Shigetaka, J. Hiroo, N. Saburo, T. Kenichi and Y. Keio, EP 0.116.865 (12.11.1986), Kanebo Ltd.
11 H. K. Beyer, G. Borbely, P. Miasnikov and P. Rozsa, Stud. Surf. Sci. Catal. 46 (1989) 635
12 N. Y. Chen, W. E. Garwood and F. G. Dwyer, Chemical Industries, Marcel Dekker Inc., New York, 1989, Vol. 36
13 N. Y. Chen, A. C. S. Symposium Series 368 (1988) 468
14 N. Y. Chen, Stud. Surf. Sci. Catal. 38 (1988) 153
15 S. L. Meisel, Stud. Surf. Sci. Catal. 36 (1988) 17
16 N. Y. Chen, Catal. Rev.-Sci. Eng. 28 (1986) 185
17 W. F. Hoelderich and E. Gallei, Chem.-Ing.-Tech. 56 (1984) 908
18 S. M. Csicsery, Zeolites 4 (1984) 202
19 P. B. Weisz, Pure + Appl. Chem. 52 (1980) 2091
20 B. P. Venuto and E. Th. Habib, Jr., Chemical Industries, Marcle Dekker Inc., New York, 1979, Vol. 1
21 A. P. Bolton, A. C. S. Symposium Series 171 (1976) 714
22 G. Parkinson and E. Johnson, Chemical Engineering, September (1989) 31
23 H. Sherry, ECN, Juli (1988) 17
24 P. B. Venuto and P. S. Landis, Adv. Catal. 18 (1968) 259
25 W. F. Hoelderich, Stud. Surf. Sci. Catal, 49A (1989) 69

26 W. F. Hoelderich in K. Tanabe et al. (Eds.), "Acid-Base Catalysis", Proceedings of the Symposium Acid-Base Catalysis, Sapporo, Japan, 1988, Kodansha Ltd., 1989, p. 1
27 H. van Bekkum und H. W. Kouwenhoven, Stud. Surf. Sci. Catal. 41 (1988) 45
28 W. F. Hoelderich, Stud. Surf. Sci. Catal. 41 (1988) 83
29 W. F. Hoelderich, M. Hesse and F. Näumann, Angew. Chem. Int. Edit. 27 (1988) 226
30 W. F. Hoelderich, Pure + Appl. Chem. 58 (1986) 1383
31 R. F. Parton, J. M. Jacobs, D. R. Huybrechts und P. A. Jacobs, Stud. Surf. Sci. Catal. 46 (1989) 163
32 W. F. Hoelderich, Stud. Surf. Sci. Catal. 46 (1989) 193
33 W. F. Hoelderich, Proceedings of the TOCAT 1, Tokyo, Japan July 1990, in press
34 N. Herron, A. D. Stucky and C. A. Tolman, J. Chem. Soc. Chem. Commun. (1986) 1521
35 C. A. Tolman and N. Herron, ACS Prep. Div. Petr. Chem. 32 (3) (1987) 798
36 N. Herron and C. A. Tolman, ACS Prep. Div. Petr. Chem. 32 (1) (1987) 200
37 N. Herron and C. A. Tolman, J. Am. Chem. Soc., 109 (1987) 2837
38 KH. M. Minachev, D. B. Tagiev, Z. G. Zul'Fugarov and V. V. Kharlamov, Heterog. Katal. 4 (1979) 505
39 J. Colonge, G. Decotes, B. Giroud-Abel and J. C. Martin, C. R. Acad. Sc. Paris, 258 (1964) 2096
40 C. D. Chang and P. D. Perkins, EP 082.613 (29.06.1983) and US 4.388.461 (14.06.1983), Mobil Oil Corp.
41 H. LeBlanc, L. Puppe and K. Wedemeyer, DE 3.332.687 (28.03.1985), Bayer AG
42 C. D. Chang and P. D. Perkins, Zeolites 3 (1983) 298
43 W. F. Hoelderich, M. Hesse and E. Sattler, in M. J. Phillips and M. Ternan (Eds.), Proceedings 9th ICC, Calgary, Canada, 1988, Vol. 1, p. 316
44 H. Speck, W. F. Hoelderich, W. Himmel, M. Irgang, G. Koppenhöfer and W. D. Mroß, Dechema-Monographs 116 (1989) 43
45 A. Miyamoto, S. Iwamoto, K. Agusa und T. Inui, in (26), p. 497
46 P. Roffia, M. Padovan, E. Moretti und G. De Alberti, EP 208.311 (14.01.1987), Montedipe S.p.A.
47 P. Roffia, M. Padovan, G. Leofanti, M. A. Mantegazza, G. De Alberti and R. G. Tauszik, EP 267.362 (18.05.1988), Montedipe S.p.A.
48 H. Sato, N. Ishii, K. Hirose and S. Nakamura, Stud. Surf. Sci. Catal. 28 (1986) 755
49 H. Sato, K. Hirose, N. Ishii und Y. Umada, EP 234.088 (02.09.1987), Sumitomo Chem. Co.
50 H. Sato, K. Hirose, M. Kitamura, H. Tojima und N. Ishii, EP 236.092 (09.09.1987), Sumitomo Chem. Co.
51 K. D. Olson, EP 251.168 (Jan. 07. 1988), UCC

R.K. Grasselli and A.W. Sleight (Editors), *Structure-Activity and Selectivity Relationships in Heterogeneous Catalysis*
© 1991 Elsevier Science Publishers B.V., Amsterdam

THE INFLUENCE OF SURFACE DEFECT SITES ON CHEMISORPTION AND CATALYSIS

JOHN T. YATES, JR., ANDRÀS SZABÒ and MICHAEL A. HENDERSON

Surface Science Center, Department of Chemistry, University of Pittsburgh, Pittsburgh, PA 15260

1. INTRODUCTION

Modern methods of surface science offer the possibility of understanding fundamental questions about the detailed behavior of adsorbed species on metallic binding sites of various types. It will be shown in this work that we have been able to characterize the surface bonding, the vibrational dynamics, and the surface reactivity of chemisorbed CO on a stepped Pt single crystal. The stepped Pt crystal contains atomic steps separated by smooth terraces, affording a convenient opportunity to witness the differing behavior of surface species on the various types of adsorption sites present. In this work, we rely on the digital ESDIAD method [ESDIAD = Electron Stimulated Desorption Ion Angular Distribution], first developed as an analog method in 1974, and recently refined by digitization [1]. We have applied this method to the study of the bonding geometry and the vibrational dynamics of chemisorbed CO on Pt(112) and to the surface reaction between CO(a) and O(a) to produce CO_2(g).

The ESDIAD phenomenon occurs because of the electronic excitation of an adsorbate species into a repulsive state, using electron impact excitation. The excited species escapes from the surface in a direction closely related to the direction of the chemical bond being broken. In many cases, positive ions are produced. The ion trajectories are intercepted by a detection system which employs microchannel plate amplification and a position sensitive detector as shown in Figure 1. In a few cases, the production of electronically excited neutral metastable species has been observed, and these trajectories may

be analyzed also in the digital ESDIAD

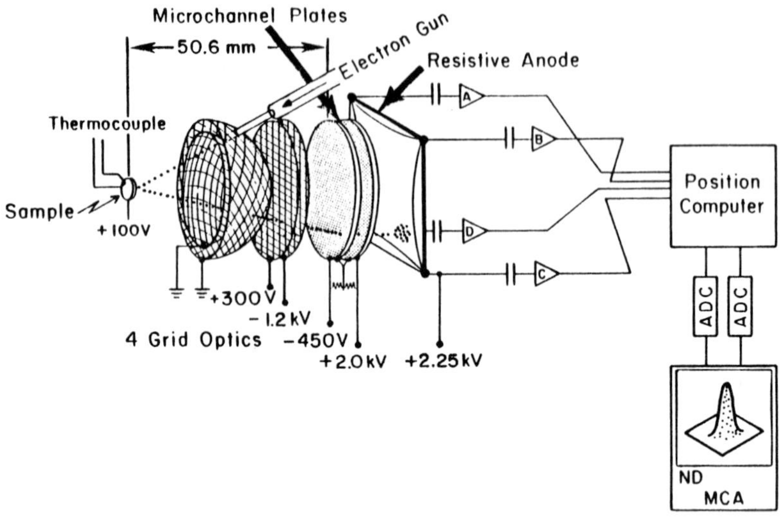

Fig. 1. Digital ESDIAD system. Positive ions or other excited species produced by electron stimulated desorption are measured by pulse counting methods and the statistical distribution of particle directions is displayed on a dual parameter multichannel analyzer [2].

apparatus. When ions and neutrals are mixed, the use of retarding potentials on the grid system may be conveniently employed for separation.

2. RESULTS AND DISCUSSION

2.1 CO Chemisorption on Pt(112).

As shown in Figure 2, the Pt(112) surface consists of three atom wide terraces of the Pt(111) structure, separated by atomic steps which have a (001) structure. The angular coordinates relative to this stepped surface are defined with respect to the normal of the macroscopic (112) crystal plane which is labeled 0° in Figure 2.

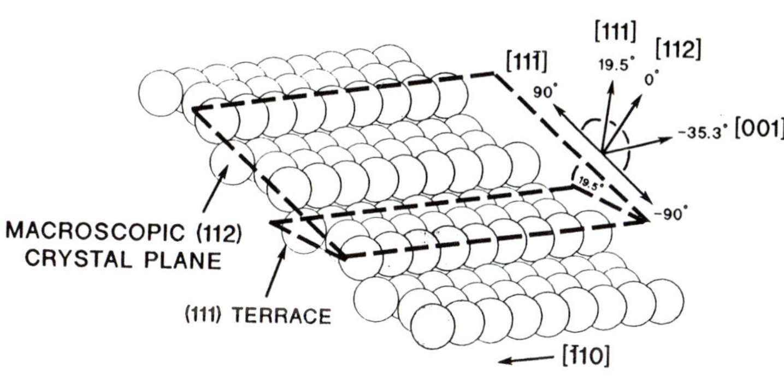

Fig. 2. Structure of the Pt(112) Surface.

CO molecules are delivered to this surface at 100 K followed by heating to 300 K to allow migration of the CO to the step sites [3]. The presence of the regular steps permits the experimenter to arrange CO molecules in linear arrays and to observe CO - CO interactions in these arrays. It is known that terminal CO is the species present on the Pt steps. It has been found that when CO layers are bombarded by electrons, an electronically excited neutral CO species is produced which is the $a^3\pi$-CO state, designated CO^* [4]. Figure 3 shows the ESDIAD patterns obtained from CO layers adsorbed to various average coverages, $\bar{\theta}$, where the CO^* angular distributions (free from image effects) are detected. Initially at a coverage of 0.19 ML, a single beam of CO^* species is observed. This beam is directed -20° in the downstairs direction and is perpendicular to the Pt step edge direction. At a coverage of 0.24 ML, the single CO^* beam is attenuated, and CO^* intensity in tilted beams at -13° and +13° is observed. Above 0.24 ML, ESDIAD patterns obtained from CO layers adsorbed to various coverages, the left and right

(± 13°) beams are attenuated, and two dominant beams at 0° and -38°, directed normal to the Pt(112) plane and in the

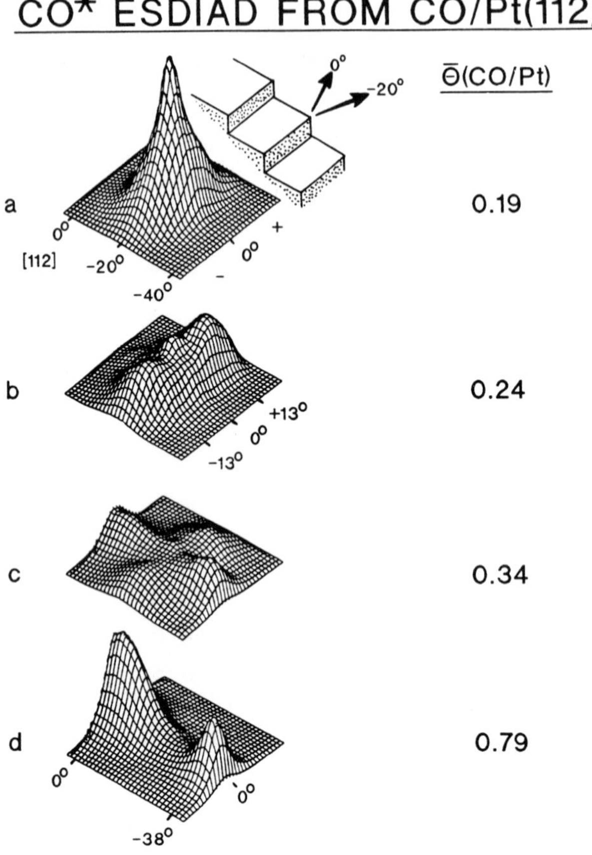

Fig. 3. CO* ESDIAD Patterns for CO on Pt(112) [5].

downstairs direction, are observed with complete extinction of all other CO* beams which were observed at lower CO coverages. These effects are thought to be due to tilting of the CO molecules as repulsive forces influence linear chains of CO species adsorbed on the step sites, and having higher and higher coverages within the chains.

A model to explain the orthogonal tilting directions selected by the CO molecules as coverage increases is shown in Figure 4. Initially, at $\overline{\theta} = 0.19$, the step sites are one-half filled, and all CO molecules are directed with the M-CO bond oriented in the -20° direction (downstairs). As three quarters filling of the step sites is approached, triplet CO groups containing left and right tilted molecules are produced. The lateral tilting occurs in the direction of the empty Pt sites in the CO linear arrays. At still higher CO coverages, in CO groups where there are no vacancy sites, the repulsive CO - CO intermolecular forces may be relieved only by forward and backward tilting of the CO molecules. The measurements shown here concern only CO molecules on the steps; other measurements on the terrace CO species have also been made [6]. At the present time, it is believed that the repulsive energy required

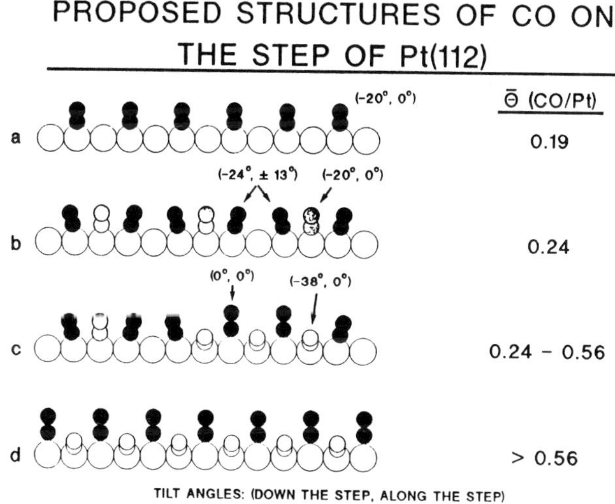

Fig. 4. Proposed Structures of CO on the Step of Pt(112) at various coverages [5].

for the tilting of the CO molecules originates mainly from steric effects (Pauli exclusion principle) rather than from dipole-dipole repulsions [7].

2.2 Dynamical Behavior of CO on Pt(112).

The digital ESDIAD method permits one to make very accurate measurements of the shape of the ESDIAD beams as a function of temperature. Since a statistical averaging of beam directions occurs in the ESDIAD measurement where as many as several million trajectories are summed up in a pattern, we have a method to observe the thermal average for bond angles, and hence to observe thermal broadening due to the increased occupancy of excited vibrational states as the temperature is increased. The ESDIAD method has also been used to observe hindered molecular rotations [8]. The broadening of an ESDIAD beam by thermal excitation will be primarily caused by the excitation of high amplitude, low frequency modes, and for terminally-bonded CO on Pt, the mode responsible for the thermal broadening will be the hindered translational mode with a frequency of about 50 cm^{-1} [9].

Figure 5 shows an ESDIAD pattern for CO^* species produced by ESD from the Pt(112) step sites. The coverage ($\overline{\theta}$ = 0.17 ML) is such that intermolecular CO - CO forces are <u>not</u> observed between CO species. The ESDIAD pattern is cut by two planes which are respectively along the step edge direction and perpendicular to the step edge direction (up-down plane). We desire to measure the shape of the cross sections in the two directions as well as the temperature dependence of the shape. This will give information on the average amplitude in two orthogonal directions for the CO frustrated translational modes.

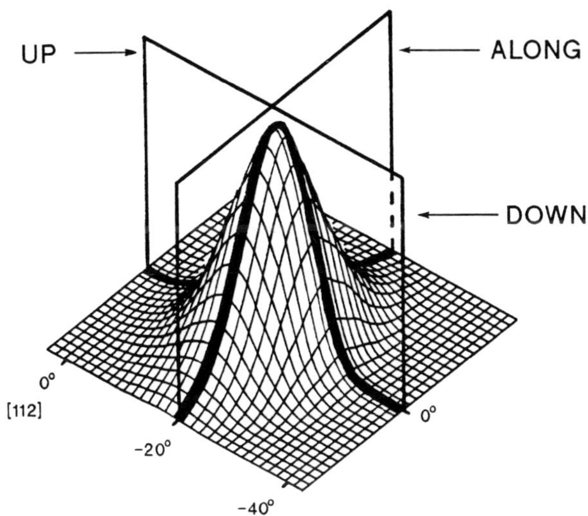

Fig. 5. CO* ESDIAD Patterns for one-half filled step sites on Pt(112) at 100 K. The two planes are parallel to and perpendicular to the step edge directions [6].

The cross section of the ESDIAD beam shapes in the two chosen directions is shown in Figure 6. It may be seen that in the up-down direction an asymmetry is observed, with larger vibrational amplitudes being seen in the "up" direction. This behavior is consistent with the asymmetry of the Pt binding site when viewed in the up-down directions.

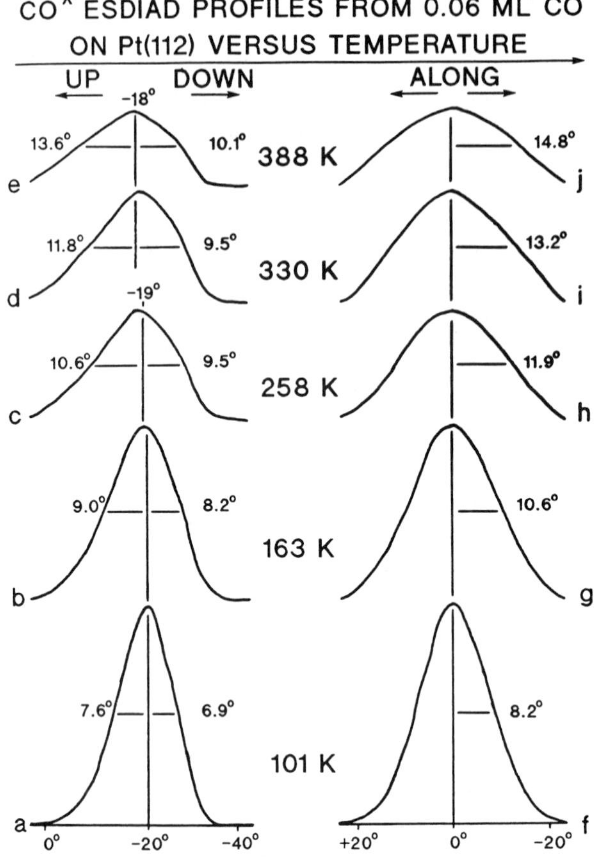

Fig. 6. CO^* ESDIAD cross sections in directions up and down, and parallel to the step directions on Pt(112) [6].

In contrast to this behavior, the cross section in the "along" direction is symmetrical, just as the symmetry in the two along directions (left and right) is identical for the CO binding sites.

These results are even more interesting when one considers the temperature dependence of the two cross sectional beam shapes. In all cases the beams expand in width as the temperature is increased, and in all cases the highest halfwidth amplitude is observed in the "along" direction, the next highest amplitude is in the "up" direction, and the lowest amplitude is in the "down" direction. These results, taken together, suggest that the freedom for vibration in the frustrated translational modes for CO chemisorbed on the step sites of Pt(112) is highest along the step edge. This means it is likely that the surface mobility of CO will be highest in this direction if the amplitudes of the frustrated translations are an indicator of the ease of surface migration [10-12].

2.3 Oxygen Chemisorption on Pt(111) and Pt(112).

The chemisorption of oxygen has been studied on both Pt(111) and Pt(112) using the ESDIAD method to image the direction of emission of O^+ from both surfaces [13]. As shown in Figure 7A, from Pt(111) O^+ ions are observed to escape from the surface in a direction peaked along the [111] direction, perpendicular to the (111) plane. In contrast to this behavior, a study of the direction of O^+ emission from Pt(112) is shown in Figure 7B. Here, the most probable direction of O^+ emission is at an angle of -38° from the [111] direction (normal to the terraces) and about - 19° from the [112] direction (normal to the macroscopic (112) crystal plane). These sections through the O^+ ESDIAD patterns clearly indicate that in contrast to normal O^+ emission from Pt(111), for Pt(112), the O^+ emission occurs dominantly in the downstairs direction. This indicates clearly that on Pt(112), at O coverages between 0.10 and 0.18 ML, the chemisorbed atomic oxygen species produced by adsorption of O_2 are localized on the atomic steps which face downstairs. Chemisorption of

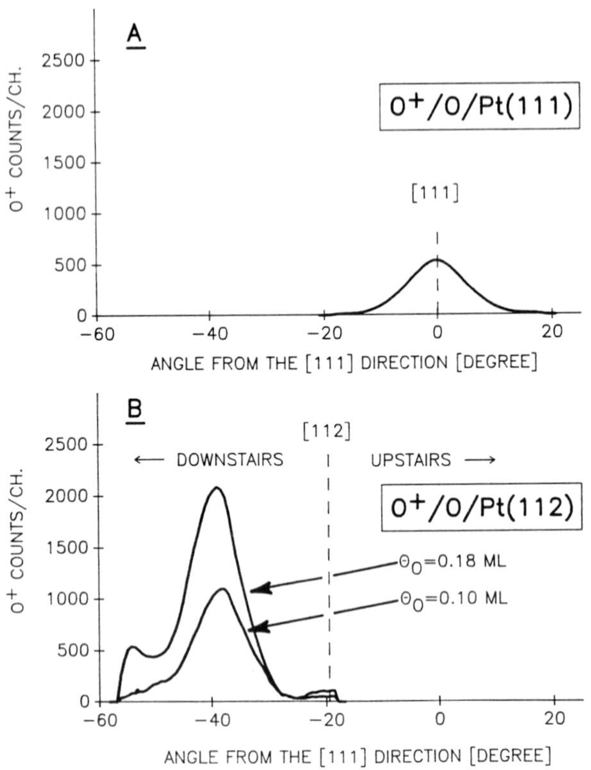

Fig. 7. Comparison of the O^+ ESDIAD Pattern for Oxygen Dissociative Adsorption on Pt(111) and Pt(112). In both cases, sections of the ESDIAD patterns are shown. Primary electron energy = 160-260 eV; crystal bias = 0 V. [13]

atomic oxygen does not occur on the (111) terrace sites at the oxygen coverages employed here. The behavior of the step sites in this regard is therefore similar for both CO(a) and O(a), with preferential adsorption on the steps. Studies of the thermal desorption behavior of oxygen from Pt(111) and Pt(112) have shown that the binding states and desorption kinetics for oxygen differ significantly for Pt(111) and Pt(112) [14,15].

2.4 CO and O Site Exchange From Coadsorption on Pt(112).

It has been found that the digital ESDIAD method may be used to obtain detailed information about the behavior of CO chemisorbed onto a Pt(112) surface containing step sites which have previously been partially filled with atomic oxygen. Figure 8 shows the CO^* ESDIAD pattern for a CO layer (0.17 ML) adsorbed at 100 K on top of the surface containing 0.18 ML of preadsorbed atomic O which was localized on the step sites as previously demonstrated in Figure 7. It is found in the case of the oxygen-covered steps, CO adsorbs on the TERRACE sites giving a CO^* ESDIAD pattern directed in the UPSTAIRS direction as shown in the upper left hand panel of Figure 8. Preadsorbed oxygen blocks the normal occupancy of the step sites by CO.

The exchange of sites by CO and adsorbed oxygen was observed upon heating this mixed layer to 230 K, as may be seen from the CO^* ESDIAD behavior in the lower left hand panel of Figure 8. Here, CO migrates to the step sites and produces a broad ESDIAD pattern which is oriented in the DOWNSTAIRS direction. This is probably accompanied by O migration onto the terraces, although we have no direct evidence for this displacement of O(a). The data given in the thermal desorption spectrum for the mixed CO(a) + O(a) layer in the right hand section of Figure 8 shows that the site exchange process is not accompanied by the production of CO_2(g) since CO_2 desorption occurs only above 250 K for this particular mixture of surface species.

The site exchange process is very useful to us in being able to prepare a mixed CO(a) + O(a) layer where the physical location of the two chemisorbed species is known. It should be emphasized that under the coverage conditions achieved in the experiments of Figure 8, CO_2 production does not occur in the temperature range 100 - 230 K, where site exchange has been observed. The specific occupancy of step sites by the CO(a) and the displacement of O(a) to the terraces provides a useful starting condition for experiments designed to determine where the CO(a) + O(a) reaction to produce CO_2(g) occurs, as will be shown in section 2.5, below.

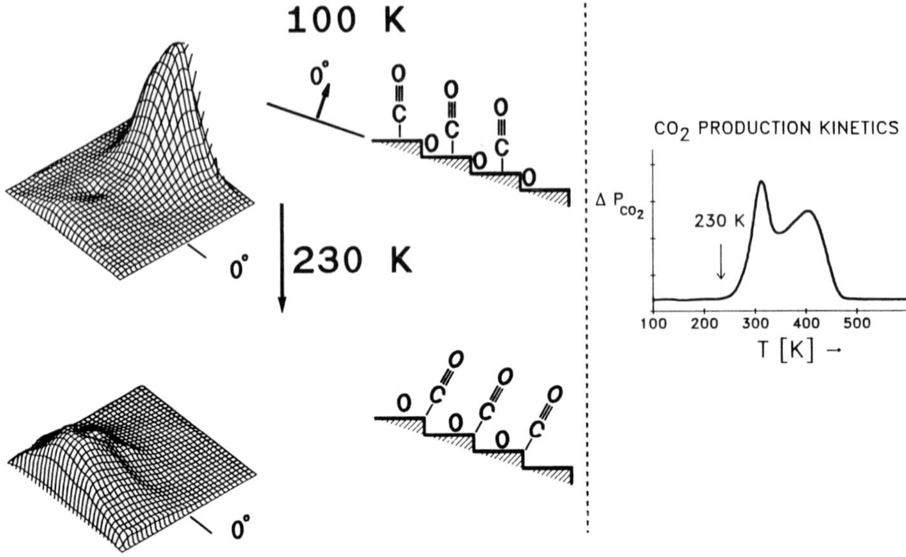

Fig. 8. CO(a) and O(a) Site Exchange as observed by ESDIAD studies of the CO^* Angular Distribution. Primary electron energy = 160 eV [13].

2.5 Detection of the Preferential Site for the CO(a) + O(a) ---> CO_2(g) Reaction.

The adsorption and reaction of adsorbed CO and adsorbed O on Pt is a classic heterogeneous catalytic reaction, studied by many others in the past [16-23].

Detailed insight into the location of the most favorable surface sites for the Langmuir-Hinshelwood reaction between CO(a) and O(a) has been obtained through the use of isotopically labeled CO species in experiments on surfaces prepared as shown in Figure 8. These experiments, and the resulting isotopic CO_2 production are shown in Figure 9. The procedure for producing the mixed CO(a) and O(a) layer is shown in the dotted box on the right hand side of each of the panels in Figure 9. In the upper panel, the surface is prepared as it was in Figure 8, with $^{12}C^{16}O$ displacing O(a) from step sites to terrace sites upon annealing at 230 K. At this point, the surface is filled with a second isotopic CO species, $^{13}C^{18}O$. This second isotopic CO species will primarily adsorb on the terrace sites of Pt(112), since these sites are only partially filled with O(a) following our postulated displacement from the step sites. **IN THIS EXPERIMENT THEN, WE HAVE ARRANGED A PARTICULAR ISOTOPIC CO SPECIES TO COEXIST WITH O(a) ON THE TERRACE SITES, WHILE A SECOND ISOTOPIC FORM OF CO IS PRESENT ON THE STEP SITES.**

Multiplexed mass spectroscopic studies of the desorption of CO_2(g) were performed on the isotopically dosed surface as shown in the upper panel of Figure 9. **THE CROSS HATCHED REGION OF THE THERMAL DESORPTION SPECTRUM INDICATES THE PREFERENTIAL PRODUCTION OF $^{13}C^{18}O^{16}O$(g) (47 amu) IN THE TEMPERATURE RANGE BETWEEN 100 K AND ABOUT 200 K.** The absence of CO_2(g) production processes when 44 amu is simultaneously monitored indicates clearly that the $^{12}C^{16}O$(a) species, present on the step sites as demonstrated in Figure 8, **DO NOT PARTICIPATE IN THIS LOWER TEMPERATURE CO_2 PRODUCTION PROCESS.** Thus, we have clear evidence that the lowest activation energy CO_2(g) production processes involve O(a) and CO(a) species which are both on the terrace sites of Pt(112).

To be certain that systematic errors do not exist in this isotopic experiment, the CO isotopic species are reversed in their order of addition in the experiment shown in the bottom panel of Figure 9. Here, $^{12}C^{16}O^{16}O$(g) is preferentially produced below 200 K, proving that the order of addition of the CO isotopes is the factor which determines the isotopic identity of the most readily produced CO_2(g) species.

It can be seen from Figure 9 that above about 200 K little memory for the order of addition of the isotopic CO species can be detected through monitoring of the isotopic CO_2 species. This is presumably because of the beginning of rapid CO site exchange between step and terrace sites above about 200 K which obscures the surface memory for the order of adsorption of the CO isotopes. Such results are consistent with other measurements of CO migration rates over terraces to steps [3], as well as with studies of CO thermal desorption from stepped Pt surfaces [24]. Oxygen site exchange during $CO_2(g)$ production is also possible.

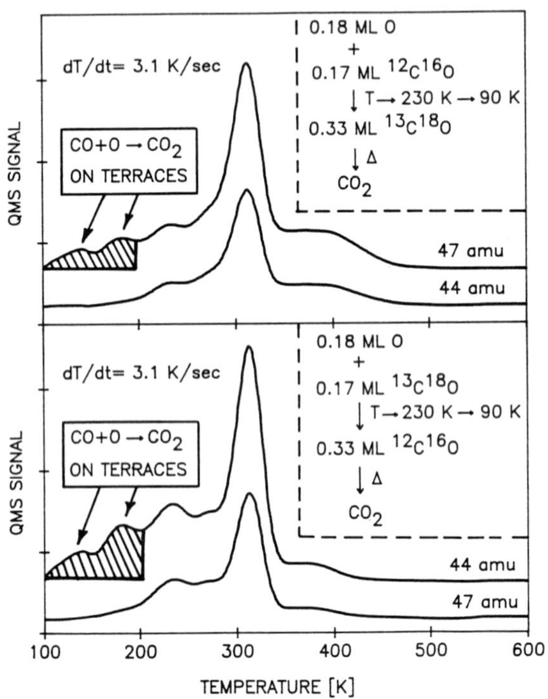

Fig. 9. Isotopic Studies of CO_2 Production from Pt(112) Using Preferential Adsorption of Isotopic CO Species on Step and Terrace Sites [13].

The overall results of these experiments are schematically summarized in Figure 10, where it is shown that labeled CO• which is preferentially adsorbed on Pt(112) terraces can react with O(a), also present on the terrace sites. Below 200 K, little or no CO from the step sites is found to react with the adsorbed terrace O(a) species. Thus for this particular combination of surface coverages of CO(a) and O(a), the involvement of the terrace sites in the production of CO_2(g) by means of the lowest activation energy pathway has been demonstrated. This may be related to the lower binding energy of CO(a), and possibly O(a), to the terrace sites compared to the step sites on Pt(112).

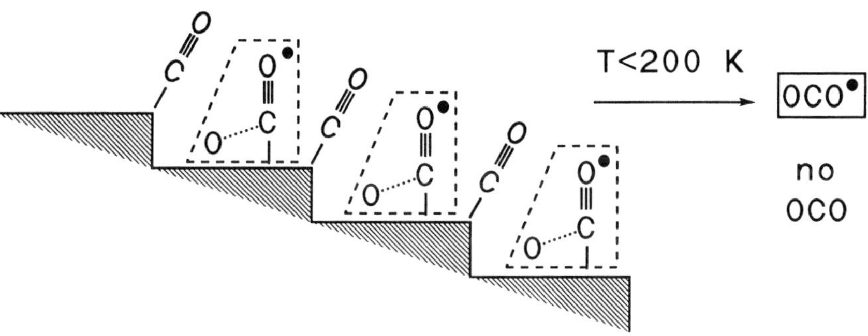

Fig. 10. Schematic Diagram of the Preferential Involvement of Terrace Sites for CO_2 Production from CO(a) and O(a) on Pt(112) [13].

3. SUMMARY

These experiments have illustrated a surface science study which has been able to determine certain details about the behavior of chemisorbed CO on a stepped Pt(112) surface. In particular, the following has been shown:

1. Preferential chemisorption of CO onto the step sites of a Pt(112) surface has been observed, in agreement with work of others.
2. The CO molecules localized on the step sites of Pt(112) undergo mutual interactional effects which cause them to tilt in particular directions (toward unfilled neighbor Pt sites). As the CO coverage is increased on the step sites, different tilt directions are selected in order to relieve the intermolecular strain between the neighboring CO molecules.
3. Under CO coverage conditions on the stepped sites where strong mutual interactions do not occur between neighbor CO species, it has been possible to monitor the relative amplitude of the low frequency, high amplitude frustrated CO translational modes. The softest mode occurs in directions parallel to the step edge.
4. It has been found that the dissociative adsorption of oxygen occurs with preferential deposition of O(a) on the step sites, giving, in ESDIAD, an O^+ beam which is oriented in the downstairs direction.
5. CO adsorption onto Pt(112) containing O(a) on the step sites occurs preferentially on the terrace sites at 100 K.
6. Upon heating the above CO(a) + O(a) layer to 230 K, site exchange between the CO(a)(terrace) and O(a)(step) occurs. This provides a convenient method for placing one isotopic type of CO on the step sites and of transferring adsorbed atomic O to the terrace sites.
7. Isotopic studies of the rate of reaction of O(a)(terrace) with CO(a)(terrace) and with CO(a)(step) have shown that below 200 K preferential reaction occurs between CO(a) and O(a) on the terrace sites. Thus, the (111) terrace sites preferentially catalyze the reaction between adsorbed CO and adsorbed O, producing CO_2 at temperatures below 200 K. Above 200 K, surface migration processes between terrace and step sites obscure experiments where particular CO isotopes are adsorbed on particular sites.

4. ACKNOWLEDGEMENT

We thank the Air Force Office of Scientific Research for support of the studies of CO adsorption on Pt(112). We thank the Department of Energy, Office of Basic Energy Sciences, for support of the studies of the site selectivity for the CO oxidation reaction.

REFERENCES

1. J.J. Czyzewski, T.E. Madey, and J.T. Yates, Jr., Phys. Rev. Lett., 32 (1974) 777; J.T. Yates, Jr., M.D. Alvey, K.W. Kolasinski, and M.J. Dresser, Nuclear Inst. and Methods in Phys. Research, B27 (1987) 147.
2. M.J. Dresser, M.D. Alvey, and J.T. Yates, Jr., Surface Sci., 169 (1986) 91.
3. J.E. Reutt-Robey, D.J. Doren, Y.J. Chabal, and S.B. Christman, Phys. Rev. Lett., 61, (1988) 2778; see also B. Poelsema, R.L. Palmer, and G. Comsa, Surface Sci., 123 (1982) 152.
4. M. Kiskinova, A. Szabò, and J.T. Yates, Jr., Surface Sci., 205 (1988) 215.
5. M.A. Henderson, A. Szabò, and J.T. Yates, Jr., J. Chem. Phys., 91 (1989) 7245.
6. M.A. Henderson, A. Szabò, and J.T. Yates, Jr., J. Chem. Phys., 91 (1989) 7255.
7. M.A. Henderson, A. Szabò, and J.T. Yates, Jr., Chem. Phys. Lett., 162 (1990) 51.
8. M.D. Alvey, J.T. Yates, Jr., and K.J. Uram, J. Chem. Phys., 87 (1987) 7221.
9. A.M. Lahee, J.P. Toennies, and Ch. Wöll, Surface Sci., 177 (1986) 371.
10. B.E. Hayden and A.M. Bradshaw, Surface Sci., 125 (1983) 787.
11. J.W. Gadzuk, J. Opt. Soc. Am. B: Opt. Phys., 4 (1987) 201.
12. R. Berndt, J.P. Toennies and Ch. Woll, J. Electr. Spect. Related Phenom., 44 (1987) 183.
13. A. Szabò, M. A. Henderson, and J.T. Yates, Jr., J. Chem. Phys., to be submitted.

14 H.R. Siddiqui, A. Winkler, X. Guo, P. Hagans, and J.T. Yates, Jr., Surface Sci., 193 (1988) L17.

15 A. Winkler, X. Guo, H.R. Siddiqui, P.L. Hagans, and J.T. Yates, Jr., Surface Sci., 201 (1988) 419.

16 I. Langmuir, Trans. Faraday Soc., 17 (1922) 671, 672.

17 J. Segner, C.T. Campbell, G. Doyen, and G. Ertl, Surface Sci., 130 (1984) 505.

18 L.S. Brown, S.L. Bernasek, J. Chem. Phys., 82 (1985) 2110.

19 R.C. Yeates, J.E. Turner, A.J. Gellman, and G.A. Somorjai, Surface Sci., 149 (1985) 175.

20 J.L. Gland and E.B. Kollin, Surface Sci., 151 (1985) 260.

21 L.F. Razon and R.A. Schmitz, Catal. Rev. Sci. Eng., 28 (1986) 89.

22 P.J. Berlowitz, C.H.F. Peden, and D. W. Goodman, J. Phys. Chem., 92 (1988) 5213.

23 M. Elswirth and G. Ertl, Phys. Rev. Lett., 60 (1988) 1526.

24 H.R. Siddiqui, X. Guo, I. Chorkendorff, and J.T. Yates, Jr., Surface Sci., 191 (1987) L813.

R.K. Grasselli and A.W. Sleight (Editors), *Structure-Activity and Selectivity Relationships in Heterogeneous Catalysis*
© 1991 Elsevier Science Publishers B.V., Amsterdam

CORRELATIONS BETWEEN STRUCTURE AND REACTIVITY OF METAL SURFACES

XUDONG JIANG and D. WAYNE GOODMAN

Department of Chemistry, Texas A&M University
College Station, Texas 77843 (U.S.A.)

ABSTRACT

Single crystal nickel and iridium have been used as model catalysts to investigate the hydrogenolysis and reactive sticking reactions of small alkanes. It has been found that the Ni(100) surface is much more reactive than the Ni(111) surface toward ethane hydrogenolysis and methane reactive sticking, and that the reconstructed Ir(110)-(1x2) surface has much higher selectivity than the Ir(111) surface for ethane production from the hydrogenolysis of n-butane. These results demonstrate the correlation between structure and reactivity of metal surfaces, and the relevance between surface science studies on single crystal model catalytic surfaces and the corresponding measurements on supported metal catalysts.

1. INTRODUCTION

An important question in catalysis is the relationship between the structure and composition of a catalytic surface and the reactivity and selectivity demonstrated by that surface. The use of oriented single crystals has been shown to be particularly informative regarding the unambiguous assessment of the effects of surface composition and geometry and provides a way leading to a microscopic understanding of the catalytic properties of various catalysts [1-4]. Although there are examples of "structure-insensitive" reaction in which the reaction proceeds at the same rate and gives the same product distribution over different facets of a metal catalyst, as shown in Fig. 1 for CO oxidation reaction on several transition metal catalysts [5, 6], many reactions have been found to be "structure-sensitive". Their reactivity and selectivity depend considerably on the surface geometry or the metallic particle size of the catalyst.

In this paper, we review some of the results of our studies on the hydrogenolysis of small alkanes over nickel [7] and iridium [8, 9], and alkane reactive sticking over nickel [10-12]. These studies were all performed on single crystal surfaces at elevated pressures. The results demonstrate the correlation between structure and reactivity of metal surfaces, and the relevance between surface science investigations on single crystal model catalytic surfaces and corresponding studies on supported metal catalysts.

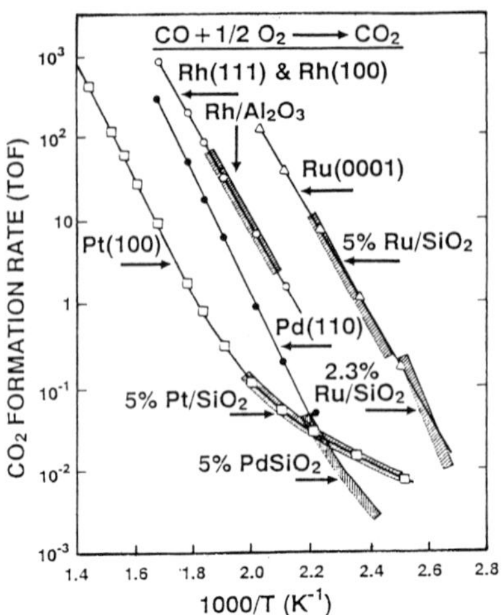

Fig. 1 Arrhenius plot for CO oxidation reaction on several transition metal single crystal surfaces and supported catalysts at a total reactant pressure P_T of 24 Torr and CO/O_2=2 [5, 6]. The data demonstrate the "structure-insensitivity" of this reaction.

2. EXPERIMENTAL

The experiments were performed in a stainless-steel, dual-chambered apparatus which has been described in detail elsewhere [2, 13]. The chambers are linked via a gate valve and each can be evacuated to $<10^{-10}$ Torr. Crystals were mounted on a retraction bellows and translated vertically between the analysis chamber and the reaction chamber. The analysis chamber is equipped with a cylindrical mirror analyzer (CMA) for Auger electron spectroscopy (AES) and a quadrupole mass analyzer for thermal desorption spectroscopy (TDS). The reaction chamber, which has a volume of ~600cm^3 and can be pressurized to several atmospheres, was operated as a batch microreactor.

The crystal temperature was monitored by either a chromel/alumel thermocouple (for nickel) or a W-5%Re/W-26%Re thermocouple (for iridium) spot-welded to the back of the crystal. The temperature of the sample was maintained during reaction by a RHK temperature programmer to ±1K.

The single crystal preparation and cleaning, reactant handling and purification, and the experimental procedures are given in detail in ref. 7-12.

In the alkane hydrogenolysis experiments, reaction products were analyzed by gas chromatography. Absolute reaction rates were calculated from the reactor volume, duration of reaction, the measured surface area and the known atomic density of each crystal surface.

In the alkane reactive sticking experiments, after reaction, the Auger ratio C(272eV)/Ni(848eV) was measured and assumed to be proportional to the concentration of carbon atoms present on the nickel surface [14-15]. This ratio was then compared to the saturation C/Ni ratio, for which the exact carbon coverage is known [16]. This enabled the calculation of initial reaction rates expressed as alkane decomposition events per site, per second.

3. RESULTS AND DISCUSSION

3.1 Hydrogenolysis of Small Alkanes on Nickel and Iridium Single Crystal Surfaces

3.1.1 Hydrogenolysis of Ethane on Nickel

The reactivity for ethane hydrogenolysis to methane on nickel has been shown to depend critically on the particular geometry of the surface. Fig. 2 shows the specific reaction rate [(product molecules)•(substrate surface atom)$^{-1}$•(second)$^{-1}$] or turnover frequency (TOF) for methane formation from ethane over Ni(100) and Ni(111) surfaces plotted in Arrhenius form [7]. It can be seen that the more open (100) surface is far more active than the close-packed (111) surface. For the Ni(100) surface, the data give an activation energy of 24 kcal/mole, which is remarkably close to the 25 kcal/mole obtained for the methanation reaction over the same surface [13]. Furthermore, the specific rates observed for both methanation and ethane hydrogenolysis on this surface are virtually identical for the same partial pressure of hydrogen. These observations strongly suggest that these two reactions over the Ni(100) surface are following the same reaction pathway and are limited by the same reaction step. As previously shown for the methanation reaction [16], ethane hydrogenolysis on this surface must involve a surface carbon formation step followed by its reduction by hydrogen. In contrast, the kinetic data shown in Fig. 2 for the Ni(111) surface give an activation energy of 46 kcal/mole, implying that a different reaction mechanism is operative.

There could be several possibilities as to the origin of the inhibited activity of the Ni(111) surface relative to the Ni(100) surface toward ethane hydrogenolysis. One possibility is the differences in the electronic structure between these two surfaces. If back-bonding from the metal to the unfilled σ^* levels of ethane is an important first step toward carbon-carbon bond scission in

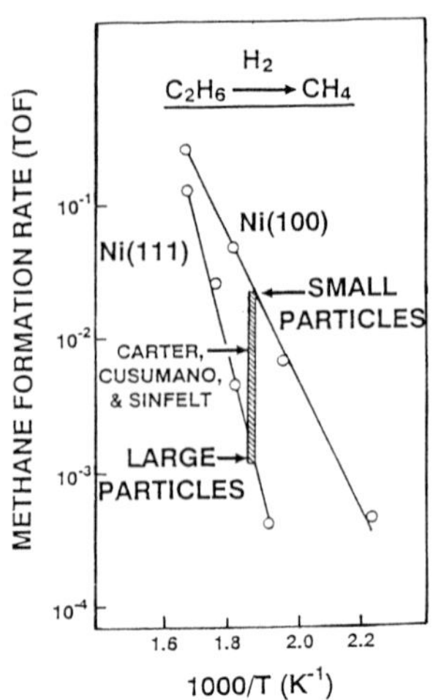

Fig. 2 Arrhenius plot for ethane hydrogenolysis reaction on Ni(100) and Ni(111) surfaces at a total reactant pressure P_T of 100 Torr and $H_2/C_2H_6=100$ [7]. Also shown is the result on supported nickel catalysts at P_T of 175 Torr and $H_2/C_2H_6=6.6$ [20].

ethane, then the (100) surface may be more active than the (111) surface since the appropriate nickel orbitals for such back-bonding are more available in the case of the (100) surface [17]. The second possibility is the differences in the spacing between high coordination bonding sites on these two surfaces. For the (100) surface the spacing between the four-fold hollow sites is approximately 2.5Å. The bond length of a carbon-carbon intermediate is expected to be from 1.3 to 1.5Å. Obviously for the Ni(100) surface, the carbon-carbon bond cannot remain intact and, at the same time, the carbon atoms bond in the preferred high coordination site [18]. However, for the Ni(111) surface, the 1.4Å spacing between the high coordination sites is ideally suited to maintaining the carbon-carbon bond intact while bonding each carbon to a three-fold hollow site. It follows then that ethane adsorbs on the (100) surface and dissociates to form a surface carbide or hydrogenated carbonaceous species. This species, in turn, hydrogenates to methane. For the (111) surface, however, we anticipate a stable adsorbed carbon-carbon species to form at relatively high surface concentrations. The rate limiting step for product formation then would be the carbon-carbon bond scission step. The surface should be virtually free of the single carbon species, which is an essential intermediate for methanation.

The (111) surfaces are encountered more prevalently in FCC

materials as the particle size is increased via successively higher annealing temperatures [19]. The results of this study then are consistent with rate measurements on supported nickel catalysts [20, 21], which show hydrogenolysis activity to be a strong function of particle size, the larger particles exhibiting the lower rates, as also shown in Fig. 2.

3.1.2. Hydrogenolysis of n-Butane and Propane on Iridium

The selectivity for ethane production from the hydrogenolysis of n-butane over iridium single crystals has been demonstrated to scale with the concentration of low-coordination-number metal surface atoms [8, 9]. Fig. 3 shows the results on Ir(110)-(1x2) and Ir(111) surfaces as well as the schematic representation of these iridium surfaces. The Ir(110)-(1x2) surface, which has a stable "missing-row" structure [22], has been found to produce ethane very selectively. This contrasts with the results for the close-packed Ir(111) surface, where only the statistical scission of C-C bonds has been observed. Although there is still some controversy regarding the surface structure of Ir(110) under reaction conditions, its selectivity for ethane production is clearly superior to Ir(111).

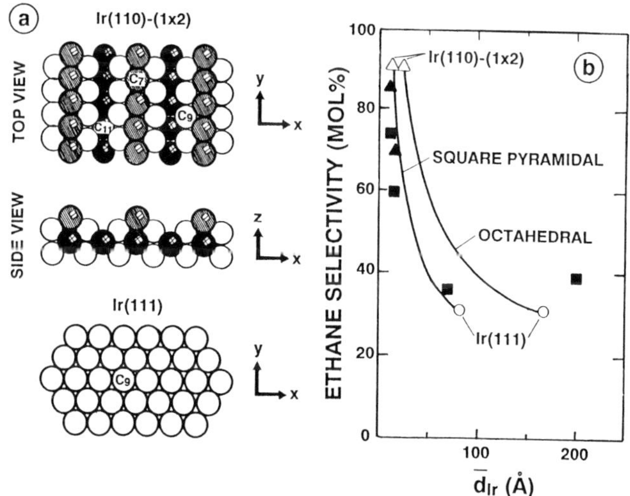

Fig. 3 Diagram showing the correlation between selectivity and structure for n-butane hydrogenolysis on iridium. (a) Schematic representation of the Ir(110)-(1x2) and Ir(111) surfaces. The z axis is perpendicular to the plane of the metal surface. C_n designates the coordination numbers of the metal surface atoms. (b) Selectivity for C_2H_6 production (mol % total products) for n-butane hydrogenolysis on iridium single crystals [8, 9] and supported iridium catalysts [23] at 475K. The effective particle size for the single crystal surfaces is based on the specified geometrical shapes. ▲, Ir/Al_2O_3; ■, Ir/SiO_2.

The results of this study correlate qualitatively with the observations made previously for selective hydrogenolysis of n-butane to ethane on supported iridium catalysts as a function of iridium particle size [23], which is also shown in Fig. 3. It can be seen that the results for Ir(110)-(1x2) model very well the small-particle limit whereas the results for Ir(111) relate more closely to the data for the corresponding large particles (>10nm). By assuming particle shapes the general behavior of declining selectivity with larger particle size can be accurately modelled, as illustrated in Fig. 3.

The stoichiometry of the surface intermediate leading to high ethane selectivity, based on kinetics and surface carbon coverages subsequent to reaction, is suggested to be a metallocyclopentane [8, 9]. The Ir(110) surface undergoes a reconstruction, described as the Ir(110)-(1x2) or "missing-row" structure, resulting in rows of the highly coordinatively unsaturated "C_7" sites, as schematically shown in Fig. 3. These sterically unhindered C_7 sites can form a metallocyclopentane species (e.g., a 1,4-diadsorbed hydrocarbon species) which has been proposed as an intermediate in the central scission of butane to ethane. Based on analogous chemistry reported in the organometallic literature [24, 25], the mechanism responsible for the hydrogenolysis of n-butane on the Ir(110)-(1x2) surface is postulated to be the reversible cleavage of the central C-C bond in this metallocyclopentane intermediate. On the other hand, butane hydrogenolysis on the Ir(111) surface appears to operate via a different mechanism. First, dissociative chemisorption of butane and hydrogen occurs followed by irreversible cleavage of the terminal carbon-carbon bond of the adsorbed hydrocarbon. Further C-C bond cleavage prior to product desorption leads to the methane and ethane observed as initial products.

For both iridium surfaces, the extent to which hydrogenolysis proceeds increases with increasing reaction temperature. This is in keeping with the general trend for increased cracking at higher temperatures for alkane reactions. The term "roll-over" has been used to describe the fall in overall activity at the high temperatures which leads to a decrease in the selectivity for the production of ethane (shown in Fig. 4) in the hydrogenolysis of propane over these two iridium surfaces. Decreasing the partial pressure of H_2 at the temperature of onset of roll-over induces the same selectivity change as observed for an increase in reaction temperature. The origin of this effect is believed to be as follows. As the reaction temperature is raised beyond a critical temperature,

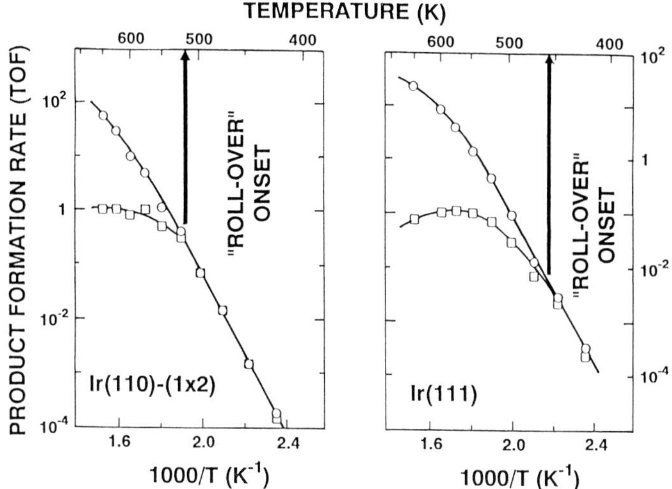

Fig. 4 Arrhenius plot for propane hydrogenolysis reaction on Ir(110)-(1x2) and Ir(111) surfaces at a total reactant pressure P_T of 101 Torr and H_2/C_3H_8=100. o, Methane; □, Ethane.

defined primarily by the hydrogen partial pressure, the hydrogen surface coverage falls below a saturation or critical coverage. The lower hydrogen coverage then reduces the efficiency of the hydrogenation of surface hydrocarbon fragments. It is shown in Fig. 4 that the roll-over onset occurs at a higher temperature on the Ir(110)-(1x2) than that on the Ir(111) surface. From previous studies, it is known that hydrogen desorbs at a higher temperature (390K at the saturation of the high temperature desorption state) from Ir(110)-(1x2) than from Ir(111) surface (255K at 250L of hydrogen) [26]. Therefore, the higher temperature of onset of roll-over on the more open Ir(110)-(1x2) surface correlates with the higher binding energy of hydrogen adatoms on this surface. This suggests that the source of the reactive hydrogen is the metal surface rather than, for example, an "active" carbonaceous overlayer.

3.2 Alkane Reactive Sticking on Nickel Single Crystal Surfaces

The reactive sticking of alkanes on nickel single crystal surfaces is strongly dependent on the surface structure. For example, methane reactivity, shown in Fig. 5 as the time-dependent carbon buildup from 1.00 Torr of methane in contact with the various low index nickel single crystal surfaces at 450K, is seen to increase in the order Ni(111)<Ni(100)<Ni(110) [10]. Initial reaction rates for the Ni(110) and Ni(100) surfaces are very similar, and are ~7 to

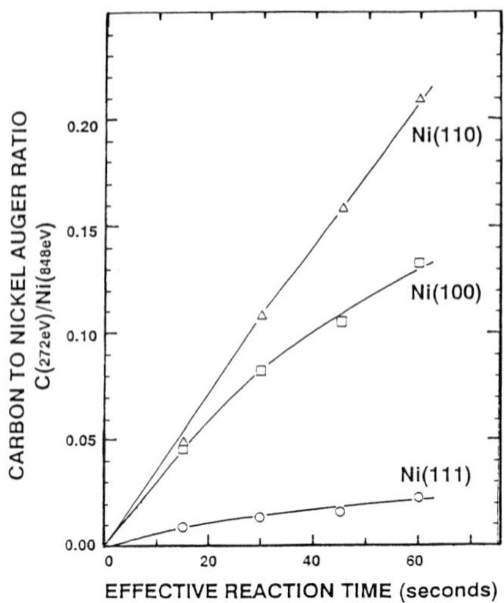

Fig. 5 Methane decomposition kinetics on low index nickel single crystal surfaces at 450K and methane pressure $P_{methane}$ of 1.00 Torr [10].

10 times greater than the initial rate for Ni(111) surfaces at 450K. However, both the Ni(100) and Ni(111) surfaces exhibit a strong coverage dependence in the methane decomposition rate, as evidenced by the deviation from linearity in the plots for these surfaces shown in Fig. 5. This behavior is in contrast with that of the Ni(110) surface, which does not exhibit the same downward curvature, possibly indicating islanding or less of a carbon coverage dependence for the methane reactivity on this surface.

Methane dissociative adsorption on sulfur-modified Ni(100) surface indicates that sulfur atoms poison this reaction by a simple site blocking mechanism [11]. The initial methane decomposition rate decreases linearly with sulfur coverage and drops to zero at a sulfur coverage of ~0.3ML. The results are consistent with a mechanism for the activated dissociative adsorption of methane on Ni(100) involving a direct process. However, dissociation of ethane, propane, and n-butane on Ni(100) is believed to proceed primarily via a trapped molecular precursor [12].

These studies on the alkane reactive sticking on nickel single crystal surfaces were carried out under the high incident flux conditions. The elevated pressures are required to produce measurable products, not because of the greater availability of higher velocity molecules, but rather because of the severe competition which is inevitably present between desorption from the precursor or adsorbed state and dissociation. Since activation energies to desorption for many reactants of interests (particularly hydrocarbons) are usually smaller than the activation energies to reaction, desorption dominates and reaction probabilities are quite small, often too small to measure at UHV conditions. For these reactants, the greater number of collisions at higher pressures simply serves to overcome this limitation.

Since the alkane dissociation rates obtained in these "thermal bath" experiments are initial rates measured in the limit of zero carbon coverage, they represent a theoretical upper limit to the rates of steam reforming of these alkanes on unpromoted nickel catalysts. Based on the kinetic experiments on sulfur-passivated nickel catalysts for carbon-free steam reforming of methane, Rostrup-Nielsen found that the kinetic data can be explained in terms of simple physical blockage by chemisorbed sulfur and that an ensemble of three nickel atoms is involved in the reforming reaction [27]. These are in complete agreement with the results obtained in the "thermal bath" experiment [11] and provide another excellent example in which surface science studies on single crystal model catalytic surfaces correlate extremely well with the corresponding measurements on supported metal catalysts.

4. CONCLUSIONS

Single crystal metal surfaces allow us to study in a systematic fashion the role of surface structure on catalytic activity and selectivity. We have shown that the reactivity for ethane hydrogenolysis to methane and alkane reactive sticking on nickel, as well as the selectivity for ethane production from the hydrogenolysis of n-butane over iridium are markedly affected by surface structure. For ethane hydrogenolysis reaction, the Ni(100) surface is much more active than the Ni(111) surface, possibly due to the different electronic structure and spacing between high coordination bonding sites on the two surfaces. For methane reactive sticking on nickel, the reactivity increases in the order Ni(111)<Ni(100)<Ni(110). For n-butane hydrogenolysis on iridium, the reconstructed Ir(110)-(1x2) surface, which has a high concentration of C_7 low-coordination-number sites, shows a marked propensity for central bond scission. On the other hand, the Ir(111) surface exhibits non-selective hydrogenolysis, yielding a statistical distribution of the products. The selective hydrogenolysis on Ir(110)-(1x2) may involve adsorption

of the n-butane as a metallocyclopentane and subsequent cleavage at the central carbon-carbon bond. All the results summarized here from studies on single crystal metal surfaces correlate very well with measurements on supported metal catalysts.

ACKNOWLEDGEMENTS

We acknowledge with pleasure the support of this work by the Department of Energy, Office of Basic Energy Sciences, Division of Chemical Sciences.

REFERENCES

1 D.W. Goodman, J. Vac. Sci. Technol., 20 (1982) 522-526.
2 D.W. Goodman, Acc. Chem. Res., 17 (1984) 194-200.
3 D.W. Goodman, Annu. Rev. Phys. Chem., 37 (1986) 425-457.
4 D.W. Goodman and J.E. Houston, Science, 236 (1987) 403-409.
5 D.W. Goodman and C.H.F. Peden, J. Phys. Chem., 90 (1986) 4839-4843.
6 P.J. Berlowitz, C.H.F. Peden and D.W. Goodman, J. Phys. Chem., 92 (1988) 5213-5221.
7 D.W. Goodman, Surf. Sci., 123 (1982) L679-L685.
8 J.R. Engstrom, D.W. Goodman and W.H. Weinberg, J. Am. Chem. Soc., 108 (1986) 4653-4655.
9 J.R. Engstrom, D.W. Goodman and W.H. Weinberg, J. Am. Chem. Soc., 110 (1988) 8305-8319.
10 T.P. Beebe, Jr., D.W. Goodman, B.D. Kay and J.T. Yates, Jr., J. Chem. Phys., 87 (1987) 2305-2315.
11 X. Jiang and D.W. Goodman, Catal. Lett., 4 (1990) 173-180.
12 A.G. Sault and D.W. Goodman, J. Chem. Phys., 88 (1988) 7232-7239.
13 D.W. Goodman, R.D. Kelley, T.E. Madey and J.T. Yates, Jr., J. Catal., 63 (1980) 226-234.
14 F.C. Schouten, E.W. Kaleveld and G.A. Bootsma, Surf. Sci., 63 (1977) 460-474.
15 F.C. Schouten, O.L.J. Gijzeman and G.A. Bootsma, Surf. Sci., 87 (1979) 1-12.
16 D.W. Goodman, R.D. Kelley, T.E. Madey and J.M. White, J. Catal., 64 (1980) 479-481.
17 M.C. Desjonquères and F. Cyrot-Lackmann, J. Chem. Phys., 64 (1976) 3707-3716.
18 M. Kiskinova and D.W. Goodman, Surf. Sci., 108 (1981) 64-76.
19 J.K.A. Clarke and J.J. Rooney, Adv. Catal., 25 (1976) 125-183.
20 J.L. Carter, J.A. Cusumano and J.H. Sinfelt, J. Phys. Chem., 70 (1966) 2257-2263.
21 G.A. Martin, J. Catal., 60 (1979) 452-459.
22 C.-M. Chan, M.A. Van Hove, W.H. Weinberg and E.D. Williams, Surf. Sci., 91 (1980) 440-448.
23 K. Foger and J.R. Anderson, J. Catal., 59 (1979) 325-339.
24 R.H. Grubbs and A. Miyashita, J. Am. Chem. Soc., 100 (1978) 1300-1302.
25 R.H. Grubbs, A. Miyashita, M. Liu and P. Burk, J. Am. Chem. Soc., 100 (1978) 2418-2425.
26 P.D. Szuromi, J.R. Engstrom and W.H. Weinberg, J. Chem. Phys., 80 (1984) 508-517.
27 J.R. Rostrup-Nielsen, J. Catal., 85 (1984) 31-43.

R.K. Grasselli and A.W. Sleight (Editors), *Structure-Activity and Selectivity Relationships in Heterogeneous Catalysis*

© 1991 Elsevier Science Publishers B.V., Amsterdam

ALKYL IODIDES ON COPPER SURFACES: C-H ACTIVATION AND COUPLING REACTIONS OF HYDROCARBON FRAGMENTS TO PRODUCE ETHYLENE

C.J. JENKS, J.-L. LIN, C.-M. CHIANG, L. KANG, P.S. LEANG, T.H. WENTZLAFF, and B.E. BENT
Department of Chemistry, Columbia University, New York, NY 10027

ABSTRACT

The dissociative adsorption of alkyl iodides onto a Cu(100) surface under ultra-high vacuum conditions has been utilized to generate and study the chemistry of adsorbed hydrocarbon fragments. It is found that iodomethane, iodoethane, and diiodomethane all react to produce, among other hydrocarbon products, ethylene. The surface reaction mechanisms, as determined by thermal desorption and isotope labelling studies, include: α-elimination from methyl groups, β-hydride elimination from ethyl groups, methylene coupling, and methylene insertion into methyl groups.

INTRODUCTION

Alkyl radicals bound to metal surfaces are postulated to be intermediates in numerous heterogeneous catalytic processes [1], yet little is known about the bonding and chemistry of these surface species. *In situ* study is difficult both because under typical reaction conditions the lifetimes of these species are short and because high pressure requirements prohibit the application of powerful, surface-sensitive electron spectroscopies. Recent studies, however, have shown that these transient intermediates can be isolated at low temperature on a number of metal surfaces under ultra-high vacuum (UHV) conditions using the facile thermal dissociation of alkyl iodides [2-11]:

SCHEME 1

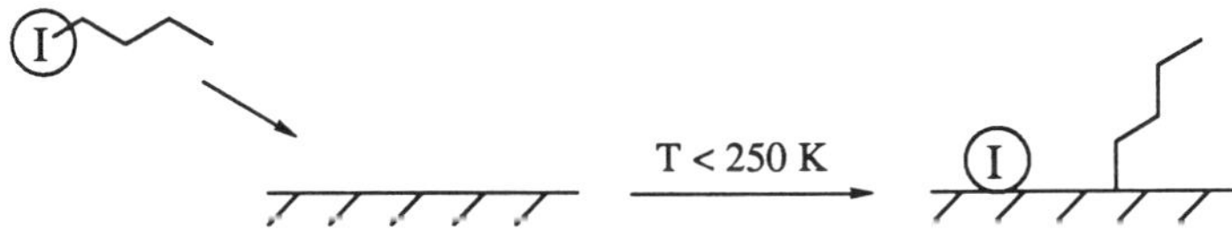

It should be emphasized that when alkyl iodides are used as alkyl precursors, the iodine atoms generated on the surface remain coadsorbed with the alkyl throughout the temperature range in which the alkyl reacts. Previous studies, however, show that the alkyl reaction pathways and temperatures are relatively insensitive to changes in alkyl iodide coverage (i.e. the coverage of coadsorbed iodine), suggesting that the effect of iodine is small compared with the dramatic effects of changing the metal (see Table 1). Also, studies on aluminum comparing aluminum alkyls and alkyl iodides as alkyl precursors show the same reaction pathways and similar rates [7,8].

Table 1 summarizes the chemistry observed for alkyls on metals. It is evident that on platinum C-H bond breaking initiates the surface reactions, whereas on silver alkyl coupling to form C-C bonds dominates. Aluminum shows β-hydride elimination chemistry similar to platinum, but at a much higher temperature (530 K vs. 230 K). In this paper, we report studies of alkyl iodides on a Cu(100) surface which show both coupling and C-H activation of the alkyl groups. We focus in particular on the ethylene-producing reactions of iodomethane, iodoethane, and diiodomethane since the diverse chemistries which convert these alkyl iodides to ethylene are illustrative of the many surface reaction pathways we observe. Specifically, we find that ethylene forms on copper (1) via β-hydride elimination of ethyl groups, (2) via coupling of methylene species, (3) via the disproportionation of methyl groups, and (4) via the insertion of methylene into methyl species followed by β-hydride elimination.

TABLE 1
Reaction Products and Temperatures for Alkyl Groups on Metals[a]

Surface	Alkyl Precursor	Reaction Products[b]	Temperature (K)[c]	Reference
Ni(100)	CH_3I	C(a), H_2(g)	<260	2
Ni(111)	"hot" CH_4	CH_2(a)	>150	3
Pt(111)	CH_3I	CH_4(g)	~290	4
	C_2H_5I	C_2H_4(a)	~230	5
Al(111)	CH_3I	CH(a)	<250	6
Al(100)	C_4H_9I	C_4H_8(g)	~540	7
	$Al(C_4H_9)_3$	C_4H_8(g)	~520	8
Ag(111)	CH_3I	C_2H_6(g)	~250	9
	C_2H_5I	C_4H_{10}(g)	~180	10
Cu film	CH_3Br	CH_4(g), C_2H_4(g)	~450	11

[a]Only systems for which experimental evidence strongly supports the formation of stable alkyl fragments on the surface prior to decomposition are tabulated.
[b]First reaction products detected.
[c]Temperature of maximum thermal desorption rate or equivalent.

EXPERIMENTAL

The reactions of alkyl iodides on Cu(100) were studied using an ultra-high vacuum (UHV) apparatus equipped with ion sputtering, low energy electron diffraction (LEED), Auger electron spectroscopy (AES), high resolution electron energy loss spectroscopy (HREELS), and a differentially pumped, quadrupole mass spectrometer. The experimental details will be published elsewhere [12], but aspects particularly significant for this work will be reviewed here.

The Cu(100) single crystal substrate used in these studies was a 0.5 cm^2 disk, 2 mm thick. It was mounted on a molybdenum resistive heating element using two small tantalum tabs which, in conjunction with liquid nitrogen cooling, allowed us to achieve temperatures between 110 K and 1100 K. The surface temperature was monitored using a chromel-alumel thermocouple wedged into a hole which had been spark-eroded into the side of the sample. The surface was routinely cleaned by a combination of sputtering at 850 K and room temperature with 1 kV Ar^+ ions followed by annealling at 890 K in UHV for 30 minutes. The alkyl iodides, obtained from Aldrich, were purified by filtration through an alumina column (basic pH), stored in glass ampules (shielded from light), and taken through several freeze-pump-thaw cycles prior to dosing onto the copper surface. An in-line alumina plug was used to remove traces of HI. Dosing was achieved by backfilling the chamber, and exposures are reported in Langmuirs (L) where 1 L = 10^{-6} torr·sec. The dosing pressures are uncorrected for differing ion gauge sensitivities.

The surface chemistry of these alkyl iodides was determined using mass spectrometry to monitor the desorbing products and isotope labelling to delineate surface reaction pathways. In temperature-programmed reaction (TPR) experiments, the alkyl iodide-dosed Cu(100) surface was positioned 1 to 2 mm from an aperture (2 mm diameter) leading to the mass spectrometer. The surface was then heated at 2.5 K/sec while recording the desorption rate at a particular mass. Between each of these experiments, the surface was flashed twice to 990 K to remove iodine.

RESULTS AND DISCUSSION

We present and discuss our results in four subsections according to adsorbate: (1) iodomethane, (2) diiodomethane, (3) diiodomethane + iodomethane-d_3, and (4) iodoethane. As mentioned previously, we focus on the ethylene produced by these compounds since in each case a different rate-determining reaction controls ethylene formation.

Iodomethane

When low coverages of iodomethane are adsorbed on Cu(100) at 110 K and the surface is heated, three hydrocarbon products are evolved: methane, ethylene, and propylene. All three are produced with peak temperatures of ~470 K. The low exposure (0 - 1 L) desorption curves for ethylene (m/e = 27) are shown in Figure 1a; similar profiles are found for methane and propylene. At higher coverages, we also observe methyl coupling to form ethane as indicated by the integrated product plots in Figure 1b. The surface chemistry in this higher coverage regime (> 1 L) will be discussed in detail elsewhere [12]. The low coverage methane and ethylene results are consistent with previous studies of bromomethane photolysis on evaporated Cu films on Ru(001) [11]; however, the reaction channels producing propylene and ethane were not previously observed.

The peak shapes in Figure 1a for ethylene evolution as well as the constant value of the peak temperature as a function of dose suggest a first order rate-determining step [13]. The fact that methane evolution also shows apparent first order reaction kinetics at the same temperature implicates bond dissociation (either C-I or C-H) as the rate-determining step.

Carbon-iodine bond dissociation can be ruled out as the rate-determining step on several accounts. First, studies of bromomethane on Cu films [11] show that even after photolytic cleavage of the carbon-bromine bond at 100 K to form adsorbed methyl groups, methane and ethylene are not produced until 460 K, analogous to the results here for iodomethane. Further, in the absence of photolysis, monolayers of bromomethane desorb molecularly from Cu films at 160 K [11]. Since iodomethane presumably has a heat of adsorption similar to bromomethane, the lack of molecular CH_3I desorption implicates carbon-iodine bond scission below 200 K. In fact, preliminary HREELS studies of CH_3I on Cu(100) indicate that the C-I bond is broken at temperatures below 150 K [14]. Furthermore, since we observe no hydrogen evolution in the temperature range where recombinative hydrogen desorption occurs on iodine-covered Cu(100) (310 K to 360 K), we infer that

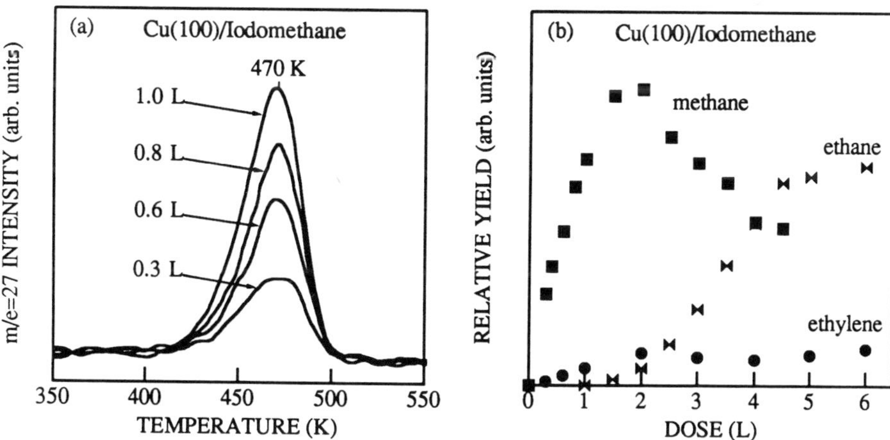

Fig. 1. (a) Rate of ethylene evolution as a function of surface temperature after exposing Cu(100) at 300 K to the indicated submonolayer doses of CH_3I. (b) Relative product yields as a function of CH_3I exposure at 300 K; relative quadrupole sensitivities were determined using mass spectral standards and correcting for differing ionization efficiencies [12].

the methyl group remains intact after C-I bond scission. These results are analogous to those for alkyl iodide dissociation on Ag(111) [9] and Al(111) [7] where C-I bond dissociation also occurs below 200 K to form adsorbed alkyls.

Rate-determining C-H bond scission in the formation of methane and ethylene from iodomethane on Cu(100) is supported by the deuterium isotope effect for this reaction. Specifically, the peak temperature for CD_3I decomposition is 15 K higher than that for CH_3I, corresponding to a k_H/k_D of about 3 at 475 K. This result, in combination with the apparent first-order reaction kinetics for methane and ethylene evolution, suggests that the following coupled surface reactions convert adsorbed methyl groups to methane and ethylene on Cu(100) and that α-elimination is the rate-determining step:

$$CH_3(a) \longrightarrow CH_2(a) + H(a) \qquad \alpha\text{-elimination} \qquad [1]$$

$$CH_3(a) + H(a) \longrightarrow CH_4(g) \qquad \text{reductive elimination} \qquad [2]$$

$$CH_2(a) + CH_2(a) \longrightarrow C_2H_4(g) \qquad \text{methylene coupling} \qquad [3]$$

Note that the reaction sequence above does not fully account for the low-coverage surface chemistry of methyl groups on Cu(100)/iodine since propylene is also produced. The surface reactions responsible for propylene formation are discussed below.

Diiodomethane

To substantiate the mechanism above, we studied the kinetics of the methylene coupling reaction (eqn. 3) utilizing diiodomethane (CH_2I_2) to form the surface-bound methylene species. Figure 2a shows ethylene desorption after diiodomethane adsorption at 140 K on Cu(100); no other C_1 - C_3 hydrocarbons products were detected. Since the ethylene peak temperature of 225 K is 40 K higher than that for desorption of molecularly adsorbed ethylene from Cu(100), we conclude that the ethylene evolution rate is reaction-limited. If we assume that CH_2I_2, like CH_3I, undergoes carbon-iodine bond scission below 200 K to produce CH_2 surface fragments, then the rate-determining step in ethylene formation is either methylene diffusion or recombination. In either case, the rate is orders of magnitude faster than ethylene evolution from CH_3I decomposition, consistent with the mechanistic scheme presented above.

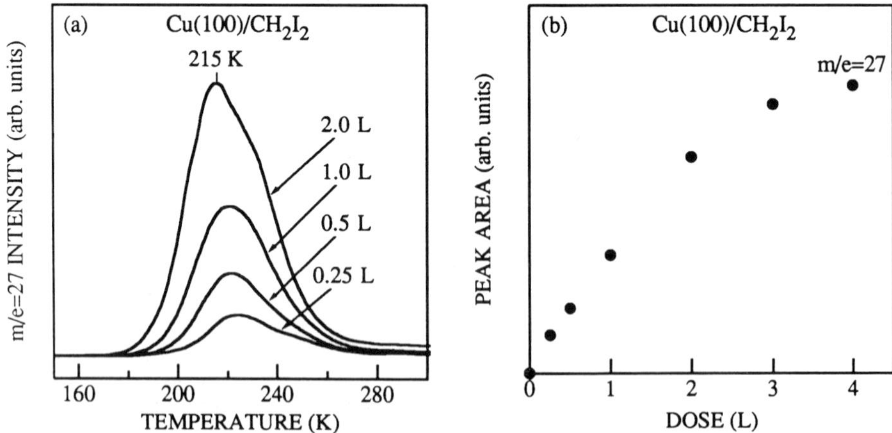

Fig. 2. (a) Ethylene evolution rate as a function of surface temperature after dosing the indicated amounts of CH_2I_2 onto Cu(100) at 120 K. (b) Relative ethylene yield as a function of dose.

Diiodomethane + Iodomethane-d_3

Coadsorption of CH_2I_2 and CD_3I was studied to investigate potential reactions between methylene and methyl species on Cu(100). In the absence of a reaction between CH_2I_2 and CD_3I, one expects a linear combination of the desorption peaks in Figures 1 and 2, that is, C_2H_4 formation from CH_2I_2 at ~215 K followed by CD_4 and C_2D_4 formation from CD_3I at ~470 K. No deuterated products are expected below 470 K. Experimentally, however, we find three deuterated products below this temperature: m/e = 30 (ethylene-d_2), m/e = 45 (propylene-d_3), and m/e = 20 (methane-d_4). Figure 3 shows the ethylene and methane products for a 0.5 L dose of CH_2I_2 followed by 1 L of CD_3I. The partially deuterated propylene desorbs at 275 K, coincident with the low temperature ethylene peak, but this reaction channel does not become substantial until the CD_3I exposure exceeds 1 L.

These experimental observations suggest that CH_2 species insert into CD_3 groups (and CH_2CD_3 groups) followed by β-hydride elimination to evolve ethylene (or propylene). The surface deuterium atoms so-generated can then react with the remaining CD_3 to produce the CD_4 observed at 335 K:

$CD_3(a) + CH_2(a) \longrightarrow CD_3CH_2(a)$		methylene insertion	[4]
$CD_3CH_2(a) \longrightarrow CH_2CD_2(g) + D(a)$		β-hydride elimination	[5]
$CD_3(a) + D(a) \longrightarrow CD_4(g)$		reductive elimination	[6]

In support of this mechanistic scheme, we have investigated reactions 5 and 6 separately. Reaction 6 was studied on a Cu(110) surface by coadsorbing deuterium and iodomethane [15]. The methane-d_1 evolution observed at 330 K is quite consistent with the CD_4 peak at 335 K in Figure 3. Reaction 5 was studied, as described below, using iodoethane as the ethyl precursor.

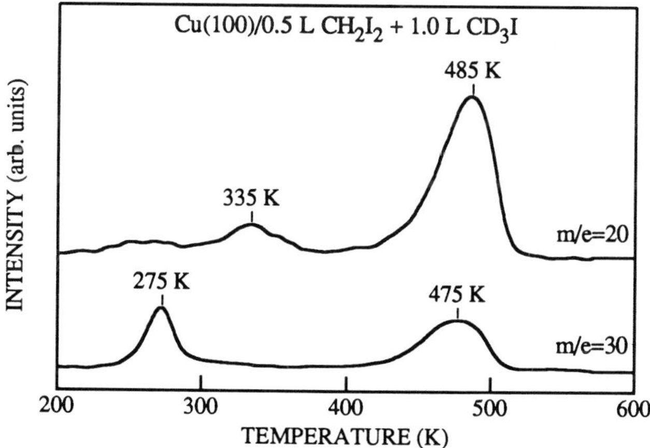

Fig. 3. Ethylene-d_2 (m/e=30) and methane-d_4 (m/e=16) produced upon heating a Cu(100) surface after exposure to 0.5 L of CH_2I_2 followed by 1.0 L of CD_3I at 140 K. The peaks at ~480 K are due to CD_3 disproportionation to ethylene-d_4 and methane-d_4 analogous to the CH_3I chemistry in Fig. 1, while the peaks at 275 K and 335 K implicate insertion of CH_2 into CD_3 as discussed in the text.

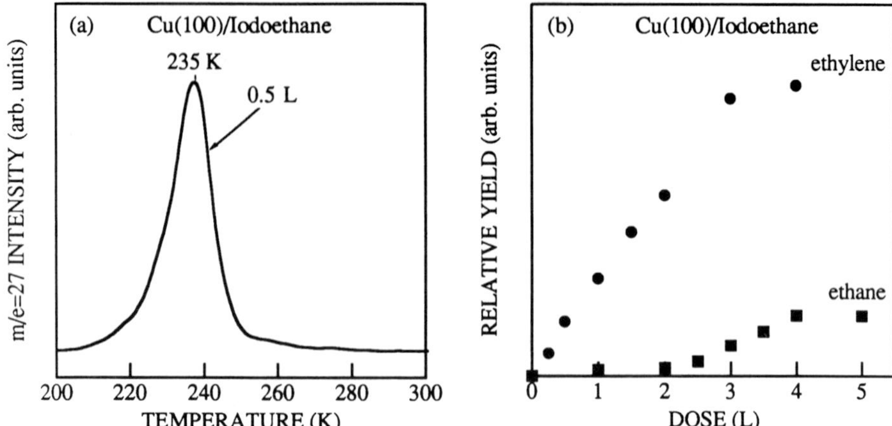

Fig. 4. (a) Ethylene formation after exposing a Cu(100) surface to 0.5 L of C_2H_5I at 120 K. (b) Relative ethylene and ethane yields as a function of C_2H_5I exposure.

Iodoethane

Figure 4a shows ethylene evolution (m/e = 27) after adsorbing 0.5 L of iodoethane onto Cu(100) at 110 K. The ethylene peak temperature of 235 K is higher than that for molecular ethylene desorption from Cu(100) (185 K), indicating that ethylene evolution is reaction-limited. Also, the peak temperature shifts to 250 K in studies with iodoethane-d_5, implicating C-H bond-breaking in the rate-determining step with a k_H/k_D of about 3 at 300 K. We have confirmed that alkene production from ethyl and longer alkyls occurs by C_β-H bond cleavage by studying propylene formation from 1-iodopropane-2,2-d_2 and finding that the sole alkene product is propylene-d_1 [15]. At higher exposures, ethane is also produced (peak temperature = 235 K) as shown by the integrated product curves in Figure 4b.

Surface Reaction Summary

Figure 5 plots the peak temperatures (T_p) for all the hydrocarbon reactions we observe on Cu(100)/iodine. The temperature intervals reflect the range of T_p as a function of coverage. For example, the peak temperature for β-hydride elimination increases from 235 K to 250 K over the range of 0.5 - 4 L, while that for methyl coupling shifts down from 470 K to 390 K as expected for a second order surface process [13].

These surface coverage effects may be particularly significant in the case of CH_3I decomposition. Specifically, while Fig. 5 shows CH_2 coupling at ~215 K and CH_2 insertion at ~275 K, these peak temperatures are for fractional surface coverages in the range of 0.1 - 1. Computer modelling studies show that during CH_3I decomposition, steady state CH_2 coverages are extremely small (fractional coverage < 0.001). Under these conditions, methylene coupling (eqn. 3) may be preempted by methylene insertion followed by β-hydride elimination (eqns. 5 and 6) as the ethylene formation pathway. In either case the net effect is the same, the only difference being whether the reaction sequence is hydrogen abstraction (α-elimination) followed by coupling or coupling (methylene insertion) followed hydrogen abstraction (β-hydride elimination).

We conclude by emphasizing several additional implications of the relative reaction rates tabulated in Figure 5. For example, it is evident that when comparable surface coverages of CH_3I and H are coadsorbed on Cu(100), reductive elimination to give methane is favored over either

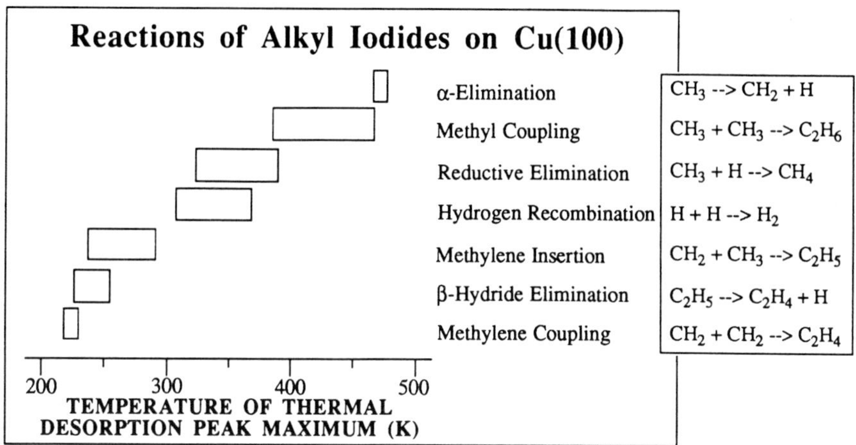

Fig. 5. Reaction pathways observed for alkyl fragments generated on Cu(100) by the dissociative adsorption of alkyl iodides. The temperature ranges given indicate the range of peak temperatures observed in TPR experiments as a function of surface coverage.

methyl group coupling or hydrogen atom recombination. Second, the methylene insertion rate is competitive with methylene coupling, and both of these reactions are much more facile than methyl coupling. Finally, the rates of α- and β- C-H bond cleavage are enormously different, the ratio being 10^9 at 300 K assuming equal preexponential factors of 10^{13} sec^{-1} [15]. Studies using molecular beam scattering and surface vibrational spectroscopy are in progress to gain further insight into the factors controlling these surface reactions.

CONCLUSIONS

We have utilized the dissociative adsorption of alkyl iodides to generate and study the chemistry of alkyl fragments on a Cu(100) surface. Our results show that monolayers of iodomethane, iodoethane, and diiodomethane on Cu(100) all decompose to evolve (among other hydrocarbon products) ethylene, but at significantly different temperatures in each case, implicating different surface reaction pathways. Surface reaction kinetics and isotope labelling studies show that β-hydride elimination at ~235 K converts iodoethane to ethylene, methylene coupling at ~215 K converts diiodomethane to ethylene, and α-elimination from adsorbed methyl groups limits ethylene formation from iodomethane to temperatures above 400 K. In addition, we find that CH_2I_2 reacts with coadsorbed CD_3I to form ethylene-d_2 and propylene-d_3 at ~275 K, implicating a mechanism involving methylene insertion into surface alkyl groups followed by β-hydride elimination to form the alkene. To our knowledge, this is the first evidence from surface chemistry studies on well-defined single crystal surfaces for the methylene insertion chain growth step proposed for Fischer-Tropsch synthesis.

ACKNOWLEDGEMENTS

Research support from the American Chemical Society (Petroleum Research Fund) and the National Science Foundation (Presidential Young Investigator Award to B.E.B.) is gratefully acknowledged. C.-M.C. thanks IBM for a predoctoral fellowship.

REFERENCES

1 (a) F. G. Gault, Gazz. Chim. Ital., 109 (1979) 255-269.
(b) J.J. Rooney, J. Molec. Catal., 31 (1985) 147-159.
(c) C. K. Rofer-DePoorter, Chem. Rev., 81 (1981) 447-474.

2 (a) X.-L. Zhou and J.M. White, Chem. Phys. Lett., 142 (1987) 376-380.
(b) X.-L. Zhou and J.M. White, Surf. Sci., 194 (1988) 438-456.

3 (a) M.B. Lee, Q.Y. Yang, S.L. Tang, and S.T. Ceyer, J. Chem. Phys., 85 (1986) 1693-1694.
(b) M.B. Lee, Q.Y. Yang, and S.T. Ceyer, J. Chem. Phys., 87 (1989) 2724-2741.
(c) S.T. Ceyer, Ann. Rev. Phys. Chem., 39 (1988) 479-510.

4 M.A. Henderson, G.E. Mitchell, and J.M. White, Surf. Sci., 184 (1987) L325-L331.

5 (a) K.G. Lloyd, A. Campion, and J.M. White, Catal. Lett., 2 (1989) 105-112.
(b) F. Zaera, Surf. Sci., 219 (1989) 453-466.
(c) F. Zaera, J. Am. Chem. Soc., 111 (1989) 8744-8745.

6 J.G. Chen, T.P. Beebe, Jr., J.E. Crowell, and J.T. Yates, Jr., J. Am. Chem. Soc., 109 (1987) 1726-1729.

7 B.E. Bent, R.G. Nuzzo, and L.H. Dubois, J. Am. Chem. Soc., 111 (1989) 1634-1644.

8 B.E. Bent, R.G. Nuzzo, and L.H. Dubois, submitted for publication.

9 X.-L. Zhou, F. Solymosi, P.M. Blass, K.C. Cannon, and J.M. White, Surf. Sci., 219 (1989) 294-316.

10 X.-L. Zhou and J.M. White, Catal. Lett., 2 (1989) 375-384.

11 B. Roop, Y. Zhou, Z.-M. Liu, M.A. Henderson, K.G. Lloyd, A. Campion, and J.M. White, J. Vac. Sci. Technol., A7 (1989) 2121-2124.

12 C.-M.Chiang, J.-L. Lin, C.J. Jenks, L. Smoliar, and B.E. Bent, in preparation.

13 P.A. Redhead, Vacuum, 12 (1962) 203-211.

14 J.-L. Lin and B.E. Bent, in preparation.

15 C.J. Jenks, C.-M. Chiang, and B.E. Bent, in preparation.

R.K. Grasselli and A.W. Sleight (Editors), *Structure-Activity and Selectivity Relationships in Heterogeneous Catalysis*
© 1991 Elsevier Science Publishers B.V., Amsterdam

EFFECTS OF HALOGENS ON OXIDATION REACTIONS OVER SINGLE CRYSTALS OF PALLADIUM

K. KLIER, G.W. SIMMONS, Y.-N. WANG AND J.A. MARCOS

Zettlemoyer Center for Surface Studies and Department of Chemistry
Sinclair Lab. #7, Lehigh University, Bethlehem, PA 18015, U.S.A.

ABSTRACT

Dissociative chemisorption of dichloromethane occurs on a Pd(100) surface at room temperature with C, Cl, and H adsorbing into 4-fold sites while H tends to diffuse underneath the surface layer with increasing CH_2Cl_2 exposures. The CH_2Cl_2-precovered surface is very reactive toward oxygen in the 350-510K temperature range. The selectivity of oxidation of the carbon fragments to CO increases with CH_2Cl_2 preexposures, i.e. Cl-C-Cl coverges. A Monte Carlo simulation of the oxidation of surface carbon by surface oxygen atoms resulted in a semiquantitative account of the experimentally observed rates of production of CO and CO_2 as a function of Cl coverage. The model explains selective oxidation to CO_2 on Cl-free palladium and to CO on partially Cl-covered surfaces in terms of restricted supply of oxygen to the carbon fragments on the surface (ensemble control).

INTRODUCTION

The introduction of trace amounts of gaseous CH_2Cl_2 had been reported to selectively promote the oxidation of methane over palladium catalysts to form formaldehyde, in addition to CO_2 and water (1,2). It is therefore of interest to determine how the dichloromethane molecules react with the palladium (100) surface and how fragments produced by their chemisorption influence the adsorption and reactivity of the reacting components, i.e. oxygen and methane. In the current study, emphasis was placed on the reactivity of surface oxygen and carbon in the presence of the chlorine promoter.

EXPERIMENTAL

The experiments were performed in ultra-high vacuum (UHV) systems equipped with Auger electron spectroscopy (AES), low energy electron diffraction (LEED), mass spectrometry/temperature programmed desorption (MS/TPD), and high resolution electron energy loss spectroscopy (HREELS) techniques and with base pressures of $\leq 2x10^{-10}$ torr. The instrumentation and Pd(100) single crystal cleaning procedures have been described elsewhere (3).

The extent of Cl_2 or CH_2Cl_2 chemisorption for a given exposure was measured by following the Auger peak-to-peak height ratio of Cl(176 eV) to that of Pd(323 eV). An auxiliary experiment with chlorine gas adsorption was carried out to calibrate the chlorine overlayers during CH_2Cl_2 adsorption. The chlorine

AES peaks were calibrated on the assumption that the well-ordered c(2x2) LEED pattern obtained after room temperature saturation of the surface with chlorine gas corresponded to a surface coverage θ_{Cl} = 0.50 ML (monolayer) (4).

For the sequential adsorption of oxygen and CH_2Cl_2, the Pd(100) surface was exposed to CH_2Cl_2 at room temperature by backfilling the UHV chamber with CH_2Cl_2 via a variable leak valve. Oxygen, however, was directed onto the surface via a multicapillary beam doser. The surface coverages of oxygen were determined by calibrating the Auger peak-to-peak height ratios of O(507 eV)/Pd(323 eV) and the areas under TPD curves against that of room temperature saturation with oxygen coverage of 0.50 ML and a well-ordered c(2x2) LEED pattern (3).

RESULTS AND DISCUSSION

Dichloromethane adsorbed dissociatively on Pd(100) surface at room temperature as revealed by HREELS spectra. Loss peaks were observed at 190, 225, and 475 cm^{-1} after exposures of ≤0.5 L Langmuir) suggesting that C, Cl and H species, respectively, adsorb into the four-fold sites. For higher exposures, i.e. ≥19 L, however, the 475 cm^{-1} peak was not seen indicating that H atoms may diffuse into the bulk of the Pd crystal with increasing CH_2Cl_2 exposure. The CH_2Cl_2 adsorption proceeded with a near unity initial sticking coefficient and reached saturation with Cl coverage of 0.22 ML. At saturation, only faint streaks in the position of (±1/2,±1/2) were observed by LEED indicating a disordered overlayer from the random distribution of the dissociated CH_2Cl_2 fragments.

From the coadsorption studies, it was found that the oxygen precovered surface with p(2x2) or c(2x2) structure completely inhibited CH_2Cl_2 adsorption while oxygen uptake on CH_2Cl_2-precovered surface was detected by both AES and HREELS. A LEED pattern with mixed c(2x2) and p(2x2) structure was observed for all of the CH_2Cl_2 precovered surfaces after being exposed to 90 L oxygen at room temperature. These results suggest that oxygen adsorption requires a smaller area of free metal sites than that of CH_2Cl_2 and oxygen can form well-ordered c(2x2) overlayers in the area between CH_2Cl_2 fragments.

Thermal desorption spectra obtained immediately after oxygen adsorption on the CH_2Cl_2-precovered surfaces gave rise to CO_2 and CO at 350-510K in addition to O_2 and Cl signals, shown in Fig. 1, indicating that the carbon fragments were very reactive toward O_2. Atomic chlorine (m/e=35) desorption peaked at 1100K and was observed in all cases corresponding to an activation energy of desorption of ca. 68 kcal/mol. HCl, H_2O, HCHO, and CH_3OH were surveyed during the TPD measurements but not found. The relative amounts of CO and CO_2 formed by surface oxidation have been calculated as a function of carbon coverage arising from the CH_2Cl_2 adsorption. The TPD spectra of CO were calibrated using results from a CO saturated Pd(100) surface corresponding to 0.50 ML, i.e. $(2\sqrt{2}x\sqrt{2})R45°$ LEED pattern (5-8). The amounts of desorbed CO_2, equivalent to the coverages of

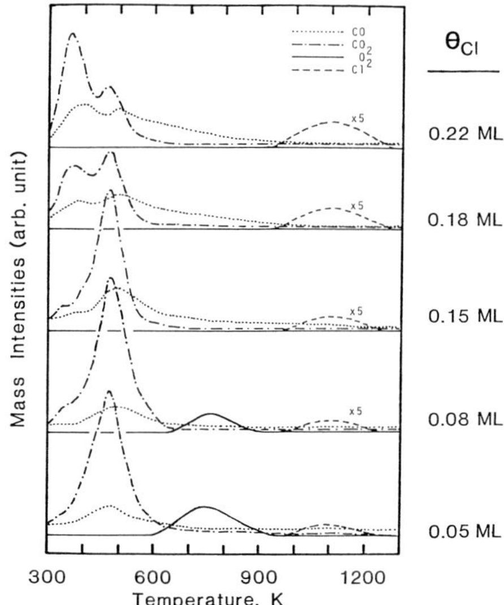

Fig. 1. Thermal desorption spectra after the first dose of 90 L of oxygen on CH_2Cl_2 -precovered Pd(100) surfaces. Note that CO, CO_2, and most of the oxygen could be desorbed, while leaving the surface chlorine, if the temperature ramp were stopped at 900K.

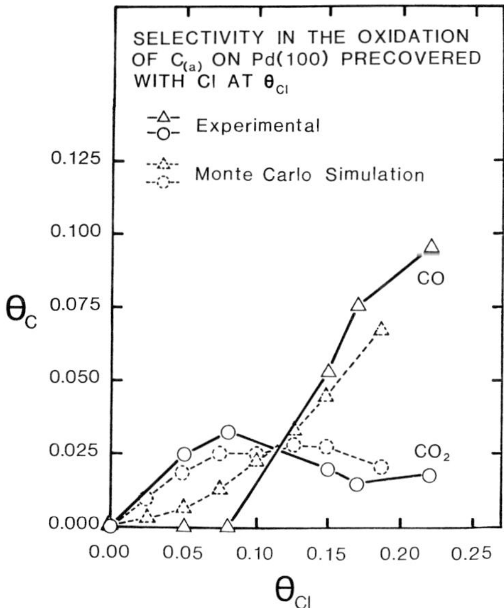

Fig. 2. Experimental and simulated CO and CO_2 selectivities over Pd(100) in terms of carbon and chlorine coverages. The experimental distributions were derived from thermal desorption yields obtained after the first oxygen dose on the CH_2Cl_2-precovered surface. The simulated distributions were obtained via the Monte Carlo method by statistically counting the oxygen atoms that reside adjacent to each carbon atom.

surface carbon species that were oxidized and evolved as CO_2, was calculated based on the assumption that the amount of CO_2 evolved from the 1.0 L CH_2Cl_2 preexposed surface was equal to the total amount of adsorbed carbon species, i.e. 0.025 ML. Corrections were made in the calibration of the CO and CO_2 distributions to account for (i) the fragmentation pattern of CO_2 yielding 12% of mass 28 and (ii) the falling baseline in the desorption spectra due to the recovery of the pumping speed during each thermal desorption run. The amounts of CO and CO_2 presented in terms of carbon coverages and chlorine coverages are shown in Fig. 2 (the solid lines), where selectivity to CO is seen to increase with increasing Cl coverage. The reaction between the carbon species, which were generated by dissociative adsorption of CH_2Cl_2, and oxygen may be of significance for catalysis as CO is produced with increasing selectivity at the expense of the CO_2 production upon increasing preadsorption of CH_2Cl_2.

Monte Carlo simulations and kinematic diffraction analysis (9) were carried out to test various proposed adsorption and the reaction mechanisms. Only C and Cl fragments were used in the model development because of the experimental evidence that hydrogen is either displaced or diffused into the bulk of Pd crystal. A 50x50 square array was randomly filled with various adsorption geometries of CCl_2 groupings. To consider a structure acceptable, the main features of the LEED patterns, namely the presence of the weak spots and streaks of the CCl_2 overlayers, the saturation chlorine coverage, and the extent of oxygen coverage on the CCl_2- precovered surface, had to agree with the experimental results. It was found that a random distribution of Cl-C-Cl fragments, having Cl and C adsorbed at next nearest neighboring (NNN) sites with 90° Cl-C-Cl angle, that excludes all the nearest neighbor (NN) adsorption and Cl-Cl NNN adsorption possibilities gave the most satisfactory fit to the experimental observations. The maximum coverage of chlorine obtained by randomly saturating the 50x50 array by the CCl_2 fragments was ≈0.19 ML compared with the experimental value of 0.22 ML. The kinematic calculation of this CCl_2 saturated surface and the postsaturation with oxygen atoms exhibited structures with a broad and fuzzy spot at (±1/2,±1/2) and a structure of strong c(2x2) + weak p(2x2), respectively, in reasonable agreement with the LEED observations.

In another series of Monte Carlo simulations, the 50x50 square array was randomly filled with the proposed CCl_2 groupings for different surface concentrations (0.025, 0.050, 0.075, 0.100, 0.125, 0.150, and 0.185 ML in terms of Cl coverage). The residual surface was then filled with oxygen atoms into an ordered c(2x2) structure. In a specific version of the model presented here, it was assumed that C and Cl atoms are immobile while oxygen atoms are mobile and may move along [01] and [10] directions on the Pd(100) surface. The rules for the CO and CO_2 formation were made on the basis of the oxygen distribution around each carbon atom: (i) If there were two or more than two oxygen atoms simultaneously present at the third nearest neighboring (3NN) sites of a carbon atom, then a CO_2 molecule was formed; (ii) If there was one or no oxygen atom present at the 3NN sites of a carbon atom, then a CO molecule was formed; (iii) If there were two oxygen atoms simultaneously present at the 3NN sites of two carbon atoms, then two CO molecules were formed; and (iv) The maximum number of oxygen atoms that could be attached to a carbon atom is two while if there were three oxygen atoms simultaneously present at the 3NN sites of two carbon atoms, then both CO and a CO_2 molecules were formed, and so on.

The resulting distributions of CO and CO_2 formed according to the above rules 1-4 from the simulated O/CCl_2 overlayers appear to be in a favorable agreement with experiment, shown in Fig. 2. This indicates that the random distribution of CCl_2 fragments, and thereby Cl atoms, does provide for the control of the access of oxygen to the surface carbon and thus directs the reactions in favor of the CO product. The controlling characteristic of carbon oxidation by chlorine is believed to be an ensemble size effect and the ensemble-controlled reaction pattern is therefore expected to be of catalytic significance in the partial oxidation process for hydrocarbons.

A full account of this work will be published in the Journal of Physical Chemistry, in press. This research was partially supported by U.S. Department of Energy Contract DE-FG02-86ER13580.

REFERENCES

(1) C.F. Cullis, D.E. Keene, and D.L. Trimm, J. Catal., **19** (1970) 378.
(2) R.S. Mann and M.K. Dosi, J. Chem. Technol. Biotechnol., **29** (1979) 467.
(3) G.W.Simmons, Y.-N. Wang, J.A. Marcos, and K. Klier (submitted to Surface Science for publication).
(4) Y.-N. Wang, J.A. Marcos, G.W. Simmons and K. Klier, J. Phys. Chem., **94** (1990) in press.
(5) J.C. Tracy and P.W. Palmberg, J. Chem. Phys., **51** (1969) 4852.
(6) A.M. Bradshaw and F.M. Hoffmann, Surf. Sci., **72** (1978) 513.
(7) R.J. Behm, K. Cristman, G. Ertl, and M.A. VanHove, J. Chem. Phys., **73** (1980) 2984.
(8) E.M. Stuve, R.J. Madix, and C.R. Brundle, Surf. Sci., **146** (1984) 155.
(9) G. Ertl and J. Kuppers, Surf. Sci., **21** (1970) 61.

R.K. Grasselli and A.W. Sleight (Editors), *Structure-Activity and Selectivity Relationships in Heterogeneous Catalysis*
© 1991 Elsevier Science Publishers B.V., Amsterdam

SMALL MOLECULE REACTIONS ON CLEAN AND MODIFIED IRON SURFACES

S.L. BERNASEK, J.-P. LU, M.R. ALBERT, and W.-S. HUNG

Department of Chemistry, Princeton University, Princeton, New Jersey 08544 USA

ABSTRACT

The connection between structure and reactivity in heterogeneous reactions can be best studied using model substrates whose structure can be characterized and controlled. Several examples of this approach to the understanding of structure-reactivity relationships in heterogeneous catalysis are presented, based on recent investigations of the Fe(100) surface. This surface exhibits a rich and complex chemistry, illustrated here by discussions of the adsorption and reaction of CO, O_2, H_2O, CH_3OH and CH_3SH on the clean and modified Fe(100) surface. An approach to the preparation of characterizable mixed metal oxide substrates based on the hydroxylated Fe(100) surface is also presented.

1. INTRODUCTION

Direct experimental information about the connection between structure and reactivity in heterogeneous reactions can best be obtained by using simplified model systems where the detailed microscopic structure of the solid substrate can be determined. A series of small molecule reaction studies which take advantage of this model system approach are described here. The model substrate common to this series of studies is the well characterized single crystal Fe(100) surface. In spite of the difficulty of working with iron surfaces, the Fe(100) surface has displayed a rich chemistry which provides a number of interesting examples of the connection between structure and reactivity in heterogeneous systems.

Iron, of course, has significant technological importance (beyond its inherently interesting chemistry) as a structural material and as an important catalytic surface. Its use in Fischer-Tropsch catalysis (ref. 1), in coal liquefaction catalysis (ref. 2), and in ammonia synthesis (ref. 3), all justify the intense study of iron surface chemistry and catalysis which has occurred over the past fifty years. Even though there has been intense study of iron as a supported catalyst, there have been relatively fewer investigations of small molecule adsorption and reactivity on the well characterized single crystal faces of iron. Its reactivity, and the subsequent difficulty of preparing a clean, well ordered single crystal surface, has restricted the number of investigations which have provided detailed structure reactivity information on the iron substrate.

In the following pages, studies of the adsorption and reaction of CO, O_2, H_2O, CH_3OH, CH_3SH, and an organometallic complex, tetrakis allyl molybdenum, on the Fe(100) surface will be described. Interaction with the clean Fe(100) surface will be considered first. Then the effect of sulfur, oxygen, and water modification of the Fe(100) surface on small molecule reactions will be described. The connection between detailed structure and observed reactivity will be emphasized throughout.

2. Experimental Procedures

The studies described here use a multiple technique approach to detail the structure and reactivity of the systems under study. Single crystal samples of high purity iron are oriented using Laue back reflection X-ray diffraction, and cut with a diamond wafering saw to expose the (100) plane. This surface is mechanically polished using successively finer abrasives, finishing with a 0.05μ Al_2O_3 slurry and a brief etch with dilute $FeCl_3$ solution. The single crystal disk, about 10 mm in diameter and 1 mm thick, is then heated in a tube furnace under a flow of 1 atm pure H_2 at about 600°C for several weeks. This procedure is effective in removing bulk sulfur from the sample, and saves considerable effort in cleaning the sample once it is mounted in the ultra high vacuum (UHV) system.

The single crystal sample is then mounted in a UHV system equipped with several surface spectroscopic probes (Figure 1). The work described here was

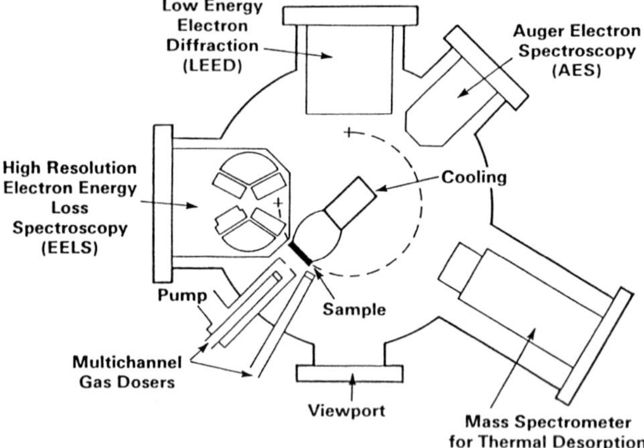

Fig. 1. Schematic diagram of typical ultra high vacuum apparatus used for this work.

carried out in several separate UHV systems, all equipped with Auger electron spectroscopy (AES), low energy electron diffraction (LEED), and a quadrupole mass spectrometer for thermal desorption measurements (TDS). In addition, various other probes were available in the different systems, primarily high resolution electron loss spectroscopy (HREELS), X-ray photoelectron spectroscopy (XPS), and in one case a synchrotron source for near edge X-ray absorption fine structure measurements (NEXAFS). Each system was also equipped with facilities for heating and cooling the sample and an ion gun for sputter cleaning. The UHV mounted sample was cleaned by cycles of high temperature ion bombardment and annealing until a contaminant free Auger spectrum, a sharp, ordered (1x1) LEED pattern, and a loss free HREELS spectrum were obtained (see Figure 2 (a)).

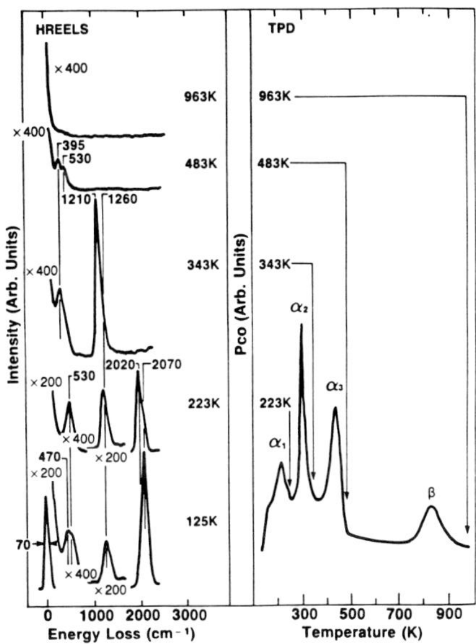

Fig. 2. HREELS data correlated with thermal desorption of CO from the CO saturated Fe(100) surface (reference 6).

3. RESULTS - CLEAN SURFACE

The adsorption and reaction of even simple small molecules on the clean Fe(100) surface can be surprisingly complex. This is well illustrated by the adsorption of CO on the Fe(100) surface (ref. 4-8). Figure 2 summarizes TDS and HREELS data for a saturation coverage of CO on the initially clean Fe(100) surface. Three molecular adsorption states, and one dissociative adsorption state, are evident from the data shown in Figure 2. Subsequent studies have shown that the state labelled α_2 is in reality two separate adstates distinguished by their adjacency to an occupied or unoccupied fourfold hollow site on this surface (ref. 7).

The most striking feature of this adsorption system is the state labelled α_3. As seen from the HREELS data of Figure 2, this is a highly perturbed molecular state with a C-O stretching frequency of ~ 1200 cm^{-1}. This state has been shown by coverage dependence, XPS, and isotopic labelling measurements to be the precursor to the dissociative (β) state (ref. 4,5,7). Its structure, identified by NEXAFS (ref. 8) and coadsorption measurements, is illustrated in Figure 3, along with structures proposed for the α_1 and α_2 states. This study

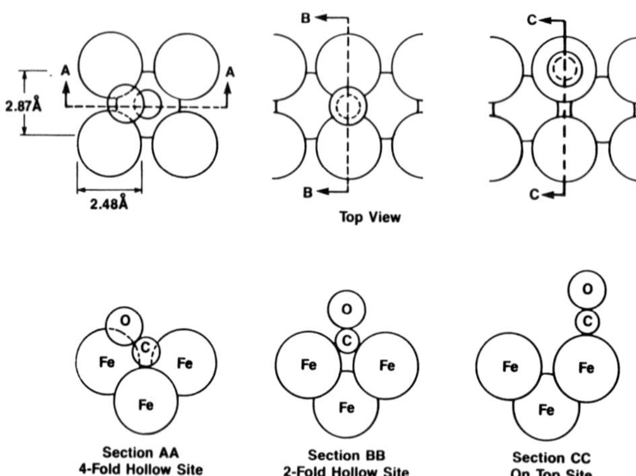

Fig. 3. Top and cross sectional views of a) CO(α_3) b) CO(α_2) and c) CO(α_1) on the Fe(100) surface (reference 6).

has identified this fourfold hollow site on the Fe(100) surface as being particularly interesting, with a rich and complex chemistry dependent on this relatively open structural site.

This theme continues with the adsorption of O_2 on this surface (ref. 9). Even at low temperature, disordered dissociative adsorption occurs on the Fe(100) surface, by way of a mobile molecular precursor. Fourfold hollow sites are occupied preferentially at low coverage, followed by bridging site occupation at higher coverage. These atomic adsorption states are identified by their characteristic HREELS spectra (Figure 4). At still higher coverages, a vibrational band near 650 cm^{-1} is observed which may be indicative of O-O bonding in a highly perturbed non-dissociated adsorbed species. Annealing the low temperature disordered overlayer to 923 K results in a well ordered p(1x1) oxygen structure with a sharp loss spectra and phonon structure indicative of oxide formation. An intermediate, complex structure is observed during the annealing process. This complex chemistry strongly affects the adsorption and decomposition of molecules such as methanol, as will be seen below.

The adsorption and subsequent decomposition of water on the clean Fe(100) surface also exhibits a complex and interesting chemistry (ref. 10). In this case, exposure of the surface held at low temperature to water results in

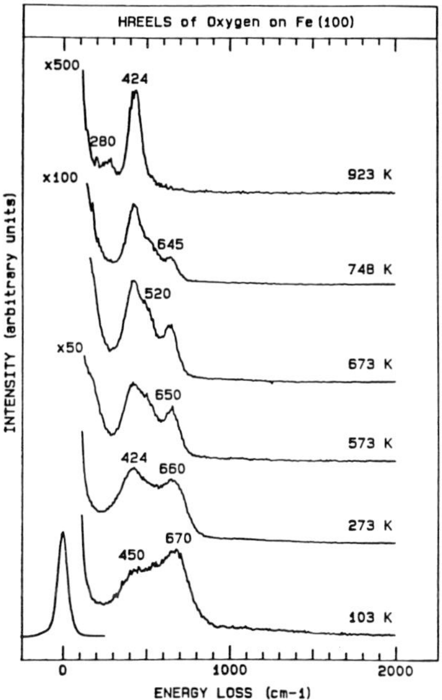

Fig. 4. HREELS for Fe(100) exposed to 10L O_2 at 103 K, and heated to the indicated temperature (reference 9).

multilayer formation and evidence for water clustering even at very low exposures. As the surface is heated (Figure 5), TDS and HREELS data suggest the formation of monolayer molecular adsorption states. Above 222 K, water

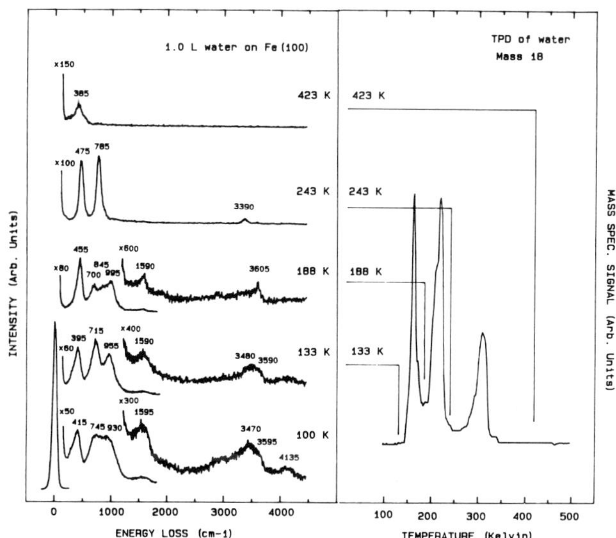

Fig. 5. HREELS data correlated with thermal desorption of H_2O from the H_2O saturated Fe(100) surface.

dissociates and an ordered hydroxylated surface is formed. This surface is characterized by a well ordered p(1x2) LEED pattern, and a HREELS spectrum with intense OH stretching and bending (~ 3300 and ~ 800 cm^{-1}) bands. Again the fourfold hollow site is the likely adsorption site for the OH species although this has not been confirmed. Based on relative HREELS intensities the OH species appears to be bent. The reactivity of this surface in interactions with molybdenum organometallic species will be discussed in the following section.

As a final example of the clean surface reactivity of the Fe(100) surface, the decomposition of methanol[11] and methanethiol (ref. 12) will be considered. The combination of HREELS, TDS and XPS has been especially effective in detailing and contrasting the differences in decomposition mechanism for these molecules on the Fe(100) surface. Thermal desorption data indicates that methanol decomposes to give CO and H_2 in a desorption limited step at 440 K. Methanethiol, however, decomposes to produce methane and H_2, leaving sulfur behind adsorbed in the fourfold hollow site. Isotope substitution measurements indicate that methyl C-H bond cleavage is the rate determining step leading to product formation in methanol decomposition, while product formation occurs via adsorbed methyl in the case of CH_3SH decomposition.

HREELS data indicates further subtle differences in the decomposition mechanisms (Figure 6). In both cases, multilayers of hydrogen bonded molecular ices are formed, with characteristic OH and SH stretching bands observed in the HREELS spectra. As the surface is warmed, desorption of intact molecules takes

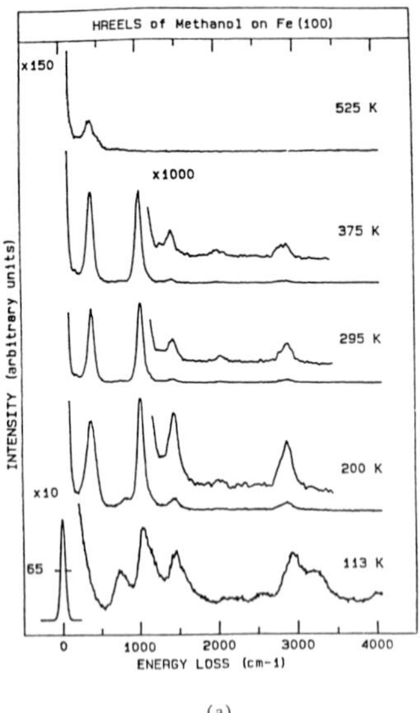

(a)

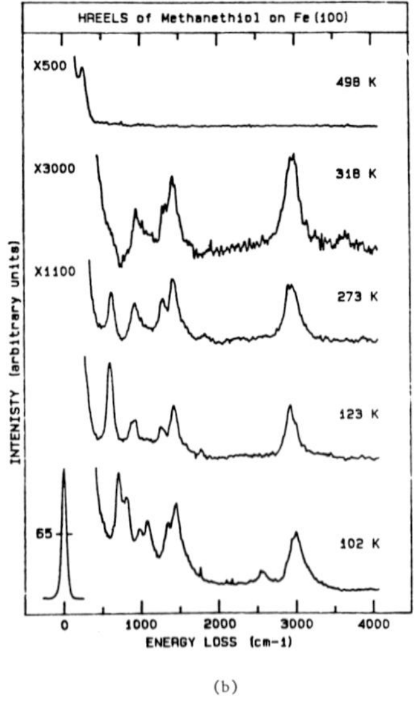

(b)

Fig. 6. HREEL spectra of a saturated overlayer of methanol (a) and methanethiol (b) adsorbed at low temperature on Fe(100) and subsequently heated to higher temperatures (references 11 and 12).

place, and SH and OH bond scission occurs. Analogous intermediates, methoxy($-OCH_3$) and thiomethoxy($-SCH_3$), are formed and appear to be bound normal to the surface. As the surface temperature is increased further, the methoxy intermediate is seen to be stable up to 443 K. At this temperature, methyl C-H bond breaking occurs and CO and H_2 desorb from the surface. The thiomethoxy intermediate appears to undergo C-S bond scission at around 250 K, forming an adsorbed methyl species which recombines with adsorbed H and desorbs as methane around 330 K. The adsorbed sulfur left behind in this decomposition reaction appear to stabilize the adsorbed methyl species resulting in methane desorption rather than further CH bond cleavage as has been observed for other substrates.

4. RESULTS - MODIFIED SURFACE

Adsorption and reaction chemistry on the Fe(100) surface is strongly affected by structural and compositional modification of the surface with preadsorbed species. Three examples of this behavior will be discussed; 1) the effect of sulfur on CO adsorption and dissociation, 2) the effect of oxygen on CH_3OH adsorption and decomposition, and 3) the reaction of tetrakis allyl molybdenum with the ordered hydroxylated iron surface.

Changes in CO adsorption behavior can be observed for the Fe(100) surface modified by pre-exposure to sulfur (ref. 5), or for the CO saturated surface post-exposed to a sulfur containing molecule such as methanethiol (ref. 13). In the case of surface modification by sulfur pre-exposure, HREELS and TPD evidence (Figure 7) suggest that sulfur has a local structural effect on the CO (α_3)

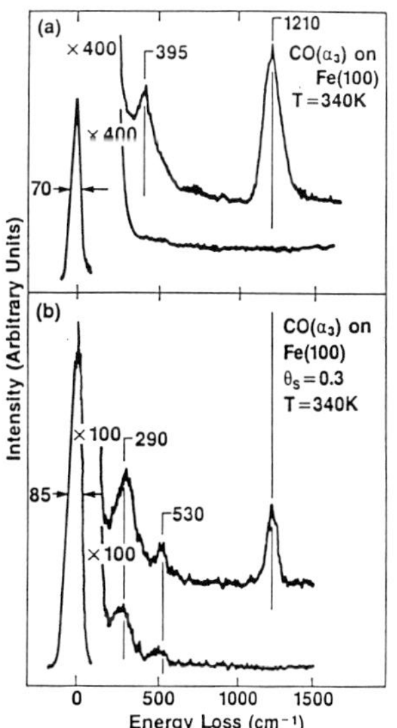

Fig. 7. HREEL spectra for (a) CO(α_3) at 340 K and (b) for CO(α_3) on presulfided (θ_S = 0.3) Fe(100) surface (reference 6).

adstate, blocking the fourfold hollow site for adsorption. Low coverages of sulfur block the CO dissociation process and prevent α_3 adsorption. The CO stretching frequency of the α_3 state is not changed at intermediate sulfur coverages (Figure 7), suggesting that the electronic nature of the Fe(100) surface is not strongly modified by sulfur pre-exposure.

Saturation of the Fe(100) surface with CO followed by heating to 383 K prepares a c(2x2)CO overlayer consisting of only α_3 adsorbed molecules. When this surface is post-exposed to methanethiol, competition for the fourfold hollow site occurs and the α_3 molecule is displaced into a new adstate with a lower desorption temperature and a higher C-O stretching frequency (Figure 8).

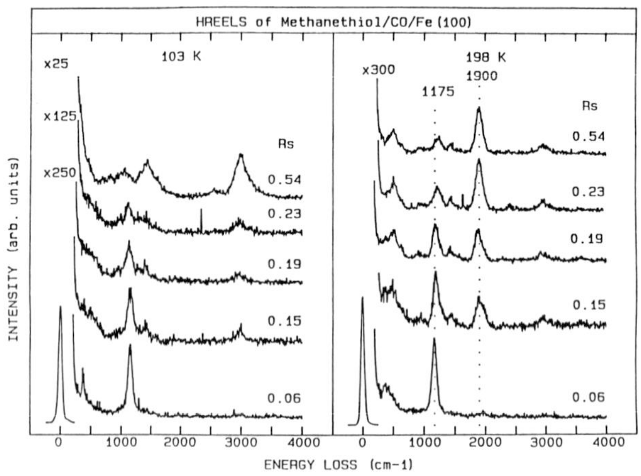

Fig. 8. HREEL spectra of $CH_3SH/CO/Fe(100)$ overlayers (left panel) and subsequently heated to 198 K (right panel) (reference 13).

Coverage dependence and temperature dependence indicate that this is again a localized effect; the CO(α_3) state is displaced in a thermally activated process without significant perturbation of the α_3 molecule. Displacement out of the α_3 state results in a decrease in dissociation of CO, as the α_3 site appears to be essential as a precursor to dissociation.

Oxygen modification of the Fe(100) surface has a profound effect on methanol decomposition chemistry (ref. 14,15). Oxygen modification affects the decomposition pathway so that formaldehyde becomes a dominant reaction product at the expense of CO and H_2 production. This behavior is illustrated for both the low and high temperature oxygen overlayers in Figure 9. As a function of oxygen coverage, it can be seen that formaldehyde production peaks as CO and H_2 production falls off on the high temperature oxygen modified surface. Above this peak H_2CO production coverage, the surface is deactivated to methanol decomposition. For the low temperature modified surface it appears that formaldehyde production continues to increase as long as the coverage of 650 cm^{-1} oxygen continues to increase.

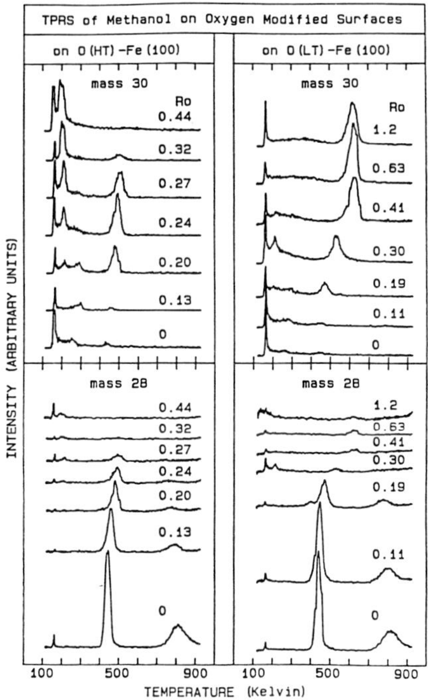

Fig. 9. Comparison of TPD data for methanol decomposition on high temperature (left) and low temperature (right) oxygen modified Fe(100) surfaces (ref. 15).

HREELS data for the methanol decomposition reaction on the oxygen modified surfaces indicates that the decomposition again proceeds via a methoxy intermediate, as was seen on the clean Fe(100) surface. However, this methoxy species is much more stable on the oxygen modified surface, with its decomposition temperature increasing proportionally to the coverage of oxygen on the surface. Isotope labelling studies again identify C-H bond scission as the rate determining step in the decomposition reaction. No evidence for a stable formaldehyde species present on the surface is obtained from the HREELS spectra.

A somewhat more complex example of a reaction on a modified iron surface is the recent investigation of molybdenum alkyl adsorption on the hydroxylated Fe(100) surface (ref. 16). This well ordered overlayer, discussed above (Figure 5), was exposed to tetrakis allyl molybdenum vapor in the ultra high vacuum chamber. The resulting reaction was monitored using HREELS, as illustrated in Figure 10.

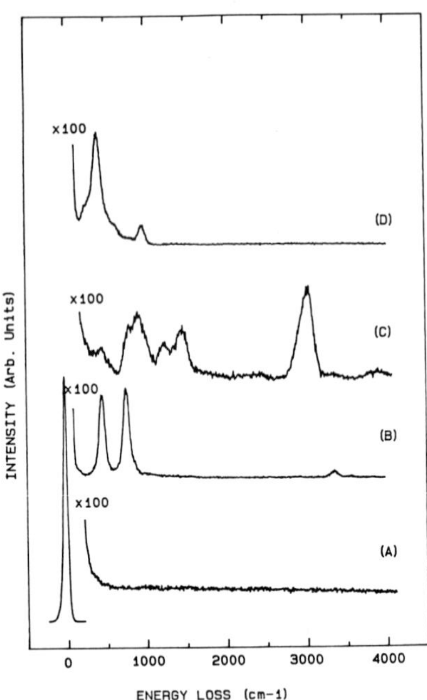

Fig. 10. HREEL spectra illustrating formation of ordered mixed oxide substrate. Traces keyed to scheme in text. (A) Clean Fe(100). (B) p(2x1)-OH structure on Fe(100). (C) Surface of (B) after exposure to tetrakis allyl molybdenum. (D) Fe-O-Mo dioxo surface obtained after reduction and oxidation of (C).

The expected chemistry is illustrated in the following scheme.

Clean Fe(100) (A) —[H_2O, 105K]→ H_2O multilayer —[243K]→ p(2x1) -OH (B) —[Mo(allyl)$_4$, 100K]→

Fe-O-Mo(allyl)$_2$ (C) —[H_2, 10^{-7} torr, 673K, 15 min]→ Fe-O-Mo- (?) —[O_2, 10^{-8} torr, 573K, 15 min]→ Fe-O-MoO$_2$ (D)

The ordered OH overlayer reacts readily with the molybdenum alkyl vapor, producing a supported Mo diallyl species. This species can be subsequently reduced and oxidized to form the iron oxide supported molybdenum dioxo species. This surface will be used as a substrate for methanol adsorption and decomposition, in order to model the behavior of mixed metal oxide catalysts.

5. DISCUSSION

The several examples presented above provide an indication of the effect of structure on the chemistry of small molecule reactions on the Fe(100) surface. The Fe(100) surface, carefully characterized using LEED, Auger spectroscopy, HREELS and other spectroscopic methods, serves as a detailed model of complex and interesting heterogeneous chemistry. Four particular points are well illustrated by the examples presented here.

The first concerns the special nature of the fourfold hollow site on the Fe(100) surface. This site is the location of the α_3 CO adstate, and as such is the site effective for CO dissociation on this surface. The α_2 (bridging) and α_1 (on-top) sites desorb from the surface, and do not serve as effective dissociation precursors. The presence of sulfur in this fourfold hollow site blocks the α_3 dissociation pathway. Competition for this adsite affects the relative amount of desorption versus dissociation when the CO(α_3) adlayer is post-exposed to a strongly binding molecule such as methanethiol.

Final product structures also affect reaction pathways on this surface, for reactant molecules of similar structure. This is well illustrated by the contrast in final products for the decomposition of methanol versus methanethiol. The stability of sulfur in the fourfold hollow site on this surface controls the formation of methane in the CH_3SH decomposition reaction. The analogous process, forming CH_4 from CH_3OH and leaving O behind in the fourfold hollow, is perhaps not as likely because of the lower stability of the atomic oxygen adlayer.

The detailed structure of this oxygen overlayer is seen to strongly affect the chemistry of methanol decomposition on the oxygen modified Fe(100) surface. The high temperature overlayer, characterized by fourfold hollow atomic oxygen and a well developed phonon structure, is essentially inactive for methanol decomposition at saturation oxygen coverages. At intermediate coverages, the fourfold hollow overlayer favors formaldehyde production over decomposition to form CO and H_2. The low temperature oxygen modified surface also actively produces formaldehyde in the methanol decomposition reaction, in proportion to the presence of the unusual molecularly bound oxygen species observed on this surface. The stability of the methoxy intermediate in the decomposition reaction is strongly affected by the oxygen modified surface structure.

Finally, it is seen that complex mixed oxide structures can be rationally constructed based on the ordered hydroxylated Fe(100) surface. The reaction of this overlayer with volatile metal alkyls could serve as a general route to the preparation of ordered mixed metal oxide overlayers which can be used as characterizable structural models for still more complex heterogeneous systems. This approach will extend the use of well characterized model substrates to probe the microscopic nature of the structure-activity relationship in heterogeneous catalysis.

ACKNOWLEDGEMENTS

This work was supported by the National Science Foundation, Division of Materials Research, and in its early stages by grants from Exxon Research and Engineering. Collaboration and discussion with Dr. John Gland and Dr. Dan Dwyer formerly at Exxon Research and Engineering, and with Professor Jeffrey Schwartz of Princeton University are gratefully acknowledged.

References

1. R.B. Anderson, The Fischer-Tropsch Synthesis, Academic Press, Orlando, Florida, 1984.
2. N. Berkowitz, The Chemistry of Coal, in Coal Science and Technology, Volume 7, Elsevier, Amsterdam, 1986.
3. A. Mittasch, Adv. Catal. 2, (1950) p. 81.
4. D.-W. Moon, D.J. Dwyer, J.L. Gland and S.L. Bernasek, J. Am. Chem. Soc. 107, (1985) p. 4363.
5. D.-W. Moon, D.J. Dwyer and S.L. Bernasek, Surface Sci. 163, (1985) p. 215.
6. D.-W. Moon, J.-P. Lu, D.J. Dwyer, J.L. Gland and S.L. Bernasek, Surface Sci. 184, (1987) p. 90.
7. J.-P. Lu, M.R. Albert, S.L. Bernasek and D.J. Dwyer, Surface Sci. 199, (1988) p. L406.
8. D.W. Moon, S. Cameron, F. Zaera, W. Eberhardt, R. Carr, S.L. Bernasek, J.L. Gland and D.J. Dwyer, Surface Sci. Lett. 180, (1987) p. L123.
9. J.-P. Lu, M.R. Albert, S.L. Bernasek and D.J. Dwyer, Surface Sci., 215, (1989) p. 348.
10. W.-S. Hung, J. Schwartz, and S.L. Bernasek, "Adsorption of H_2O on Fe(100): Formation of an Ordered Hydroxylated Surface," manuscript in preparation.
11. M.R. Albert, J.-P. Lu, S.L. Bernasek, and D.J. Dwyer, Surface Sci., 221, (1989) p. 197
12. M.R. Albert, J.-P. Lu, S.L. Bernasek, S.D Cameron, and J.L. Gland, Surface Sci., 206, (1988) p. 348
13. J.-P. Lu, M.R. Albert and S.L. Bernasek, "Effects of Post-dosed Species on Preadsorbed CO on Fe(100): Adsorption Site Conversion Caused by Site Competition", J. Phys. Chem., submitted.
14. J.-P. Lu, M.R. Albert, S.L. Bernasek, and D.J. Dwyer, Surface Sci., 218, (1989) p. 1
15. J.-P. Lu, M.R. Albert, S.L. Bernasek, and D.J. Dwyer, "Decomposition of Methanol on Oxygen Modified Fe(100) Surfaces 2: Preadsorbed Oxygen as Poison, Selectivity Modifier and Promoter," Surface Sci., submitted.
16. W.-S. Hung, J. Schwartz and S.L. Bernasek, "Reaction of Tetrakisallyl Molybdenum with the Hydroxylated Fe(100) Surface: Formation of an Orderd Mixed Metal Oxide Substrate," manuscript in preparation.

R.K. Grasselli and A.W. Sleight (Editors), *Structure-Activity and Selectivity Relationships in Heterogeneous Catalysis*

© 1991 Elsevier Science Publishers B.V., Amsterdam

STRUCTURE-ACTIVITY AND STRUCTURE-SELECTIVITY RELATIONS FOR REACTIONS OF CARBOXYLIC ACIDS ON TiO_2 (001) SURFACES

H. IDRISS, K. S. KIM, AND M. A. BARTEAU

Center for Catalytic Science and Technology
Department of Chemical Engineering
University of Delaware
Newark, DE 19716

ABSTRACT

The reactions of carboxylic acids were investigated on a TiO_2 (001) single crystal surface which exhibits two stable faceted structures upon annealing at different temperatures. The {011}-faceted structure contains only Ti^{+4} cations which are five-fold coordinated, and the {114}-faceted structure exposes Ti^{+4} cations which are four-, five-, and six-fold coordinated. It was observed that the dissociation, reduction, and dehydration of carboxylic acids require only one degree of coordinative unsaturation of the surface cations and are therefore relatively insensitive to surface structural transformations, since both faceted structures of the TiO_2 (001) surface contain five-fold coordinated cations. However, the bimolecular ketonization of acetates to acetone and of acrylates to divinylketone was only observed on the {114}-faceted structure and is, therefore, a structure-sensitive reaction. This reaction requires two coordination vacancies on a common cation to accommodate the pair of carboxylate species to be coupled. This latter result suggests that the active site for bimolecular ketonization is analogous to that on Ziegler-Natta catalysts for the oligomerization and polymerization of olefins.

INTRODUCTION

Metal oxides are important materials for a variety of catalytic reactions. In general one would like to identify and characterize surface intermediates in these processes and to relate their formation and conversion to the structure of the surface sites on these materials. However while structure sensitivity in catalysis by metals is widely investigated (ref. 1), less effort has been devoted to understanding reactions on metal oxides at the molecular level or to recognizing the different ways that structure sensitivity can arise in catalysis by oxides and by metals.

Several different strategies have been used to probe structure-sensitivity in catalysis by oxides (also referred to as crystal-face anisotropy (ref. 2) or catalytic anisotropy (ref. 3)). Modification of the bulk crystallographic structure of the oxides by structural promotion or by synthesis techniques leading to metastable bulk structures (ref. 4) allows the manipulation of surface structures. Variation of crystallite morphology, especially on materials with highly anisotropic bulk structures such as MoO_3, is a common strategy (ref. 5, 6). Changes in the relative population of the different crystallographic planes exposed may be correlated with changes in catalytic activity or selectivity.

The range of structural variations that one may produce by the above approaches is often limited. Moreover, confusion may arise in attempting to establish structure-reactivity relations for reactions in which selectivity is a strong function of conversion, as in sequential oxidation steps. The selective oxidation of propylene is one such example. While it has been reported that the basal plane of MoO_3 unselectively oxidizes propylene and the apical faces oxidize it selectively to acrolein (ref. 7, 8) it was also contended in other work (ref. 9) that acrolein is formed on the basal plane. Alcohol dissociation is another example where structure-reactivity studies are not in agreement. Bowker et al. (ref. 10) have determined that decomposition of alcohols occurs only on the polar surfaces of zinc oxide, while Djega-Mariadassou et al. (ref. 11) concluded that this reaction is structure insensitive.

Surface science studies of oxide single crystals offer an attractive approach to establishing structure-reactivity relationships for oxides, as they have for metals. However, low index planes of oxides frequently exhibit a greater tendency to undergo thermal rearrangement and faceting than do those of metals. This has often led to the choice of oxide surfaces with the greatest average coordination number, as these tend to exhibit greater thermal stability. Fortunately, if the structures of faceted surfaces are well resolved, it is possible to exploit this instability to permit the examination of surface reactions with changes in surface structure of a single oxide sample. Thus manipulation of surface structure and composition by high temperature pretreatment under thermodynamically controlled conditions may help in understanding the activity and selectivity of metal oxides in catalysis.

Titanium dioxide has been suggested to be a good model system on which to establish structure-reactivity relations, due in part to the ease with which surface defects may be detected on this material (ref. 12, 13). We have previously examined the reactions of methanol on $TiO_2(001)$ single crystal surfaces (ref. 14). The principal products of methanol decomposition included methane, formaldehyde and dimethylether, however the selectivity towards these products depended on the structure. The {011}-faceted $TiO_2(001)$ single crystal surface, a stable structure produced by thermal treatment at ca. 700-750K, exhibited the highest selectivity for methane formation, while dimethylether was observed only on the {114}-faceted $TiO_2(001)$ surface. Ether formation via disproportionation of pairs of methoxides requires surface cations with a pair of coordination vacancies to accommodate them. These four-fold oxygen-coordinated Ti^{+4} cations are available only on the {114}-faceted surface. Of the $TiO_2(001)$ surfaces examined, this is the only one active for ether synthesis via alkoxide coupling (ref. 14, 15).

In this work we present the results for the decomposition and reaction of saturated and unsaturated carboxylic acids (acetic acid and acrylic acid) by temperature programmed desorption (TPD) on a $TiO_2(001)$ single crystal subjected to different pretreatments. These included sputtering with argon ions and annealing under ultrahigh vacuum. These results illustrate the variety of reactions which occur on oxide surfaces and provide insights into the surface characteristics required to produce different products.

EXPERIMENTAL

Experiments were performed in two different UHV systems which have been described in detail previously (ref. 16, 17). TPD experiments were conducted in a PHI model 548 surface analysis system equipped with a UTI 100C quadrupole mass spectrometer. XPS experiments were conducted in a PHI model 550 ESCA/Auger surface analysis system. The TiO_2 single crystal was prepared from a rutile crystal boule (99.9% Atomergic Chemetals Corp.) aligned to within 0.5° of the (001) orientation by the Laue method. The crystal mounting and cleaning procedures have been described previously (ref. 14). The ratio of the peak-to-peak height of the O(KLL) Auger transition at 510 eV to that of the Ti(LMM) transition at 380 eV was ca. 1.2 for the sputtered surface and increased to ca. 1.65-1.70 for the fully annealed crystal (annealed at ca. 950K for 20 minutes); this latter ratio is typical of fully oxidized TiO_2 surfaces (ref. 18). A typical TPD experiment consisted of initially dosing the crystal to saturation with the reactant (acetic acid or acrylic acid) through a variable leak valve equipped with a stainless steel dosing needle. After dosing, the chamber was allowed to pump down to a background pressure of ca. 1×10^{-9} torr. Heating was then initiated and the desorption flux was monitored by the mass spectrometer which was multiplexed with an IBM PC. Typically 8, but occasionally as many as 20 masses were monitored simultaneously as a function of temperature during a single TPD experiment. The computer was also used to control the heating rate of the crystal (1.2 K/sec.)

TPD experiments were conducted on a $TiO_2(001)$ single crystal surface on which different structures were produced by thermal faceting. On the ideal (001) surface all Ti^{+4} cations are four-fold oxygen coordinated, whereas the cations in the bulk are six-fold coordinated. This high degree of coordinative unsaturation renders the structure thermally unstable, and faceting occurs upon annealing. Previous work in this laboratory (ref. 19) reproduced the LEED patterns of the faceted surfaces reported and assigned by Firment (ref. 20). The sputtered surface was a disordered surface which exhibited no LEED pattern. Upon annealing to ca. 700-750K a {011}-faceted structure was formed, while the {114}-faceted surface was formed by annealing the crystal to 900K or above. Those two faceted structures were the only two ordered structures (ref. 19). The ideal {114}-faceted surface exposes four-, five-

and six-fold coordinated cations, while the {011}-faceted surface exposes only five-fold coordinated titanium cations (ref. 20).

RESULTS

Acetic acid:

TPD following adsorption of acetic acid at 300K on the {011}-faceted TiO_2 (001) surface indicated two desorption states. The lower temperature state consisted of the coincident desorption of acetic acid and water at ca. 390K. The higher temperature state included acetic acid at 590K, CO at 600K and ketene at 610K. Unlike our results for acetic acid adsorption on titania powder (ref. 21), no acetone was observed. However, on the {114}-faceted TiO_2(001) surface, TPD after acetic adsorption did produce acetone at 580K. This represents an important difference in the reaction pathways available to acetic acid on the {114}-faceted versus the {011}-faceted TiO_2(001) surfaces. Figure 1 illustrates the product selectivity for acetic acid TPD on the TiO_2(001) surface as a function of the prior annealing temperature of the surface. On the {011}-faceted surface the selectivity for ketene was 64%. Further oxidation and faceting of the surface produced by prior annealing temperatures above 750K gradually decreased the selectivity for ketene while an additional product, acetone, was produced with increasing selectivity. The selectivities for ketene and acetone were 47% and 13% respectively for the 950K-annealed surface that exhibits the {114}-faceted structure. The selectivity for CO and adsorbed carbon, which was ca. 5% on the {011}-faceted structure increased gradually to ca. 10% for the {114}-faceted structure. The total activity of the surface for dissociative adsorption of acetic acid remained relatively constant for surfaces annealed at 850K and below, but decreased by ca. 20% for surfaces annealed above this temperature.

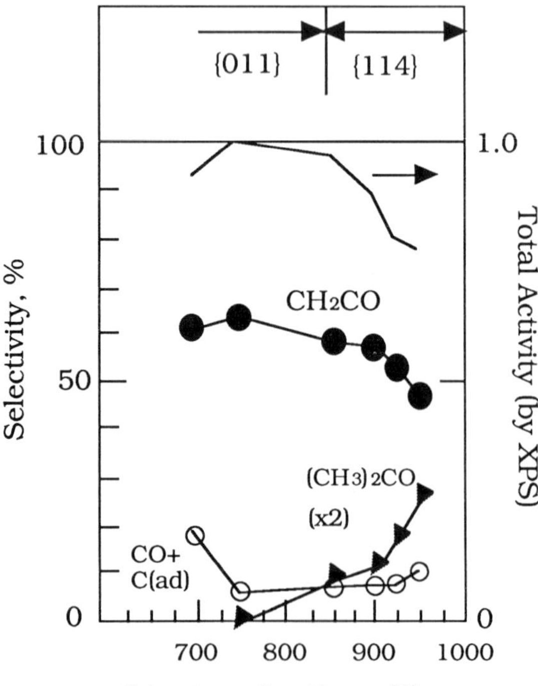

Fig. 1. Activity and selectivity of faceted TiO_2(001) surfaces for acetic acid decomposition.

Acrylic Acid:

In order to extend this study to unsaturated acids, we have also examined the reaction of acrylic acid (CH_2=CHCOOH), a molecule containing both olefinic and carboxyl groups. TPD following adsorption of acrylic acid on the TiO_2(001) surface at room temperature was carried out as a function of the prior annealing temperature of the surface. As in the case of acetic acid TPD, two desorption states were observed on the {011}-faceted structure. The first consisted mainly of acrylic acid desorption at ca. 380K accompanied by water. The second state consisted of acrylic acid (ca. 600K), ethylene (ca. 600K), carbon monoxide (ca. 580K), carbon dioxide (550-600K) and water (ca. 650K). The coupling of two acrylate species to divinylketone was not observed on this {011}-faceted surface. This result is in agreement with that for acetic acid TPD on the same faceted surface. However, two unexpected products were observed. Acrolein (m/e 29, 56) and butadiene (m/e 54, 39) both exhibited desorption peaks at ca. 580-600K.

The influence of further prior annealing of the surface to 950K to form the {114}-faceted structure was also investigated in the study of acrylic acid decomposition. Figure 2 displays the TPD spectrum following adsorption of acrylic acid at room temperature on the {114}-faceted structure. As on the {011}-faceted structure, two sets of desorption peaks were observed. The first consisted of a large peak for acrylic acid (m/e 72) accompanied by water at ca. 390-400K, while the second state consisted of the same products observed during acrylic acid TPD on the {011}-facetted surface plus the coupling product of two acrylate species, divinylketone, at ca. 590-600K. The formation of divinylketone on the {114}-

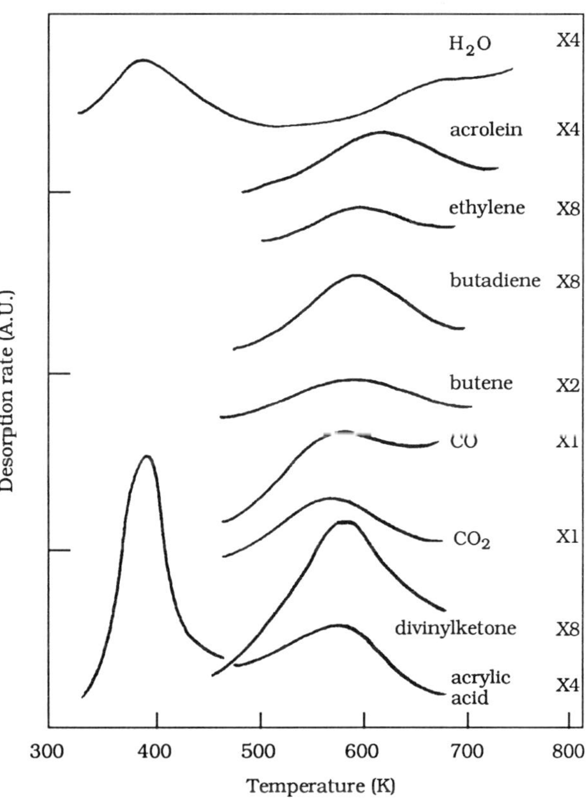

Fig. 2. TPD spectrum for acrylic acid on the {114}-faceted TiO_2(001) surface (previously annealed at 950K). The spectra illustrated are uncorrected for mass spectrometer sensitivity.

faceted surface from acrylic acid is similar to that of acetone from acetic acid on the same faceted surface. This result indicates that the carboxyl group rather than the vinyl group of the reactant (acrylic acid) interacts preferentially with the surface.

The selectivity of the desorbed products and their peak temperatures during acrylic acid TPD on the {114}-faceted stucture are presented in table 1. As indicated, the selectivity (on a carbon basis) to divinylketone was ca. 15%, very close to that of acetone formed from acetic acid on this surface (13%), while that of acrolein was 7%. Quantitative determination of the selectivity to divinylketone and acrolein as a function of prior annealing temperature is displayed in figure 3. On the low temperature annealed surface and the {011}-facetted $TiO_2(001)$ surface the selectivity to acrolein was more than 20%. Further prior annealing of the surface dramatically decreased the selectivity to acrolein to 7%, while divinylketone, first observed on the 750K annealed surface, increased to 15% on the 950K-annealed surface. This figure may be interpreted as follows: increasing (by prior annealing) the oxygen concentraton at the surface, demonstrated by the increase in the O/Ti peak to peak ratio by AES (ref. 14), decreased the extent of the reduction reaction of surface acrylates to acrolein. Creation of four-fold coordinated Ti^{+4} cations by

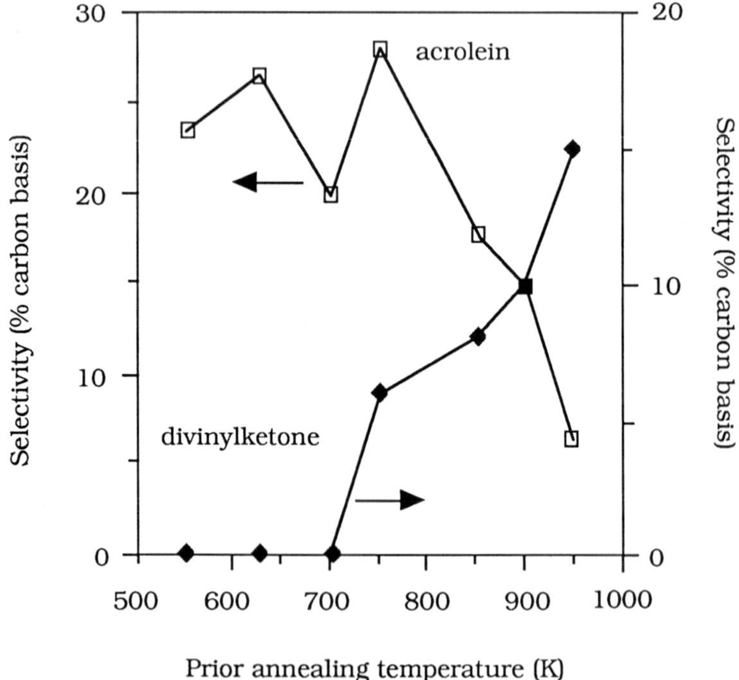

Fig. 3. Acrolein and divinylketone selectivities from acrylic acid TPD on TiO_2 surfaces as a function of prior annealing temperature.

TABLE 1

Product distribution for acrylic acid TPD from the {114}-faceted TiO_2(001) surface

Product	Peak Temperature	Selectivity (moles of product/ mole of acrylic acid adsorbed)	Selectivity (% of the total carbon in each product)
CH_2=CHCOOH	390-400K	0.21	21
CH_2=CHCOOH	570-580K	0.09	9
butene	570-600K	0.04	5
butadiene	570-580K	0.07	9
CO_2	550-580K	0.25	8
acrolein	590-600K	0.07	7
CO	550-600K	0.30	10
ethylene	570-580K	0.15	10
divinylketone	590-600K	0.09	15
surface carbon		0.19	6

thermal faceting to produce the {114}-faceted structure provided the sites for coupling of pairs of adsorbed acrylate species to give divinylketone.

DISCUSSION

The results of the study of the reactions of acetic acid and acrylic acid on TiO_2 (001) single crystal surfaces demonstrate that the selectivity is sensitive to the surface structure as well as the oxidation state of the surface.

On the low temperature annealed surface and the {011}-faceted TiO_2(001) surface, acetic acid undergoes unimolecular dehydration to ketene and acrylic acid is reduced to acrolein. On the {114}-faceted surface the selectivity to ketene in the case of acetic acid TPD and that of acrolein in the case of acrylic acid TPD decrease, while the bimolecular ketonization of two acrylates to divinylketone and of two acetates to acetone takes place.

Carboxylic acids are first dissociated to form carboxylate species on the surface of titania, as demonstrated XPS results on TiO_2(001) single crystals (ref. 19) and by FT-IR on titania powders (ref. 21). Dissociation of carboxylic acids on metal oxides requires only accessible cation-anion site pairs on the surface; the conjugate base of the parent acid binds at a coordination vacancy of the metal cation upon dissociation (ref. 22, 23). In the case of acetic acid decomposition, the identities of the products desorbed during TPD on polycystalline powders were consistent with those observed on the single crystal surfaces in this study. However the structural dependence of the reactions of acetate species could not be determined with polycrystalline samples. The single crystal results, then, resolved the nature of the sites required by each of the reactions observed on the powder.

Dissociation, net unimolecular dehydration, and reduction of carboxylic acids appear to require, according to this work, only a single coordination vacancy on the part of the surface cations. This observation is in agreement with results from the ZnO(0001) surface, a surface which exposes cations with a single coordination vacancy (ref. 24). This surface also gives rise to unimolecular decomposition and dehydration of carboxylic acids, but no evidence for bimolecular ketonization of carboxylates was observed.

The bimolecular ketonization of carboxylates is clearly a structure-sensitive reaction. It requires a pair of coordination sites on a common cation, a requirement met only by the four-fold oxygen coordinated Ti cations of the {114}-faceted TiO_2(001) in this study.

The site required for the ketonization reaction of carboxylates is directly analogous to that for Ziegler-Natta polymerization of terminal olefins (ref. 25). This reaction requires a pair of vacant coordination sites on a surface Ti cation, one vacancy at which to bind the growing polymer and the other to accommodate the olefin monomer to be added to it. The formation of divinylketone indicates that the vinyl group of acrylate species remains intact and may migrate from one adsorbed species to another without being hydrogenated. The resulting product is a highly unsaturated ketone. The correspondence of the site requirement for olefin polymerization with a classical Ziegler catalyst and carboxylic acid ketonization on TiO_2(001) single crystal surfaces indicates that the active sites of the oligomerization and polymerization reactions can be successfully probed by surface science techniques.

In summary, the dissociation, reduction and dehydration of carboxylic acids do not appear to be structure sensitive reactions on TiO_2(001) single crystal surfaces, since all require only one degree of coordinative unsaturation of the surface cations. Bimolecular ketonization of carboxylates, in contrast, requires two coordination vacancies on a common cation (four-fold coordinated Ti^{+4} sites). Highly unsaturated ketones can be formed, including divinylketone from acrylates. The surface site coordination requirements for carboxylate ketonization appear to be analogous to those for olefin polymerization.

ACKNOWLEDGEMENT

We gratefully acknowledge the support of the National Science Foundation(Grant CBT-8714416).

REFERENCES

1 M. Boudart and G. Djega-Mariadassou, Kinetics of Heterogeneous Catalytic Reactions , Princeton University Press, Princeton, N.J., 1984 and references therein.
2 S. T. Oyama, Bull. Chem. Soc. Japan, 61 (1988) 2585.
3 J. Ziolkowski, J. Catalysis, 80 (1983) 263.
4 E. M. McCarron, J. Chem. Soc. Chem. Commun., (1986) 336.
5 J. M. Tatibouet and J. E. Germain, J. Catalysis, 72 (1981) 375.
6 J. M. Tatibouet, J.E. Germain and J. C. Volta, J. Catalysis, 82 (1983) 240.
7 J. C. Volta and J. M. Tatibouet, J. Catalysis, 93 (1985) 467.
8 J. C. Volta and J. L. Portefaix, Appl. Catalysis, 18 (1985) 1.
9 K. Brückman, R. Garbowski, J. Haber, A. Mazurkiewics, J. Sloczynski and T. Wilkowski, J. Catalysis, 104 (1987) 71.
10 M. Bowker, H. Houghton, K. C. Waugh, T. Giddings and M. Green, J. Catalysis, 84 (1983) 252.
11 G. Djega- Mariadassou, L. Davignon and A. R. Marques, J. Chem. Soc. Faraday Trans. I, 78 (1982) 2447.
12 W. J. Lo, Y. W.Chung and G. A. Somorjai, Surface Sci., 71 (1978) 199.
13 W. Göpel, G. Rocker and R. Feierabend, Phys. Rev. B, 28 (1983) 3427.
14 K. S. Kim and M. A. Barteau, Surface Sci., 223 (1989) 13.
15 K. S. Kim and M. A. Barteau, J. Mol. Catalysis, in press.
16 J. M. Vohs and M. A. Barteau, Surface Sci., 176 (1986) 91.
17 R. Martinez and M. A. Barteau, Langmuir, 1 (1985) 684.
18 G. B. Hoflund, H. L. Yin, A. L. Jr. Grogan, D. A. Asbury, H. Yoneyama, O. Ikeda and H. Tamura, Langmuir, 4 (1988) 346.
19 K. S. Kim and M. A. Barteau, J. Catalysis, 125 (1990) in press.
20 L. E. Firment, Surface Sci., 116 (1982) 205.
21 K.S. Kim and M.A. Barteau, Langmuir, 4 (1988) 945.
22 T. W. Root and T. M. Duncan, J. Catalysis, 101 (1986) 527.
23 A. Zecchina, S. Coluccia and C. Mortera, Appl. Spect. Rev., 21 (1985) 259.
24 J. M. Vohs, PhD Dissertation, University of Delaware, Newark, DE (1988).
25 B. C. Gates, J. R. Katzer and G. C. A. Schuit, Chemistry of Catalytic Processes, McGraw-Hill, New York, 1979.

R.K. Grasselli and A.W. Sleight (Editors), *Structure-Activity and Selectivity Relationships in Heterogeneous Catalysis*
© 1991 Elsevier Science Publishers B.V., Amsterdam

SURFACE CHEMISTRY MODELLING OF ZIEGLER-NATTA CATALYSIS

P.R. Watson, J. Mischenko III and S.M. Mokler
Department of Chemistry and Center for Advanced Materials Research, Oregon State University, Corvallis, Oregon 97331, USA

ABSTRACT

This study investigates the possibility of modelling $TiCl_3$ Zeigler-Natta catalysts by building chloride layers on clean surfaces of Ti metal. The basal (0001) and lateral (10-10) surfaces were used to simulate the conventionally assumed inactive and poylerization-active surfaces of $TiCl_3$. LEED and Auger data that a coincidence lattice of Cl forms on Ti(0001) which to some extent resembles the basal plane of $TiCl_3$, while on Ti(10-10) chlorine atoms sit in the troughs. Olefins are decomposed by clean Ti, are are not adsorbed on fullt chlorided surfaces. On partially chlorided surfaces there are indications that some olefin may bond molecularly.

1. INTRODUCTION

The prototypical Ziegler-Natta olefin polymerization catalyst consists of small crystallites of the extremely water-sensitive material titanium trichloride [1]. This combination of material properties has made direct investigation of reactions occurring at the catalyst surface difficult to pursue. As a result, despite considerable technical advances in the development of efficient catalysts, basic questions concerning the nature of the reactive sites and the mechanistic steps involved in the polymerization remain unanswered. The bulk of experimental data has been obtained by indirect methods such as following the variations in products formed from olefins with different structures. The accepted scenario is that olefin adsorption occurs at defects and edges of the titanium chloride crystallites [1], and polymerization proceeds via an insertion mechanism [2]. The surface structure, extrapolated from the bulk, of the basal (0001) and $(10\bar{1}0)$ surfaces are shown in Figure 1. The $(10\bar{1}0)$ surface must have some chlorine vacancies present (not shown) to achieve charge balance as has been considered in detail elsewhere [1].

We here investigate the potential of suitably designed chlorided surfaces of titanium metal to serve as simple models for this type of catalyst. Previous studies of the Ti/Cl_2 system are rather few in number [3-7]. Such surfaces possess the advantage that they are can be prepared in reproducible forms amenable to the variety of surface science tools now available to characterize the

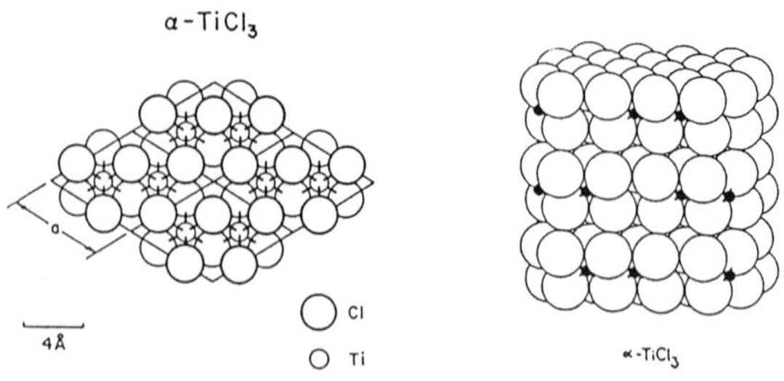

Fig. 1. Surfaces of α-$TiCl_3$, assuming bulk termination a) basal (0001) and edge ($10\bar{1}0$) surfaces.

composition, structure, and bonding at surfaces. Here we present a series of experiments on the construction and reactivity of two such surfaces - the basal (0001) and ($10\bar{1}0$) planes of Ti shown in Figure 2 - which when chlorided, are to act as models of the basal and edge planes of $TiCl_3$.

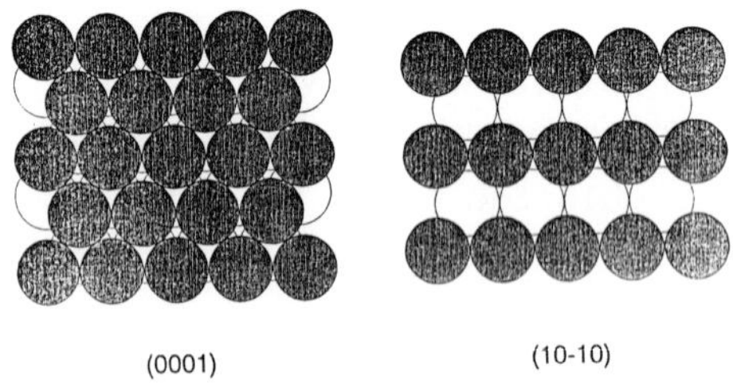

Fig. 2. Ideal surface structures of the a) (0001) and b) ($10\bar{1}0$) surfaces of Ti metal.

2. EXPERIMENTAL

These experiments were performed in a standard ion-pumped UHV chamber equipped with 4-grid LEED/Auger optics, a quadrupole mass spectrometer, and an ion-sputter gun. Olefins could be introduced via a bakeable leak valve. Bulk Ti has a phase transition at 885°C [8] and studies of the Ti($10\bar{1}1$) surface [9] revealed a lower value for the transition temperature of 800°C at this surface. Accordingly the sample was never heated above 750°C to avoid damage.

The chamber also contained an in-situ molecular chlorine source based on the design of Spencer et al [10] using a solid state Ag/AgCl/Pt,Cl_2 cell. It allows for high dosing levels without a corresponding increase in background pressure. We calculate [7,11] a chlorine dose equivalent to the passage of roughly 1600-2000 microcoulombs (4-5% efficiency) of charge through the cell will produce a monolayer coverage of Cl given a constant unit sticking probability.

Initial Auger spectra of the Ti metal samples showed substantial amounts of sulfur and carbon impurities. Following the work of Shih et al [12], we found that many cycles of ion-sputtering at elevated temperatures (650°C) and short thermal treatments at 720°C produced a surface with levels of C,S and O contamination of the order of 1%. Clean Ti surfaces are very reactive to background gases [9,12] and we found that detectable levels of C and O built up on the surface within an hour of the last flash. However, once the bulk had been depleted of S, we found that a short (15 mins) hot bombardment cycle followed by a flash to 720°C would recover the clean surface. To avoid contamination problems all experiments were carried out as quickly as possible after a cleaning cycle. Samples containing substantial amounts of surface chlorine were much less reactive.

3. RESULTS

3.1 Cl adsorption - Auger data

Figure 3 shows the increase of the Cl(183eV) Auger signal from the Ti(0001) surface, as a function of chlorine exposure at 30, 300, and 600°C. The equivalent data for the Ti(10$\bar{1}$0) surface are very similar. The room temperature data closely resemble that of Cox et al [7] for Ti foil. The increase in the chlorine signal is rapid and essentially linear over a wide range, while the Ti signal decrease is less marked. Within experimental error there does not appear to be large differences in the rate of uptake at the two highest temperatures. The data for both the Cl and Ti Auger signals at all temperatures saturate at essentially the same value

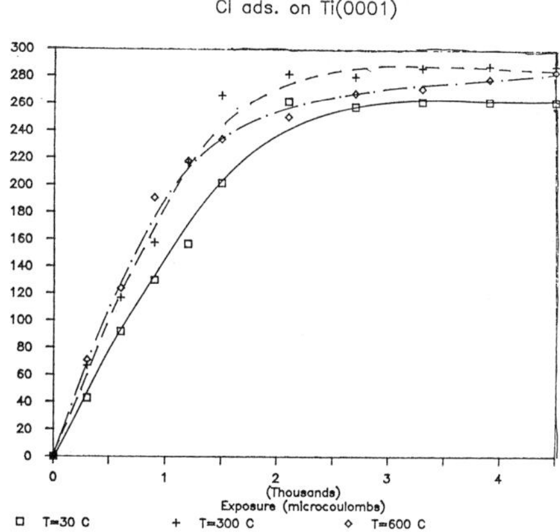

Fig. 3. Uptake of chlorine on Ti(0001) at various temperatures.

of the about 2000 microcoulombs chlorine exposure, although careful inspection of the high temperature data reveals that the signal may still be slowly changing above this exposure level. At saturation exposures we find a value of the ratio of Auger signals of Cl(183)/Ti(388) = 0.95 +/- 0.3 in reasonable agreement with the value of 9.1 at 300°C of Smith [3].

We found that adsorbed chlorine was susceptible to electron stimulated desorption (ESD) with an approximate ESD cross-section of $2x10^{-20}$ cm^2.

Because of the temperature limit imposed by the phase transition, we were only able to perform TDS experiments up to 750°C. We were unable to detect desorption of chlorine or titanium halides within this temperature range.

3.2 Cl adsorption - LEED data

Careful examination of the LEED patterns observed after adsorption of chlorine on the clean Ti(0001) surface at room temperature showed no change from the original (1x1) pattern (Figure 4a) except for a general increase in the diffuse background. Heating the chlorinated surface at temperatures of up to 650°C for periods of up to 30 minutes did not alter the Auger spectrum significantly, but did result in the appearance of a new LEED pattern shown in Figure 4b. An hexagonal arrangement of closely-spaced spots has appeared around the primary Ti spots. These new spots are very sharp and have a spacing of the superstructure spots varies of 1/16th of the Ti(0001) substrate spot spacing. The alignment of the new diffraction features is in registry with those of the substrate.

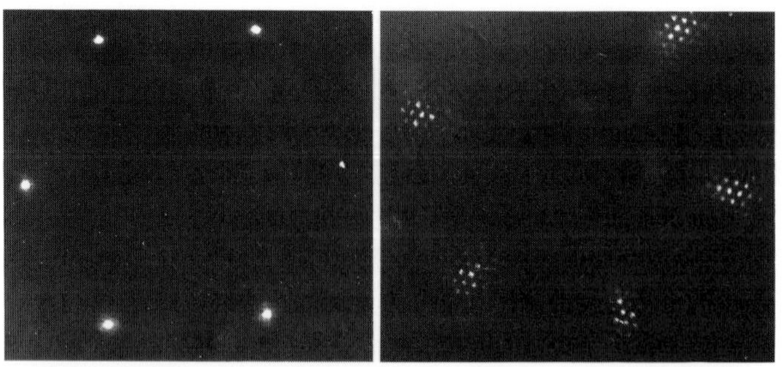

a) Clean Ti(0001)   b) after Cl2/650C

Fig. 4. LEED patterns from a Ti(0001) surface a) when clean and b) after chlorination and annealing.

We found that the clean Ti(101-0) surface always exhibited a (1x1) pattern - in contrast to an earlier report claiming superstructure LEED patterns due to inherent surface reconstructions [6]. When chlorinated this surface shows a number of superstructure LEED patterns of the (nxm) type, in particular a well-defined (3x1) phase.

3.3 Olefin adsorption

When a clean Ti surface of either crystallographic direction is exposed to olefins we observe a build of "carbidic" carbon in the Auger signal. On heating no hydrocarbon species are seen to desorb, although on occasion hydrogen desorption may be observed. The appearance or otherwise of hydrogen on heating seems to depend critically on the sample history. The surface carbon that is deposited can only be removed by a complete new cleaning cycle.

For the Ti(0001) surface only one ordered phase forms on chlorination - the (16x16) phase shown in Figure 4b. On this surface we were unable to detect any adsorption of ethylene or propylene. For the Ti(10$\bar{1}$0) surface we were able to produce a number of stable ordered Cl phases. The amount of carbon that appears on the surface correlates inversely with the amount of Cl present on the surface, Figure 5. The shape of the carbon peak again suggests that a carbidic type of carbon is present on the surface. Surfaces with a low Cl coverage show no hydrocarbon thermal desorption, whereas surfaces with a high Cl content show little olefin adsorption. Preliminary experiments with surfaces that are partially chlorinated have shown irreproducible signs of hydrocarbon thermal desorption.

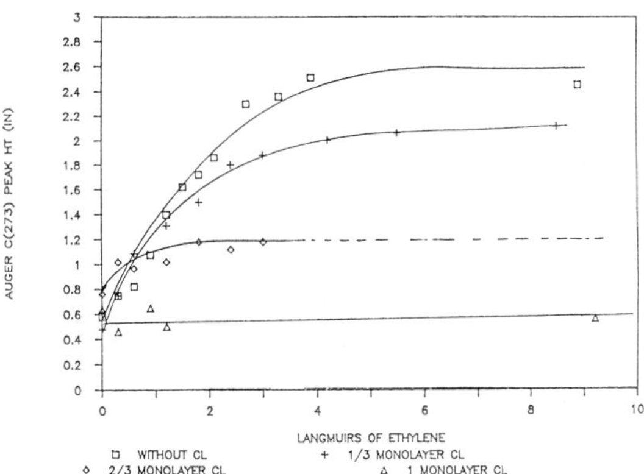

Fig. 5. Change in Auger C signal on uptake of ethylene on Ti(10$\bar{1}$0) as a function of preexisting Cl coverage on the surface.

4. DISCUSSION

4.1 Chlorine Adsorption Behavior

The shape of the Auger uptake curves in Figure 3 show that the sticking probability remains essentially constant over a large range of adsorbate coverages. Similar behavior has been noted for many systems, in particular for chlorine adsorbed on Y and Hf [7], Ta(100) and (110) [13], Cu and Ag(111) [14], and Rh(111) [11]. In common with these authors we view chlorine adsorption as non-activated with a long-lived precursor adsorption state on the surface [15].

A frequent question that occurs in halogen adsorption on metals is whether adsorption proceeds via formation of a chemisorbed monolayer, possibly followed by a corrosion reaction, or whether true halide formation occurs from the outset.

Khan [6] argued on the basis of RHEED patterns for $TiCl_3$ formation on Ti(10T0). However, work from this laboratory casts doubt on this assignment. We find [17] that we cannot reproduce the series of superstructures at various temperatures from the clean surface that he reported and attributed to surface reconstructions. It is likely that these structures were due to sulfur impurities which may have affected the halogen adsorption data.

Anderson and Gani [4] found that, at room temperature, chlorine adsorption on a Ti film was extremely rapid up to one monolayer. Further adsorption occurred slowly at pressures greater than 10^{-5} torr. Smith [3], working with Ti(0001), found from ellipsometry measurements that adsorption proceeded beyond monolayer coverage at 10^{-7} torr and 300°C. However, both work-function and Auger data showed saturation effects that might be attributed to formation of a chemisorbed monolayer.

Cox et al [7], using Ti foil, found weak evidence for high-temperature (800°C+) $TiCl_x$ (x unknown) desorption at low exposures. At high exposures $TiCl_3$ desorption grew indefinitely between 900 and 1000°C. We however were unable to detect desorption from a Ti(0001) surface below 750°C. The lack of change in the Auger spectrum on heating at 700°C also argues against desorption processes occurring below that temperature.

Our Auger data (Figure 3) show that the Cl Auger signal increases smoothly and saturates at about 2000 microcoulombs chlorine dose. A high sticking probability close to unity seems to be the rule for halogen adsorption on metal surfaces [16], and this amount of cell current would correspond closely to that expected (1600-2000 microcoulombs) for adsorption of one monolayer of Cl with unit sticking probability. Such a situation would also be consistent with dissociative adsorption. This close agreement is no doubt fortuitous, depending as it does upon some doubtful assumptions, but it does suggest that the knee of the Auger uptake curve corresponds to completion of one chemisorbed monolayer of chlorine adatoms.

Beyond the knee of the uptake curve, both the Cl signal and the Ti signal remain unchanged with further exposure to chlorine at 30°C. Further incorporation of chlorine into halide layers may be occurring at higher temperatures, as evidenced by small changes in both the Cl and Ti signals. If such growth is occurring, then it is taking place with a very low sticking probability, in agreement with the earlier studies [3,4].

4.2 Cl surface structures

Unlike many metal/halogen systems, we did not observe low-coverage LEED patterns on the (0001) surface that may help to fix the monolayer coverage point. We did however observe the complex (16x16) pattern seen in Figure 4b upon annealing a saturated surface. This pattern resembles that seen by Bowker and Waugh [18] for chlorine adsorbed on Ag(111) and for Ag on Cu(111) seen by Bauer [19]. This type of pattern can be explained in a straight-forward manner in terms of a coincidence lattice, although alternative explanations using antiphase domains are possible.

We interpret the (16x16) pattern as due to multiple diffraction interference effects between beams originating in a (0001) Ti layer and a coincident dense overlayer of Cl that has unit vectors in the same directions as the substrate.

The Ti-Ti interatomic distance in the (0001) close-packed plane is 2.94Å. This means that the repeat distance for the superstructure lattice is 47.04Å. Only a small number of possible chlorine radii are possible for a close-packed Cl layer to produce a coincidence over the superstructure repeat distance. We propose a coincidence lattice with 13 close-packed Cl atoms per 16 Ti with a Cl radius of 1.81Å. We prefer this value for the Cl radius at it corresponds to the Van der Waals radius of chlorine, and, furthermore, the Cl-Cl distance of 3.62Å equals that deduced for a (4x4) coincidence structure observed on Rh(111) [11]. Similar Cl-Cl distances have been reported for such structures on Ag and Cu [14].

This structure is in fact compatible with the growth of at

least a pseudo-$TiCl_3$ lattice. $TiCl_3$ is a layer compound with Ti situated in octahedral holes in Cl-Ti-Cl layers separated by a Van der Waals gap. The surface of $TiCl_3$ would correspond to a layer of Ti topped by a Cl layer as shown in Figure 1. The Cl-Cl distance in $TiCl_3$ is 3.6Å resulting in a close-packed Cl layer identical to that proposed for the (16x16) LEED structure above. The Ti-Ti distance in $TiCl_3$ is also 3.6Å relative to 2.94Å in the Ti(0001) surface. Only 2/3 of the octahedral holes in the $TiCl_3$ structure are occupied, and hence the atom density in the Ti layer of $TiCl_3$ ($6.5x10^{14}$) is substantially less than that in the Ti(0001) surface ($1.5x10^{15}$).

In the case of the Ti(10$\bar{1}$0) surface, the most probable explanation for the series of (nxm) structures that form as the chlorine coverage increases is for the Cl atoms to adsorb in the troughs of this surface visible in Figure 2b. An example of such a structure is shown in Figure 6 for the (3x1) phase, comprising a row filled Cl overlayer. We might compare this structure with the bulk structure of the corresponding α-$TiCl_3$ surface shown in Figure 1b. While the density of Ti atoms in the constructed surface is too high, there is a similarity in the way in which the adsorbed Cl atoms hide the underlying Ti atoms, thus possibly making unique types of adsorption sites.

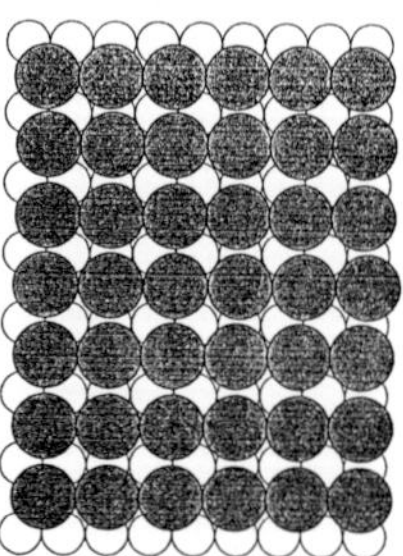

Fig. 6. Row filling model for the Ti(10$\bar{1}$0) (3x1)-Cl LEED phase.

4.3 Olefin adsorption and reaction

The clean Ti surfaces are very reactive towards gas phase olefins. The appearance of a carbidic Auger signal and lack of desorption products on heating argue for the complete fragmentation of the hydrocarbon molecule. In the case of Ti(0001) we have observed ordered LEED patterns consistent with the growth of an epitaxial TiC(111) layer [20]. The uncertain appearance of hydrogen in the gas phase on heating is probably a reflection of its well-known solubility in Ti. When the sample has been extensively heated H from newly dissociated hydrocarbon will dissolve in the bulk of the sample and not desorb on heating. When the subsurface is saturated with dissolved hydrogen some irreproducible desorption of hydrogen may occur on heating depending critically upon the sample history.

Surfaces that are completely chlorinated, such as the Ti(0001) 16x16-Cl surface, are inert towards olefin adsorption. In as much as we would wish such a surface to model the real catalysts, this observation is in accord with the observation that the Ziegler-Natta activity of α-$TiCl_3$ is related to the amount of Ti exposed at edges and defects in the basal planes, which are themselves unreactive [1,2].

The various structures observed on the chlorinated Ti($10\bar{1}0$) surfaces afford us the opportunity to study the effect of a gradual modification by adsorbed halogen of the intrinsic high reactivity of the metal. Figure 5 clearly shows that the amount of hydrocarbon adsorbed is proportional to the amount of free metal surface accessible. A highly chlorinated surface again adsorbed little olefin. Once again the Auger peak shapes suggests that the C on the surface is carbidic in nature in all cases. It appears that the chlorine is principally exerting a blocking effect to prevent adsorption, rather than substantially modifying the reactivity of the accessible metal surface. The preliminary desorption results do, however, give an indication that some molecular adsorption may be taking place at a small number of favorable sites, possibly that at edges of chlorine islands.

5. ACKNOWLEDGEMENTS

This work was supported in part by an M.J. Murdock Charitable Trust Grant of the Research Corporation and by the Dow Chemical Company. Acknowledgement is also made to the Donors of the Petroleum Research Fund, administered by the American Chemical Society, for partial support of this research.

REFERENCES

1 J. Boor Jnr., Zeigler-Natta Catalysts and Polymerizations, Academic Press, New York, 1979.
2 E.J. Arlman and P. Cossee, J. Catal., 3 (1964) 99.
3 T. Smith, J. Electrochem. Soc., 119 (1972) 1398.
4 J.R. Anderson and M.S.J. Gani, J. Phys. Chem. Solids, 23 (1962) 1087.
5 J.R. Anderson and N. Thompson, Surface Sci., 28 (1971) 84
6 I.H. Khan, Surface Sci., 48 (1975) 537.

7 M.P. Cox, J.S. Foord, R.M. Lambert and R.H. Prince, Surface Sci., 129 (1983) 375.
8 G.V. Samsonov, Handbook of the Physicochemical Properties of the Elements, Plenum, New York, 1968.
9 Y. Fukuda, G.M. Lancaster, F. Honda and J.W. Rabalais, Phys. Rev., B18 (1978) 6191.
10 N.D. Spencer, P.J. Goddard, P.W. Davies, M. Kitson and R.M. Lambert, J. Vac. Sci. Technol., A1 (1983) 1554.
11 M.P. Cox and R.M. Lambert, Surface Sci., 107 (1981) 547.
12 H.D. Shih, F.Jona, D.W. Jepsen and P.M. Marcus, J. Phys. C: Solid State Phys., 9 (1976) 1405.
13 Z.T. Stott and H.P. Hughes, Surface Sci., 126 (1983) 455.
14 P.J. Goddard and R.M. Lambert, Surface Sci. 67 (1977) 180
15 P. Kisliuk, J. Phys. Chem. Solids, 5 (1958) 78.
16 M. Grunze and P.A. Dowben, Appl. Surface Sci., 10 (1982) 209.
17 J. Mischenko III and P.R. Watson, Surface Sci., 209 (1989) L105.
18 M. Bowker and K.C. Waugh, Surface Sci., 134 (1983) 639.
19 E. Bauer, Surface Sci., 7 (1964) 351.
20 J. Mischenko and P.R. Watson, Solid State Commun., 63 693 (1987).

R.K. Grasselli and A.W. Sleight (Editors), *Structure-Activity and Selectivity Relationships in Heterogeneous Catalysis*

© 1991 Elsevier Science Publishers B.V., Amsterdam

MATHEMATICAL DESCRIPTION OF HETEROGENEOUS MATERIALS - EFFECT OF THE BRANCHING DIRECTION -

J. W. Beeckman

W. R. Grace & Co.-Conn.
7379 Route 32
Columbia, MD 21044

1. ABSTRACT

This paper elaborates on an approach to describe complex interconnected networks in two-dimensional space. The approach is based on the mathematical formulation of a travel process throughout such a structure according to a given branching algorithm. Analytical solutions are developed that allow to predict the global properties of these networks. Monte Carlo simulations are shown to be in good agreement with the theoretical predictions.

2. INTRODUCTION

Many commercial processes employing heterogeneous catalysts are hampered in performance due to the limited rate of mass transport of the reactants through the porous structure of the catalyst. It is therefore important to develop models that can portray the complex topology of the catalyst pore structure. Many mathematical models with a greater or lesser degree of sophistication are currently available in the literature [Wakao et al.,(ref. 1); Reyes et al.,(ref. 2); Mann et al.,(ref. 3); Bhatia,(ref. 4);Froment et al.(ref. 5)] In this paper, a theoretical approach is outlined that allows to investigate the influence of the pore branching frequency and the pore orientation on the overall properties of the pore structure.

3. APPROACH

In this paper, a heterogeneous catalyst is structurally viewed as a strongly interconnected maze of straight pores. The pores are assumed to represent the voids between the catalyst grains or crystallites and many possible routes are available to connect any two arbitrarily chosen points in the porous structure. The approach used in this paper to describe such a maze is based on the mathematical formulation of a travel process throughout the catalyst pore structure. Starting at a single pore mouth at the boundary of a catalyst particle, one travels inward until the pore branches in two or more pores joining in the branch node. Each of these pores is then travelled through simultaneously until they in turn branch independently and the process is repeated. Certain travel directions will inevitably run back into the already travelled through portion of the pore structure and at these reconnection nodes the travel is halted.

4. THEORY

4.1 Branching directions

The branching algorithm dictates the branching frequency and the branching angle. In this paper, the branching frequency is characterized by the branching probability, ν, and will be assumed constant. The branching direction is characterized by the angle between each of the emanating pores and the branched pore. It is further assumed that upon branching only two pores are generated.

4.2 Mathematical formulation

The case of 90 degree symmetrical branching is chosen here as an example. It is clear that four directions of travel are possible in this case. Each travel direction is characterized by a probability density function $\phi_i(x,y,t)dxdy$ that at time t, a traveller moving in direction i is located in the area dxdy. Define directions 1 and 3 as respectively in the same direction and the opposite direction of the x-axis and directions 2 and 4 as respectively in the same direction and the opposite direction of the y-axis.The following equations than describe for instance the travel in direction 1:

$$\frac{1}{\xi}\frac{\partial\phi_1}{\partial t} + \frac{\partial\phi_1}{\partial x} = -\nu\phi_1 - \xi\phi_1\int_0^t(\phi_2 + \phi_4)dt' + \nu(\phi_2 + \phi_4) \qquad [1]$$

with initial conditions :

$$\phi_1(x,y,0) = \phi_{1,0}(x,y) \qquad [2]$$

Similar equations hold for the other three travel directions and all need to be solved simultaneously. In the following, the variable $s=\xi t$ is used to simplify the notation. Figures 1a through 1d illustrate the travel process by giving the sum of the densities of all travellers as a function of time. At time zero (not shown), the travellers are all located in a small narrow peak at the center of the plane with a total density of $4cm^{-2}$. As can be observed, the total density of travellers initially increases strongly with time, but then decreases at the center due to reconnection. The final picture can be compared to a crater with walls moving outward. At the inside base of the crater is now left a strongly interconnected network of paths of travel. Figures 2a through 2d show such networks for different branching angles clearly showing the complex topology and strong connectivity of these structures.
It can further be shown for 90 degree branching, that the total number of nodes N generated by branching and reconnection equals $8\nu^2$. This relation was checked by Monte Carlo simulation and is given in Fig. 3. The simulations (trials) were performed in a 4cm by 4cm base square with a single traveller starting at the bottom portion of the square in a randomly chosen direction with $\nu=2cm^{-1}$. When the network was established, the number of nodes in a 2cm by 2cm square located at the center of the base square were counted and these are represented as dots in Figure 3. The cumulative average of the number of nodes is also given and converges to within 3 percent of the theoretically predicted value.

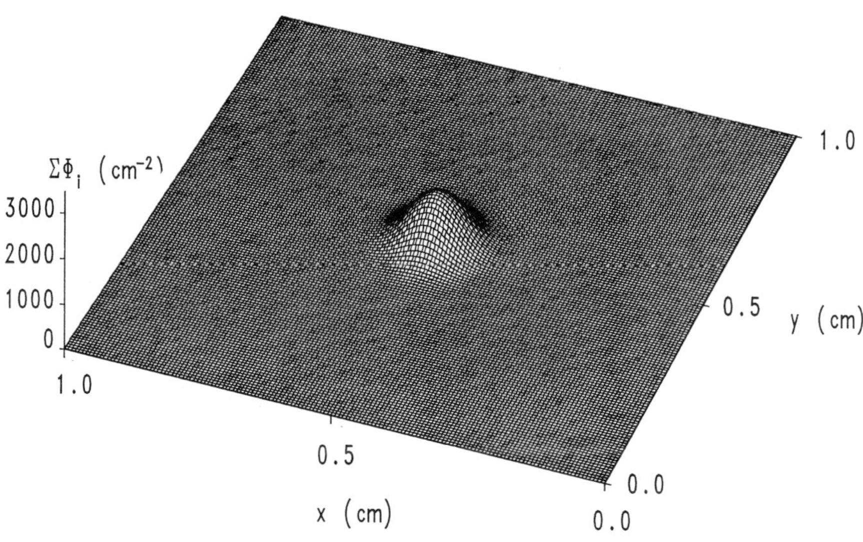

Fig. 1a. Total density of travellers for s=0.12 cm with ν=50cm^{-1}

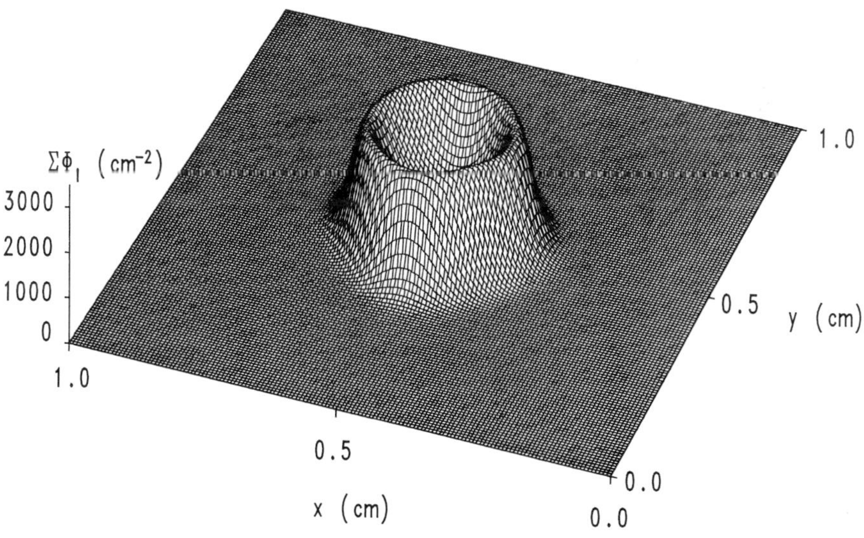

Fig. 1b. Total density of travellers for s=0.21 cm with ν=50cm^{-1}

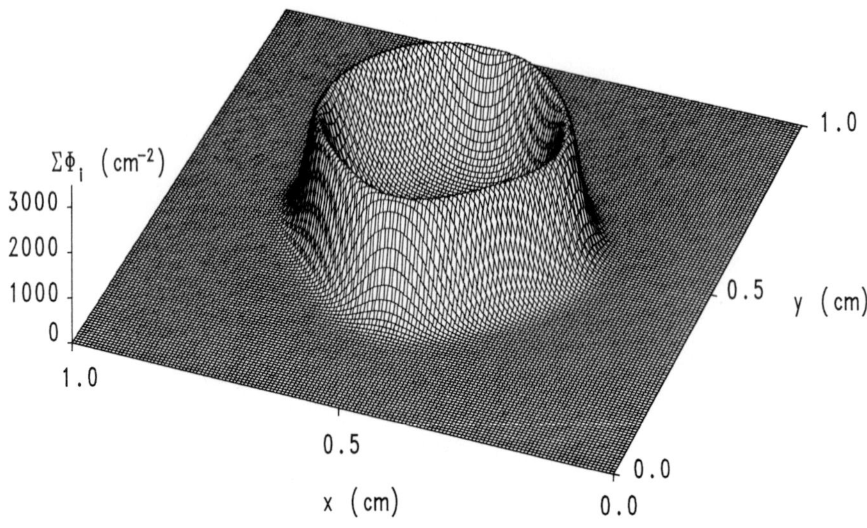

Fig.1c. Total density of travellers for s=0.28 cm with ν=50cm^{-1}

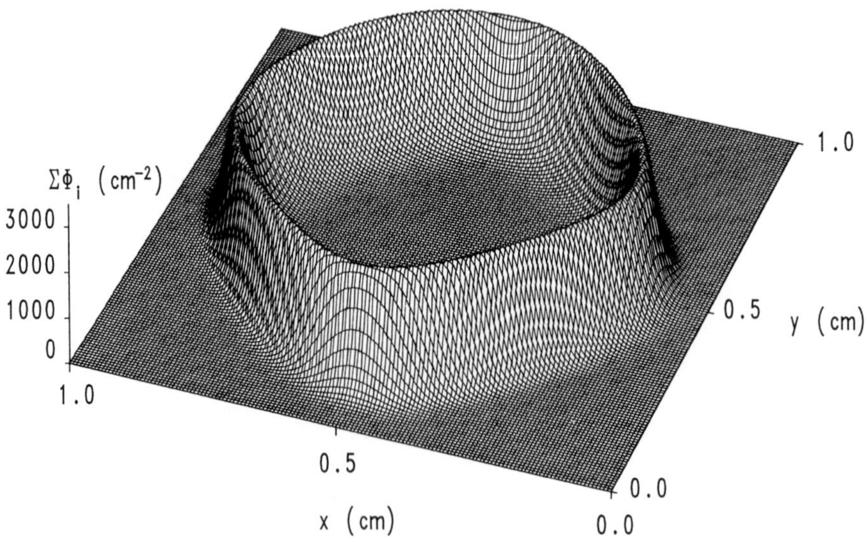

Fig.1d. Total density of travellers for s=0.38 cm with ν=50cm^{-1}

Fig. 2a. 90 degree symmetrical branching, ν=40cm^{-1}, 1cm by 1cm square

Fig. 2b. 85 degree symmetrical branching, ν=40cm^{-1}, 1cm by 1cm square

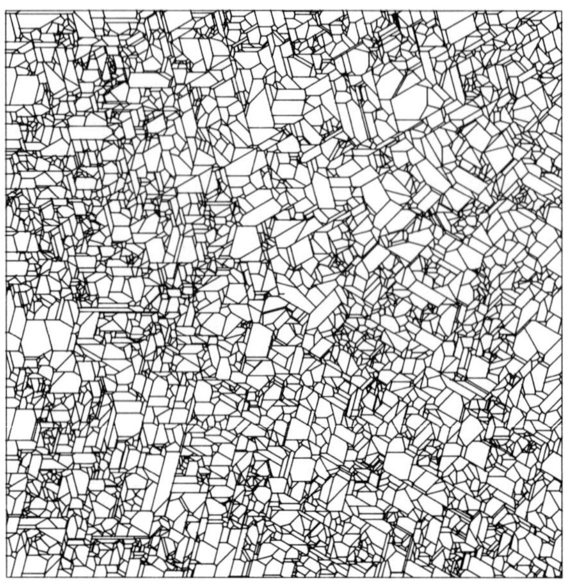

Fig. 2c. 75 degree symmetrical branching, $\nu=40cm^{-1}$, 1cm by 1cm square

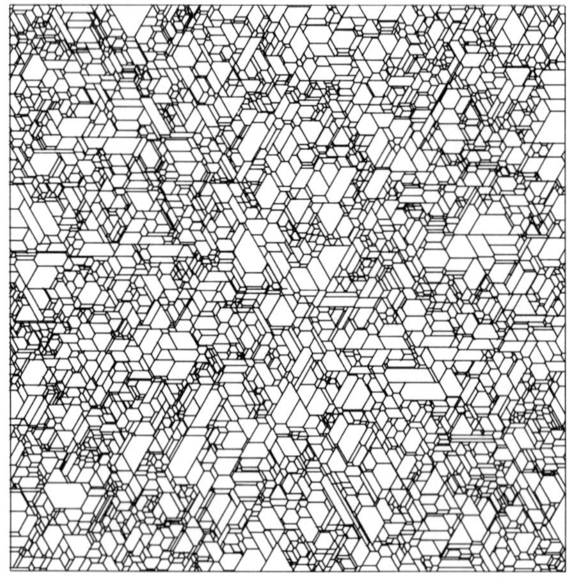

Fig. 2d. 60 degree symmetrical branching, $\nu=40cm^{-1}$, 1cm by 1cm square

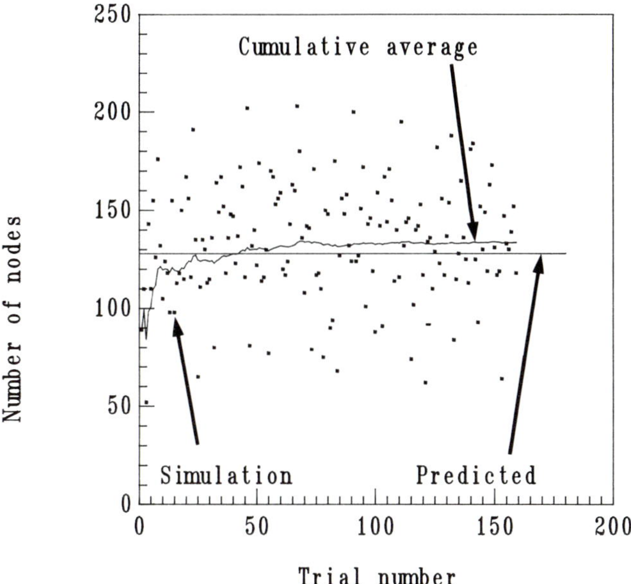

Fig. 3. Comparison of theory and Monte Carlo simulation

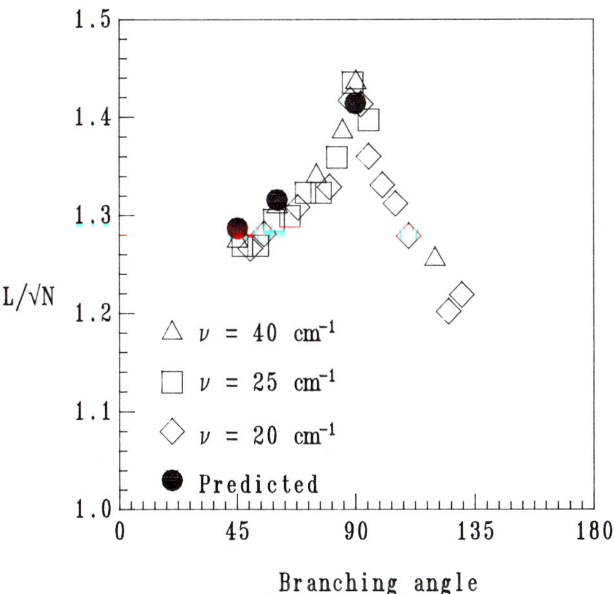

Fig. 4. Dependence of $L/\sqrt{N}$ on the branching angle

4.3 Mathematical treatment for sparsely located travellers at t=o

The treatment is here done for a 45 degree branching direction yielding obviously 8 possible travel directions. It is assumed that all initial starting point locations are random and sparse. It is further assumed that the initial travel direction associated with each starting point is randomly chosen from the 8 possible directions and that therefore no travel direction is favored. ϕ can than be expressed as follows :

$$\frac{d\phi}{ds} = \nu\phi - 2\phi \int_0^s \phi ds - 4\sin(\frac{\pi}{4})\phi \int_0^s \phi ds \qquad [3]$$

where the branching probability ν is assumed constant.
Equation [3] can be solved analytically which then allows to calculate the following overall properties of the network:

The pore density P defined as the number of pores with a given orientation crossing a unit length perpendicular to that orientation:

$$P = \frac{2\nu}{1 + \sqrt{2}} \qquad [4]$$

The specific length L of the network defined as the sum of the lengths of all pore segments per unit area:

$$L = \frac{8\nu}{1 + \sqrt{2}} \qquad [5]$$

The number of nodes N defined as the total number of nodes formed by branching or reconnection per unit area:

$$N = \frac{16\nu^2}{1 + \sqrt{2}} \qquad [6]$$

The average length p of a pore segment, i.e.: the average distance between two nodes:

$$p = \frac{1}{3\nu} \qquad [7]$$

An application of the theory is now straightforward: assume with 45 degree branching that all pore segments have a width of 100 Angstrom and that further it is desired that 50% of the area is covered by pore segments. Neglecting the overlapping of some pore segments, the branching probability ν can be calculated from [5] yielding $\nu = 1.5\ 10^5 cm^{-1}$. The average length of a pore segment is then calculated from [7] yielding p = 220 Angstrom, i.e.: the average length of a pore segment equals approximately 2.2 pore diameters. It can further be calculated that the average number of nodes amounts to $1.5\ 10^{11} cm^{-2}$ indicative of the strong interconnectivity.

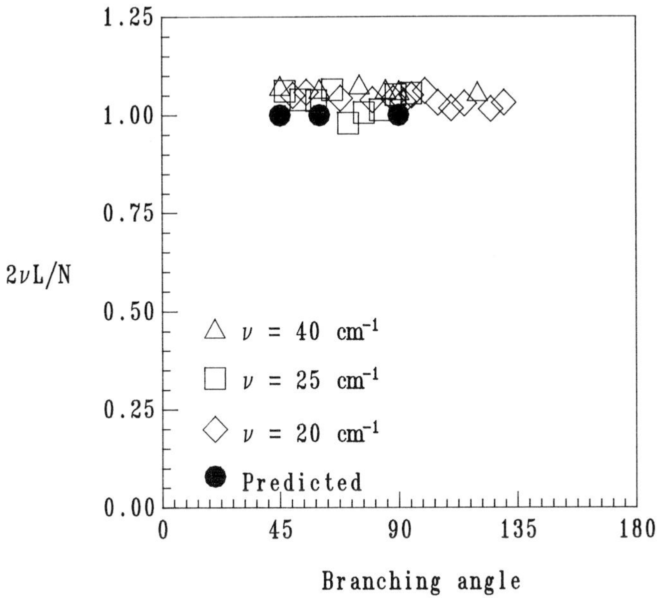

Fig. 5. Dependence of $2\nu L/N$ on the branching angle

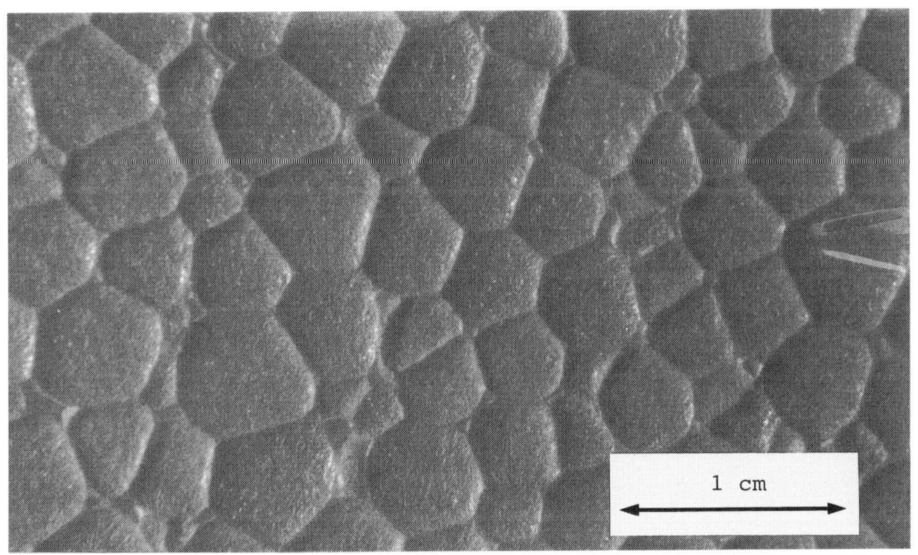

Fig. 6. Picture of Styrofoam

Figure 4 shows the ratio of the specific length L to the square root of the number of nodes N as a function of the branching angle. Open symbols are obtained through computer simulation while solid symbols are theoretical predictions. It can be observed that this ratio is independent of the branching probability and that the theory is in good agreement with the simulations. Figure 5 gives the ratio of the branching probability times the specific length to the number of nodes and clearly shows that this ratio is both independent of the branching angle and the branching probability. Also here, the theory compares well with the simulations.

5. APPLICATION

Figure 6 shows a picture of styrofoam, a commonly used packaging material and shows that most nodes have a coordination number of three with almost straight boundaries between the different particles. Measurements on the network gave : $L=5.32cm^{-1}$, $N=16.75cm^{-2}$ and p=0.19cm. The ratio of the specific length L to the square root of the specific number of nodes equals 1.30 and Fig. 4 then shows that a 60 degree branching angle should be selected (in agreement with many of the angles observed in Fig. 6). The theory applied for 60 degree branching then allows to calculate the branching probability in three different ways:

$$\nu = \frac{N}{2L} = 1.57cm^{-1} \qquad [8]$$

$$\nu = \frac{L}{2\sqrt{3}} = 1.53cm^{-1} \qquad [9]$$

$$\nu = \sqrt{\frac{N}{4\sqrt{3}}} = 1.55cm^{-1} \qquad [10]$$

The numerical values obtained for ν are quite close and the predicted value for the average distance between two nodes calculated from [7] using $\nu=1.55cm^{-1}$ yields p=0.21cm, close to the experimentally observed value.

6. CONCLUSION

A mathematical approach is presented that allows to calculate the properties of large ensembles of chaotically arranged pore segments from the branching algorithm.

LITERATURE CITED

1 N. Wakao and J. M. Smith, Chem. Eng. Sci.,17,(1976)825-834.
2 S. Reyes and K. F. Jensen, Chem. Eng. Sci.,40,(1985)1723-1734.
3 R. Mann, M. N. Almeida and M. N. Mugerwa, Chem. Eng. Sci., 41,10,(1986)2663-2671.
4 S. K. Bhatia, Chem. Eng. Sci.,41,5,(1986)1311.
5 G. F. Froment and K. B. Bischoff, Chemical Reactor Analysis and Design, 2nd edn., John Wiley & Sons, Inc., New York (1990),p151.

ACKNOWLEDGMENT

I thank the IBM Corporation for their support with the computer simulations on the IBM-3090 Supercomputer.

NOMENCLATURE

Arabic

L : specific length of a network, (cm^{-1})
N : specific number of nodes in a network, (cm^{-2})
p : average distance between two nodes, (cm)
P : number of pores crossing a unit length perpendicular to the pore direction, (cm^{-1})
s : defined in text, (cm)
t : time, (sec)
x : coordinate, (cm)
y : coordinate, (cm)

Greek

ν : branching probability, (cm^{-1})
ϕ : probability density function, (cm^{-2})
ξ : speed of travel, (cm/sec)

Subscripts

1 : direction
2 : direction
3 : direction
4 : direction
i : direction
0 : initial condition

AUTHOR INDEX

STUDIES IN SURFACE SCIENCE AND CATALYSIS

Advisory Editors: B. Delmon, Université Catholique de Louvain, Louvain-la-Neuve, Belgium
J.T. Yates, University of Pittsburgh, Pittsburgh, PA, U.S.A.

Volume 1 **Preparation of Catalysts I.** Scientific Bases for the Preparation of Heterogeneous Catalysts. Proceedings of the First International Symposium, Brussels, October 14–17, 1975
edited by **B. Delmon, P.A. Jacobs and G. Poncelet**

Volume 2 **The Control of the Reactivity of Solids.** A Critical Survey of the Factors that Influence the Reactivity of Solids, with Special Emphasis on the Control of the Chemical Processes in Relation to Practical Applications
by **V.V. Boldyrev, M. Bulens and B. Delmon**

Volume 3 **Preparation of Catalysts II.** Scientific Bases for the Preparation of Heterogeneous Catalysts. Proceedings of the Second International Symposium, Louvain-la-Neuve, September 4–7, 1978
edited by **B. Delmon, P. Grange, P. Jacobs and G. Poncelet**

Volume 4 **Growth and Properties of Metal Clusters.** Applications to Catalysis and the Photographic Process. Proceedings of the 32nd International Meeting of the Société de Chimie Physique, Villeurbanne, September 24–28, 1979
edited by **J. Bourdon**

Volume 5 **Catalysis by Zeolites.** Proceedings of an International Symposium, Ecully (Lyon), September 9–11, 1980
edited by **B. Imelik, C. Naccache, Y. Ben Taarit, J.C. Vedrine, G. Coudurier and H. Praliaud**

Volume 6 **Catalyst Deactivation.** Proceedings of an International Symposium, Antwerp, October 13–15, 1980
edited by **B. Delmon and G.F. Froment**

Volume 7 **New Horizons in Catalysis.** Proceedings of the 7th International Congress on Catalysis, Tokyo, June 30–July 4, 1980. Parts A and B
edited by **T. Seiyama and K. Tanabe**

Volume 8 **Catalysis by Supported Complexes**
by **Yu.I. Yermakov, B.N. Kuznetsov and V.A. Zakharov**

Volume 9 **Physics of Solid Surfaces.** Proceedings of a Symposium, Bechyňe, September 29–October 3, 1980
edited by **M. Láznička**

Volume 10 **Adsorption at the Gas–Solid and Liquid–Solid Interface.** Proceedings of an International Symposium, Aix-en-Provence, September 21–23, 1981
edited by **J. Rouquerol and K.S.W. Sing**

Volume 11 **Metal-Support and Metal-Additive Effects in Catalysis.** Proceedings of an International Symposium, Ecully (Lyon), September 14–16, 1982
edited by **B. Imelik, C. Naccache, G. Coudurier, H. Praliaud, P. Meriaudeau, P. Gallezot, G.A. Martin and J.C. Vedrine**

Volume 12 **Metal Microstructures in Zeolites.** Preparation – Properties – Applications. Proceedings of a Workshop, Bremen, September 22–24, 1982
edited by **P.A. Jacobs, N.I. Jaeger, P. Jirů and G. Schulz-Ekloff**

Volume 13 **Adsorption on Metal Surfaces.** An Integrated Approach
edited by **J. Bénard**

Volume 14 **Vibrations at Surfaces.** Proceedings of the Third International Conference, Asilomar, CA, September 1–4, 1982
edited by **C.R. Brundle and H. Morawitz**

Volume 15 **Heterogeneous Catalytic Reactions Involving Molecular Oxygen**
by **G.I. Golodets**

Volume 16 **Preparation of Catalysts III.** Scientific Bases for the Preparation of Heterogeneous Catalysts. Proceedings of the Third International Symposium, Louvain-la-Neuve, September 6–9, 1982
edited by **G. Poncelet, P. Grange and P.A. Jacobs**

Volume 17 **Spillover of Adsorbed Species.** Proceedings of an International Symposium, Lyon-Villeurbanne, September 12–16, 1983
edited by **G.M. Pajonk, S.J. Teichner and J.E. Germain**

Volume 18 **Structure and Reactivity of Modified Zeolites.** Proceedings of an International Conference, Prague, July 9–13, 1984
edited by **P.A. Jacobs, N.I. Jaeger, P. Jirů, V.B. Kazansky and G. Schulz-Ekloff**

Volume 19 **Catalysis on the Energy Scene.** Proceedings of the 9th Canadian Symposium on Catalysis, Quebec, P.Q., September 30–October 3, 1984
edited by **S. Kaliaguine and A. Mahay**

Volume 20 **Catalysis by Acids and Bases.** Proceedings of an International Symposium, Villeurbanne (Lyon), September 25–27, 1984
edited by **B. Imelik, C. Naccache, G. Coudurier, Y. Ben Taarit and J.C. Vedrine**

Volume 21 **Adsorption and Catalysis on Oxide Surfaces.** Proceedings of a Symposium, Uxbridge, June 28–29, 1984
edited by **M. Che and G.C. Bond**

Volume 22 **Unsteady Processes in Catalytic Reactors**
by **Yu.Sh. Matros**

Volume 23 **Physics of Solid Surfaces 1984**
edited by **J. Koukal**

Volume 24 **Zeolites: Synthesis, Structure, Technology and Application.** Proceedings of an International Symposium, Portorož-Portorose, September 3–8, 1984
edited by **B. Držaj, S. Hočevar and S. Pejovnik**

Volume 25 **Catalytic Polymerization of Olefins.** Proceedings of the International Symposium on Future Aspects of Olefin Polymerization, Tokyo, July 4–6, 1985
edited by **T. Keii and K. Soga**

Volume 26 **Vibrations at Surfaces 1985.** Proceedings of the Fourth International Conference, Bowness-on-Windermere, September 15–19, 1985
edited by **D.A. King, N.V. Richardson and S. Holloway**

Volume 27 **Catalytic Hydrogenation**
edited by **L. Červený**

Volume 28 **New Developments in Zeolite Science and Technology.** Proceedings of the 7th International Zeolite Conference, Tokyo, August 17–22, 1986
edited by **Y. Murakami, A. Iijima and J.W. Ward**

Volume 29 **Metal Clusters in Catalysis**
edited by **B.C. Gates, L. Guczi and H. Knözinger**

Volume 30 **Catalysis and Automotive Pollution Control.** Proceedings of the First International Symposium, Brussels, September 8–11, 1986
edited by **A. Crucq and A. Frennet**

Volume 31 **Preparation of Catalysts IV.** Scientific Bases for the Preparation of Heterogeneous Catalysts. Proceedings of the Fourth International Symposium, Louvain-la-Neuve, September 1–4, 1986
edited by **B. Delmon, P. Grange, P.A. Jacobs and G. Poncelet**

Volume 32 **Thin Metal Films and Gas Chemisorption**
edited by **P. Wissmann**

Volume 33 **Synthesis of High-silica Aluminosilicate Zeolites**
by **P.A. Jacobs and J.A. Martens**

Volume 34 **Catalyst Deactivation 1987**. Proceedings of the 4th International Symposium, Antwerp, September 29–October 1, 1987
edited by **B. Delmon and G.F. Froment**

Volume 35 **Keynotes in Energy-Related Catalysis**
edited by **S. Kaliaguine**

Volume 36 **Methane Conversion**. Proceedings of a Symposium on the Production of Fuels and Chemicals from Natural Gas, Auckland, April 27–30, 1987
edited by **D.M. Bibby, C.D. Chang, R.F. Howe and S. Yurchak**

Volume 37 **Innovation in Zeolite Materials Science.** Proceedings of an International Symposium, Nieuwpoort, September 13–17, 1987
edited by **P.J. Grobet, W.J. Mortier, E.F. Vansant and G. Schulz-Ekloff**

Volume 38 **Catalysis 1987**. Proceedings of the 10th North American Meeting of the Catalysis Society, San Diego, CA, May 17–22, 1987
edited by **J.W. Ward**

Volume 39 **Characterization of Porous Solids**. Proceedings of the IUPAC Symposium (COPS I), Bad Soden a. Ts., April 26–29, 1987
edited by **K.K. Unger, J. Rouquerol, K.S.W. Sing and H. Kral**

Volume 40 **Physics of Solid Surfaces 1987.** Proceedings of the Fourth Symposium on Surface Physics, Bechyne Castle, September 7–11, 1987
edited by **J. Koukal**

Volume 41 **Heterogeneous Catalysis and Fine Chemicals.** Proceedings of an International Symposium, Poitiers, March 15–17, 1988
edited by **M. Guisnet, J. Barrault, C. Bouchoule, D. Duprez, C. Montassier and G. Pérot**

Volume 42 **Laboratory Studies of Heterogeneous Catalytic Processes**
by **E.G. Christoffel**, revised and edited by **Z. Paál**

Volume 43 **Catalytic Processes under Unsteady-State Conditions**
by **Yu. Sh. Matros**

Volume 44 **Successful Design of Catalysts.** Future Requirements and Development. Proceedings of the Worldwide Catalysis Seminars, July, 1988, on the Occasion of the 30th Anniversary of the Catalysis Society of Japan
edited by **T. Inui**

Volume 45 **Transition Metal Oxides**. Surface Chemistry and Catalysis
by **H.H. Kung**

Volume 46 **Zeolites as Catalysts, Sorbents and Detergent Builders.** Applications and Innovations. Proceedings of an International Symposium, Würzburg, September 4–8, 1988
edited by **H.G. Karge and J. Weitkamp**

Volume 47 **Photochemistry on Solid Surfaces**
edited by **M. Anpo and T. Matsuura**

Volume 48 **Structure and Reactivity of Surfaces.** Proceedings of a European Conference, Trieste, September 13–16, 1988
edited by **C. Morterra, A. Zecchina and G. Costa**

Volume 49 **Zeolites: Facts, Figures, Future.** Proceedings of the 8th International Zeolite Conference, Amsterdam, July 10–14, 1989. Parts A and B
edited by **P.A. Jacobs and R.A. van Santen**

Volume 50 **Hydrotreating Catalysts.** Preparation, Characterization and Performance. Proceedings of the Annual International AIChE Meeting, Washington, DC, November 27–December 2, 1988
edited by **M.L. Occelli and R.G. Anthony**

Volume 51 **New Solid Acids and Bases.** Their Catalytic Properties
by **K. Tanabe, M. Misono, Y. Ono and H. Hattori**

Volume 52 **Recent Advances in Zeolite Science.** Proceedings of the 1989 Meeting of the British Zeolite Association, Cambridge, April 17–19, 1989
edited by **J. Klinowski and P.J. Barrie**

Volume 53 **Catalyst in Petroleum Refining 1989.** Proceedings of the First International Conference on Catalysts in Petroleum Refining, Kuwait, March 5–8, 1989
edited by **D.L. Trimm, S. Akashah, M. Absi-Halabi and A. Bishara**

Volume 54 **Future Opportunities in Catalytic and Separation Technology**
edited by **M. Misono, Y. Moro-oka and S. Kimura**

Volume 55 **New Developments in Selective Oxidation.** Proceedings of an International Symposium, Rimini, Italy, September 18–22, 1989
edited by **G. Centi and F. Trifiro**

Volume 56 **Olefin Polymerization Catalysts.** Proceedings of the International Symposium on Recent Developments in Olefin Polymerization Catalysts, Tokyo, October 23–25, 1989
edited by **T. Keii and K. Soga**

Volume 57A **Spectroscopic Analysis of Heterogeneous Catalysts. Part A: Methods of Surface Analysis**
edited by **J.L.G. Fierro**

Volume 57B **Spectroscopic Analysis of Heterogeneous Catalysts. Part B: Chemisorption of Probe Molecules**
edited by **J.L.G. Fierro**

Volume 58 **Introduction to Zeolite Science and Practice**
edited by **H. van Bekkum, E.M. Flanigen and J.C. Jansen**

Volume 59 **Heterogeneous Catalysis and Fine Chemicals II.** Proceedings of the 2nd International Symposium, Poitiers, October 2–6, 1990
edited by **M. Guisnet, J. Barrault, C. Bouchoule, D. Duprez, G. Pérot, R. Maurel and C. Montassier**

Volume 60 **Chemistry of Microporous Crystals.** Proceedings of the International Symposium on Chemistry of Microporous Crystals, Tokyo, June 26–29, 1990
edited by **T. Inui, S. Namba and T. Tatsumi**

Volume 61 **Natural Gas Conversion.** Proceedings of the Symposium on Natural Gas Conversion, Oslo, August 12–17, 1990
edited by **A. Holmen, K.-J. Jens and S. Kolboe**

Volume 62 **Characterization of Porous Solids II.** Proceedings of the IUPAC Symposium (COPS II), Alicante, May 6–9, 1990
edited by **F. Rodríguez-Reinoso, J. Rouquerol, K.S.W. Sing and K.K. Unger**

Volume 63 **Preparation of Catalysts V.** Proceedings of the Fifth International Symposium on the Scientific Bases for the Preparation of Heterogeneous Catalysts, Louvain-la-Neuve, September 3–6, 1990
edited by **G. Poncelet, P.A. Jacobs, P. Grange and B. Delmon**

Volume 64 **New Trends in CO Activation**
edited by **L. Guczi**

Volume 65 **Catalysis and Adsorption by Zeolites.** Proceedings of ZEOCAT 90, Leipzig, August 20–23, 1990
edited by **G. Öhlmann, H. Pfeifer and R. Fricke**

Volume 66 **Dioxygen Activation and Homogeneous Catalytic Oxidation.** Proceedings of the fourth International Symposium on Dioxygen Activation and Homogeneous Catalytic Oxidation, Balatonfüred, September 10–14, 1990
edited by **L.I. Simándi**

Volume 67 **Structure–Activity and Selectivity Relationships in Heterogeneous Catalysis.** Proceedings of the ACS Symposium on Structure–Activity Relationships in Heterogeneous Catalysis, Boston, MA, April 22–27, 1990
edited by **R.K. Grasselli and A.W. Sleight**

Volume 68 **Catalyst Deactivation 1991.** Proceedings of the fifth International Symposium, Evanston, IL, June 24–26, 1991
edited by **C.H. Bartholomew and J.B. Butt**

UNIVERSITY OF SURREY LIBRARY